W9-AZJ-107

EXPLORING SPACE
THE HIGH FRONTIER

JONES & BARTLETT
LEARNING

World Headquarters

Jones & Bartlett Learning
40 Tall Pine Drive
Sudbury, MA 01776
978-443-5000
info@jblearning.com
www.jblearning.com

Jones & Bartlett Learning Canada
6339 Ormindale Way
Mississauga, Ontario L5V 1J2
Canada

Jones & Bartlett Learning International
Barb House, Barb Mews
London W6 7PA
United Kingdom

Jones & Bartlett Learning books and products are available through most bookstores and online booksellers. To contact Jones & Bartlett Learning directly, call 800-832-0034, fax 978-443-8000, or visit our website www.jblearning.com.

Substantial discounts on bulk quantities of Jones & Bartlett Learning publications are available to corporations, professional associations, and other qualified organizations. For details and specific discount information, contact the special sales department at Jones & Bartlett Learning via the above contact information or send an email to specialsales@jblearning.com.

Editorial Credits

High Stakes Writing, LLC, Editor and Publisher: Lawrence J. Goodrich
Department of the Air Force Editor: Naomi L. Mitchell
Primary Writers: Ruth Walker, Mary M. Shaffrey
Editor: Katherine Dillin

Production Credits

Chief Executive Officer: Ty Field
President: James Homer
SVP, Chief Operating Officer: Don Jones, Jr.
SVP, Chief Technology Officer: Dean Fossella
SVP, Chief Marketing Officer: Alison M. Pendergast
SVP, Chief Financial Officer: Ruth Siporin
SVP, Business Development: Christopher Will
VP, Design and Production: Anne Spencer
VP, Manufacturing and Inventory Control: Therese Connell
Publisher, Higher Education: Cathleen Sether
Senior Production Editor: Susan Schultz
Production Editor: Wendy Swanson
Assistant Photo Researcher: Carolyn Arcabascio
Text and Cover Design: Anne Spencer
Composition: Mia Saunders Design
Illustrations: Morales Studios
Cover Images: Rover: Courtesy of NASA; Nebula: Courtesy of NASA/JPL-Caltech/T. Megeath (University of Toledo) & M. Robberto (STScI); Space Shuttle, Astronaut: Courtesy of NASA.
Lesson Opener Credits: **Page 6**, Courtesy of NASA/JPL; **Page 20**, Courtesy of Library of Congress, Prints & Photographs Division [reproduction number cph 3c15362]; **Page 32**, Courtesy of Library of Congress, Prints & Photographs Division [reproduction number cph 3c03175]; **Page 50**, © Photos.com; **Page 62**, © Photodisc; **Page 76**, Courtesy of Hinode JAXA/NASA/PPARC; **Page 94**, Courtesy of NASA/ESA/The Hubble Heritage Team, STScI/AURA/J. Bell, Cornell University/M. Wolff, Space Science Institute; **Page 114**, Courtesy of NASA/JPL/Space Science Institute; **Page 132**, © Datacraft/age fotostock; **Page 148**, © a. v. ley/ShutterStock, Inc.; **Page 162**, Courtesy of NASA/ESA/Space Telescope Science Institute/A. Aloisi; **Page 182**, © Photos.com; **Page 198**, Courtesy of NASA/Sean Smith; **Page 216, 230, 250**, Courtesy of NASA; **Page 268**, Courtesy of NASA/Bill Ingalls; **Page 282**, Courtesy of NASA; **Page 302**, © Alan Freed/ShutterStock, Inc.; **Page 320, 338, 358, 374**, Courtesy of NASA; **Page 394**, Courtesy of NASA/JPL; **Page 416**, Courtesy of NASA; **Page 426**, Courtesy of JPL/NASA; **Page 442**, Courtesy of NASA/Sandra Joseph and Kevin O'Connel; **Page 454**, Courtesy of NASA/Jack Pfaller; **Page 476**, Courtesy of NASA; **Page 494**, Courtesy Mars Exploration Rover Mission/JPL/NASA; **Page 512**, Courtesy of NASA; **Page 528**, © foto.fritz/ShutterStock, Inc.
Printing and Binding: Courier Companies
Cover Printing: Courier Companies

Library of Congress Cataloging-in-Publication Data
Exploring space : the high frontier.
    p. cm.
  ISBN 978-0-7637-8961-9 (case bound)
  1. Outer space—Exploration—History. 2. Astronomy—History. I. Jones & Bartlett Learning.
  TL790.E97 2010
  520—dc226048
                        2010021594

6048

Printed in the United States of America
14 13 12   10 9 8 7 6 5 4 3

Contents

Contents

UNIT 3 Manned and Unmanned Spaceflight 246

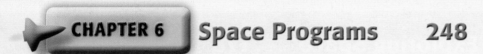

CHAPTER 6 Space Programs 248

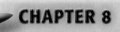

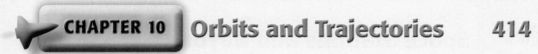
Contents

Preface

Exploring Space: The High Frontier is the third-year science course in the high school sequence of Aerospace Science courses for Air Force Junior ROTC. This textbook has been completely revised and includes the latest information available in space science and space exploration. The course begins with the study of the space environment from the earliest days of interest in astronomy and early ideas of the heavens, through the Renaissance, and on into modern astronomy. It provides an in-depth study of the Earth, Sun, stars, Moon, and Solar System, including the terrestrial and the outer planets. It discusses issues critical to travel in the upper atmosphere—such as orbits and trajectories—unmanned satellites, and space probes. It investigates the importance of entering space and discusses manned and unmanned missions, focusing on concepts surrounding spaceflight, space vehicles, launch systems, and safety.

This text is intended as a course in space for young high school students and to complement materials taught in high school math, physics, and other science-related courses. The basic concepts of spaceflight can be found in this course. Each unit and chapter also contains full-color diagrams and pictures "worth a thousand words." Included in the textbook are famous quotes at the beginning of each unit and chapter that relate to space or astronomy. Other features in each lesson include: a "Quick Write" that you can use as writing assignments; a "Learn About" box that tells you what you'll learn from the lesson; a list of vocabulary words that you should study; "Star Points" that highlight specific and interesting facts; and many biographies and profiles. Each lesson closes with "Checkpoints" which will allow you to review what you have learned. An "Applying Your Learning" section at the end of each lesson presents discussion questions that will give you a chance to use what you have learned and is another way to reinforce your understanding of the lesson's content.

The text has four units, each divided into chapters that contain a number of lessons.

"Unit 1: The Space Environment" discusses the vastness of the galaxy and the universe. The section on prehistoric and classical astronomy focuses on constellations as patterns of stars, the Sun's motion among the stars and around the Earth, the Greek Earth-centered model of the Solar System, and Ptolemy's model. "Astronomy and the Renaissance" covers Copernicus, the Sun-centered model, and Kepler's Laws of Planetary Motion. The following lesson discusses Galileo and the telescope, Newton's Laws of Motion and Gravity, and Einstein and relativity. The unit explains the special characteristics of the Earth and Moon, examining the Earth's interior, atmosphere, and magnetic field. It reviews theories of the Solar System's formation and discusses the terrestrial planets, the outer planets, dwarf planets, comets, asteroids, and the Kuiper Belt. Finally, the unit takes a look at the Milky Way Galaxy and what lies beyond it—at the Sun's location in the Milky way, other galaxies, and phenomena such as black holes.

"Unit 2: Exploring Space" looks at why humans should explore space and reviews the challenges of entering space. It examines the historical benefits of space exploration and the physical and psychological effects of spaceflight on human beings. The unit covers the US vision for space exploration and the current costs of doing so. It addresses in detail how the International Space Station will advance space exploration and the long-term goal of a manned flight to Mars. This unit also looks more deeply into how space exploration encourages the study of science, technology, engineering, and mathematics.

"Unit 3: Manned and Unmanned Spaceflight" delves into the space programs of several nations, the space shuttle program, space stations, and the many unmanned space probes. The unit begins with discussion of the US manned space program, including the history and accomplishments of Projects Mercury, Gemini, and Apollo. It also discusses in detail the achievements of the Soviet and Russian manned space programs, as well as those of the Chinese, Indian, European, and Japanese programs. It reviews the US shuttle program and the development of the original six orbiters. It also examines the lessons learned from the *Challenger* and *Columbia* accidents and highlights the changes NASA has made to reduce the possibility of another accident. In addition, it focuses on space stations and humanity's future in space. It reviews the *Salyut*, *Skylab*, and *Mir* stations, and the International Space Station—showing how nations are working together to construct and maintain it. The unit also examines humans' future in space, focusing on the planned return trip to the Moon, the plans for a Moon outpost, and a manned mission to Mars. The final chapter covers missions to the Sun, Moon, Venus, and Mars, the Hubble space telescope, and missions to comets and the outer planets.

"Unit 4: Space Technology" focuses on the science and technology of spaceflight. You'll learn about orbits and trajectories, and how they work; about maneuvering and traveling in space; and about momentum, gravitational force, orbital velocity, and orbit height, eccentricity, and inclination. The unit also covers the history and principles of rocket science, including how rockets operate and how force, mass, and acceleration apply to rockets. The unit examines solid and liquid propellant rocket engines, the evolution of rocket technology, the early use of rockets and the first rocket scientists. You'll learn about different types of launch vehicles, launch sites, launch windows, and how NASA prepares for a launch. The unit examines the use of robots in space and how NASA uses robots to explore the Solar System in ways that humans cannot. It discusses how NASA constantly works to improve its robots and keep up with new advances in technology. You'll read about the Mars Rover Expedition, the rovers *Spirit* and *Opportunity*, and the challenges of a flight to Mars. Finally, you'll read about how society has benefited from NASA technology and research—both in everyday life and in significant medical advances— and about how private companies are gearing up for their own space missions.

At the end of the text you will find a glossary defining all the vocabulary words and telling you which page each term appears on. You'll also find an index organized by subject at the end of the text, as well as a list of references.

This book has been prepared especially for you, the cadet—to increase your knowledge and appreciation of space. If space as a subject interests you, you'll find you've already embarked on your journey upon completion of this course. Students like you are the engineers, technicians and scientists of the future. The future of space exploration will soon be in your hands. Are you ready to take up the challenge? Everyone involved in the production of this book hopes it helps prepare you to do so.

Lawrence J. Goodrich and Naomi Mitchell

Acknowledgements

This new edition of *Exploring Space: The High Frontier* was based in part on suggestions received from AFJROTC instructors around the world. The Jeanne M. Holm Center for Officer Accessions and Citizen Development (Holm Center) Curriculum Directorate team involved in the production effort was under the direction of Charles Nath III, PhD, Director of Curriculum for the Holm Center at Maxwell Air Force Base, Alabama, and Roger Ledbetter, Chief, AFJROTC Curriculum. Special thanks go to Naomi Mitchell, an instructional systems specialist for Holm Center Curriculum and the primary Air Force editor and reviewer. We commend Naomi for her persistent efforts, commitment, and thorough review in producing the best academic materials possible for AFJROTC units worldwide.

We are deeply indebted to our subject matter experts, academic consultants, and reviewers: Gregory Vogt, PhD, NASA's Senior Project Manager at the Center for Educational Outreach, Baylor College of Medicine, Houston, Texas, for his meticulous review and expertise on important topics which helped make the book complete; Mr. Joseph M. Schuh of Ares Airborne and Orbiter Electrical Engineering at the Kennedy Space Center, Florida, for also taking the time to provide his expertise in the development of this textbook; Colonel John Gurtcheff (retired) AFJROTC Unit SC-873, Crestwood HS, Sumter, South Carolina, for thoroughly reviewing the material; and Kimberly Combs-Hardy, PhD, Chief of Educational Technology, Holm Center Curriculum, for her advice and suggestions throughout the development.

We would also like to express our gratitude to the High Stakes Writing, LLC, team for all its hard work in publishing this new book. That team consisted of contractors at High Stakes Writing, LLC—Lawrence Goodrich; Katherine Dillin; Ruth Walker, and Mary M. Shaffrey—subcontractors from Perspectives, Inc.—Philip G. Graham, PhD; Emily Davis; Aaron Paula Thompson; and Suzanne M. Perry—numerous personnel from Jones and Bartlett Learning—including Christopher Will; Anne Spencer; Susan Schultz; and Wendy Swanson—and Mia Saunders of Mia Saunders Design.

The AFJROTC mission is to develop citizens of character dedicated to serving their nation and communities. Our goal is to create materials that provide a solid foundation for producing members of society able to productively fulfill their citizenship roles. We believe this is another course that will continue the precedent set with the previous curriculum materials. All the people identified above came together on this project and combined their efforts to form one great team providing "world class" curriculum materials to all our schools.

UNIT 1

This cloud of dust and gas called the Orion Nebula sits 1,500 light-years from Earth. It's part of the constellation Orion and home to hundreds of infant stars, which appear as orange-yellow specks of light.

Courtesy of NASA/JPL-Caltech/T. Megeath, University of Toledo/M. Robberto, STSci

The Space Environment

Unit Chapters

The constellation Orion includes Orion's belt, three stars hanging in the sky from left to right named Alnitak, Alnilam, and Mintaka. These blue stars are larger and hotter than the Sun. The Orion Nebula is the fuzzy bright cloud just "south" of the belt.

The History of Astronomy

"Philosophy is written in this grand book—
I mean the Universe—which stands
continually open to our gaze, but it cannot
be understood unless one first learns to
comprehend the language and interpret
the characters in which it is written."

Galileo

Quick Write

_____

_____

What does Eratosthenes' experience in measuring the Earth tell you about what's needed for scientific discoveries?

Learn About

- the celestial sphere
- the Greek Earth-centered model
- Ptolemy's model

You have perhaps never heard of Eratosthenes (Eh-ra-TAHS-thin-ees) of Cyrene (276–195 BC). But this Greek mathematician and astronomer was the first person to measure the Earth's approximate size. His work was important not only for its results but for its illustration of scientific method.

Eratosthenes was a surveyor working for the Pharaoh, or king of Egypt. He was in charge of the scrolls at the Library of Alexandria, at that time the largest in the world. He had in his collection a map of the Kingdom of Egypt, with distances paced off by his surveyors. By ancient standards, it was quite a collection of data. He was the best mapmaker of his time, and he had invented a system of latitude and longitude.

Scientists don't know much about his life. But his work surely kept him and his staff out and about all over Egypt, measuring, observing, making notes, and asking questions.

Somehow he became aware of an apparently insignificant fact: In the town of Syene (today's Aswan), far up the Nile and well south of Alexandria, there was a remarkable well. When the summer solstice occurred each year, the Sun would rise so high in the sky that at noon, people could see its reflection in the water at the bottom of that well. That meant the Sun was directly overhead.

Not so on Midsummer Day in Alexandria, however. Eratosthenes measured the Sun there and observed that at noon, it was seven degrees off the vertical. He understood that the world was round—a circle, to think in terms of the simplest math for a moment.

The Sun struck Syene and Alexandria differently because they were at different points on that circle. The arc—represented by the distance between the two cities as points on that circle—was seven degrees. (Think of the arc as the outer edge of a slice of pie.) A circle has 360 degrees. So the distance from Syene to Alexandria was a little less than one-fiftieth, or 2 percent, of the whole way around the globe.

Eratosthenes knew the distance between Syene and Alexandria because his surveyors had paced it off. Once he realized that path was nearly 2 percent of the way around the entire globe, all he had to do was multiply that distance by a number a little greater than 50, and—voilà!—he knew the size of the Earth. To give his figures in modern measurements, he calculated its circumference to be about 25,937.5 miles (41,500 km). He also figured its radius to be 4,125 miles (6,600 km). He came amazingly close to the correct value for the Earth's radius of 3,986.25 miles (6,378 km).

Eratosthenes had measured the Earth without even having to leave home. It was an early demonstration of science as we know it.

Vocabulary

- celestial sphere
- north celestial pole
- south celestial pole
- constellation
- ecliptic
- zodiac
- retrograde motion
- parallax
- epicycle

The Celestial Sphere

If you ever get a chance to watch the stars for a few hours at a time—around a campfire, for instance—at almost any middle latitude on Earth, you can see how some stars rise in the east and set in the west, just like the Sun. Stars above the poles move in concentric circles in the sky. The center of these circles is a spot in the sky above the Earth's north and south poles.

This time-exposure photo of the northern sky shows stars making arcs around a single point. Polaris, the North Star, is almost, but not quite, at that point. It appears as a short, bright streak at the top of the image.

Courtesy of NOAO/AURA/NSF

CHAPTER 1   The History of Astronomy

It's also easy to observe that the stars stay in the same patterns every night. It may take only a few minutes of observing one evening after another before you start to recognize the arrangements of stars in the sky.

The ancients observed this, too—and many of them had more time to see and watch than you do. They came up with the idea that the Earth was surrounded by a large sphere that rotated around it (Figure 1.1). The stars were somehow glued to this sphere and moved with it around the Earth.

Today people know that there is no actual sphere with stars glued to it rotating around the Earth. But astronomers still use the term celestial sphere to refer to *an imaginary sphere of heavenly objects that seems to center on the observer.* It's a useful concept for describing and predicting the motions of stars and other objects in the sky. Astronomers speak of the north celestial pole and the south celestial pole as well. These refer to *the points on the celestial sphere directly above Earth's poles.*

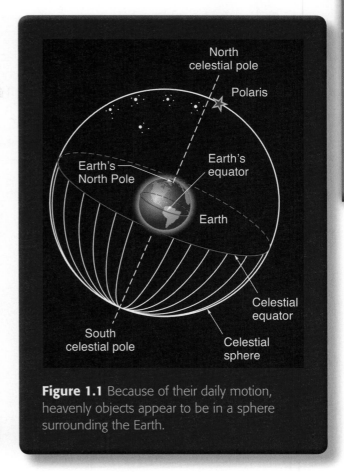

Figure 1.1 Because of their daily motion, heavenly objects appear to be in a sphere surrounding the Earth.

Constellations—Patterns of Stars

A constellation is *an area of the sky containing a group of stars in a pattern.* The search for patterns is very human, and it's very much a part of science, too. The ancients recognized shapes in the night sky and gave them names drawn from their myths.

As early as 2000 BC the Sumerians had identified several constellations, including a bull and a lion. But one of the best-known constellations is Ursa Major—Latin for "great bear" or "big bear." Ancient peoples across North America, Europe, Asia, and Egypt identified this shape as a bear. You may be more familiar with a part of it: the Big Dipper (Figure 1.2).

According to Greek legend, the bear had once been a nymph—a beautiful young goddess—who had attracted the attention of Zeus, father of the gods. His jealous wife changed her into a bear. And then, to protect the bear from hunters, Zeus grabbed her by her unusually long tail and flung her into the sky.

The Sun's Motion Across the Sky

Like the other stars, the Sun seems to revolve around the Earth. Except in the Arctic and Antarctic Circles—where it may not rise or set for several days in winter or summer—the Sun rises in the east and sets in the west. It is much closer than

Constellations become familiar patterns to anyone who watches the sky regularly. But their patterns aren't completely fixed. They're all in motion, with some stars moving toward Earth, others moving away. These pictures show how the Big Dipper continues to change shape over thousands of years.

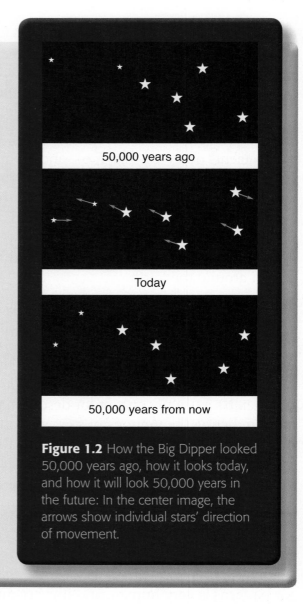

50,000 years ago

Today

50,000 years from now

Figure 1.2 How the Big Dipper looked 50,000 years ago, how it looks today, and how it will look 50,000 years in the future: In the center image, the arrows show individual stars' direction of movement.

the other stars, though. And so whereas sunrise and sunset happen every day (even when it's cloudy!), the more distant stars have a much longer cycle (Figure 1.3).

The Sun takes about 365.25 days to revolve around the celestial sphere. This cycle matches the cycle of seasons on Earth and the length of the year. That extra quarter day is the reason for adding an extra day to the calendar in leap years.

The Sun's apparent rotation around the Earth does not follow the Earth's celestial equator. Rather, it swings above and below that imaginary line. The ecliptic is the name for *the Sun's apparent path among the stars around the Earth*. (An eclipse can occur only when the Moon is on or very near this line; more on that later [Figure 1.4].)

The zodiac is *the group of constellations the Sun passes through on its apparent path along the ecliptic*. Long before people had calendars, they learned to "read" the change of season in the stars. The constellations of the zodiac were the stars they looked for.

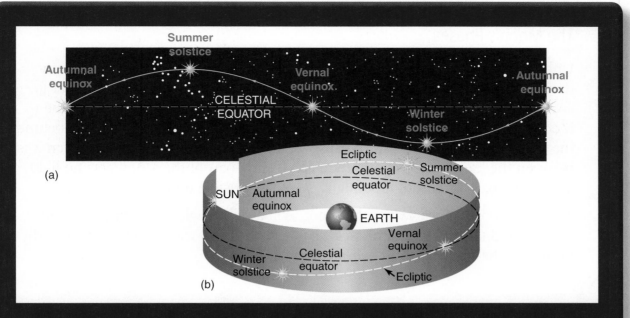

(a)

(b)

Figure 1.3 These two pictures show how the Sun cycles past the other, more distant stars as seen from Earth.

Solar Eclipses

The Sun rises every day on Earth. But sometimes something gets in the way: the Moon. When the Moon is in just the right position, it can block the view of the Sun from Earth— at least from parts of the Earth. If you are in just the right place, you may experience a total solar eclipse, in which the Sun is blacked out, with only its atmosphere visible. This is one of the world's most spectacular natural phenomena. The reason only some people can view a given total solar eclipse is that the Moon's shadow on Earth is relatively narrow. It may stretch in a thin band that is thousands of kilometers long across the Earth's surface, but it seldom exceeds 250 miles (about 400 km) in width.

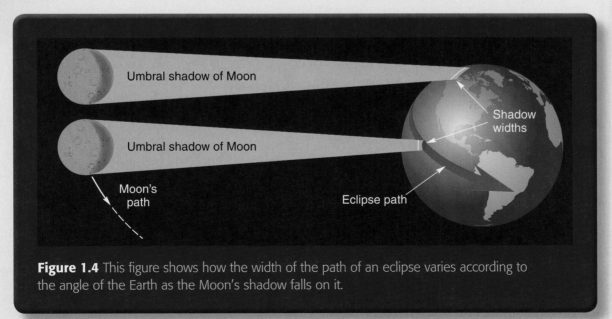

Figure 1.4 This figure shows how the width of the path of an eclipse varies according to the angle of the Earth as the Moon's shadow falls on it.

The Moon's Phases

Unlike the other heavenly objects you've read about so far, the Moon actually does orbit the Earth. Earth's gravity holds the Moon in such a firm grip that it always keeps the same face turned toward the Earth. That is the secret of the Moon's phases (Figure 1.5). The Sun lights the Moon. When the Moon is behind the Earth, it gets a full splash of sunshine and appears as a full moon. When it is between the Earth and the Sun, the Earth-facing part of the Moon gets no light, and so the palest sliver of moonlight is all that's visible on Earth.

The half moons appear when the Moon is at a right angle to the line between the Sun and the Earth. The phases when the Moon appears as a half are actually known as the first and third *quarters*, respectively.

Lunar Eclipses

Earlier in this lesson, you read about how occasionally the Moon blocks the Sun to create a solar eclipse. Sometimes the Earth comes between the Moon and the Sun to create a lunar eclipse.

Don't be tempted to believe an eclipse happens with every full moon. For one thing, the Earth and the Moon are both too small and too far apart for that to happen. The diagrams on page 11 are not all to scale. For an eclipse to happen, you'd need an alignment of Sun, Earth, and Moon that would be like aligning a grapefruit with a ping-pong ball 12 feet away, and then with a third object in the far distance.

The most important factor affecting lunar eclipses is that the Moon orbits the Earth at a tilt relative to the Earth's orbit around the Sun. When the Earth is in certain positions, its shadow cannot fall on the Moon. Lunar eclipses can only occur about twice a year.

Observing Planetary Motion

The planets are another major class of objects in the sky. Five are visible to the naked eye: Mercury, Venus, Mars, Jupiter, and Saturn. *Planet* comes from a Greek word meaning "wanderer." How did they get that name? They were observed in retrograde motion—*backward motion*, in other words. The planets moved eastward for a while, and then would start moving westward, or so it seemed.

Scientists today know that the planets all travel in continuous motion. This would appear counterclockwise if you were hovering over the Solar System from some point near the north celestial pole. Scientists also know that planets move in elliptical, not circular, orbits. And this is the key point: The planets revolve around the Sun, not the Earth. What appears as a loop-the-loop in the movements of Mars, for instance, is the result of the planet being on the same side of the Sun as Earth, and relatively close. Earth moves faster than Mars in its orbit and overtakes Mars, making it look as if Mars is going backward for a time (Figure 1.6).

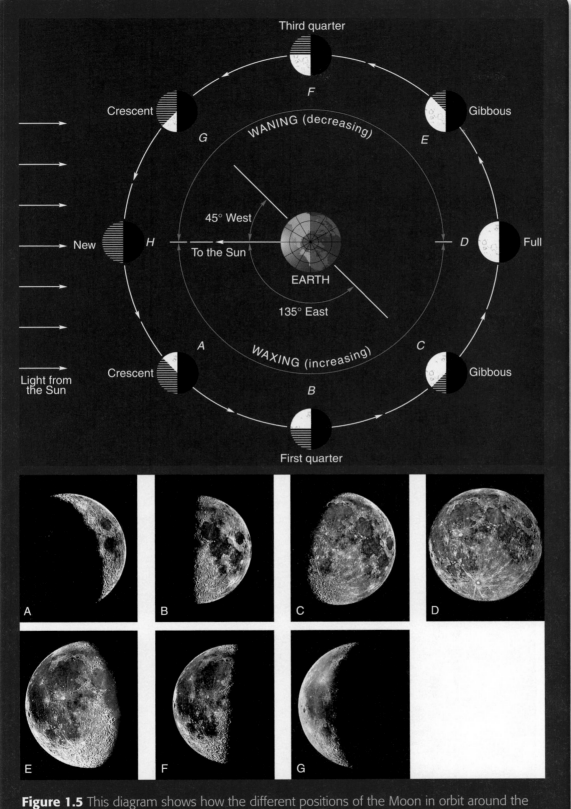

Figure 1.5 This diagram shows how the different positions of the Moon in orbit around the Earth create phases of the Moon. The photos show what the phases look like in the sky. Note the different terms for these phases.

© J. Gatherum/ShutterStock, Inc.

So it is with the planets farther than Earth from the Sun. They appear at some points to observers on Earth to be changing course. But in reality they are continuing in the same direction—around the Sun, not the Earth. This would be one of the most difficult lessons for astronomers to grasp.

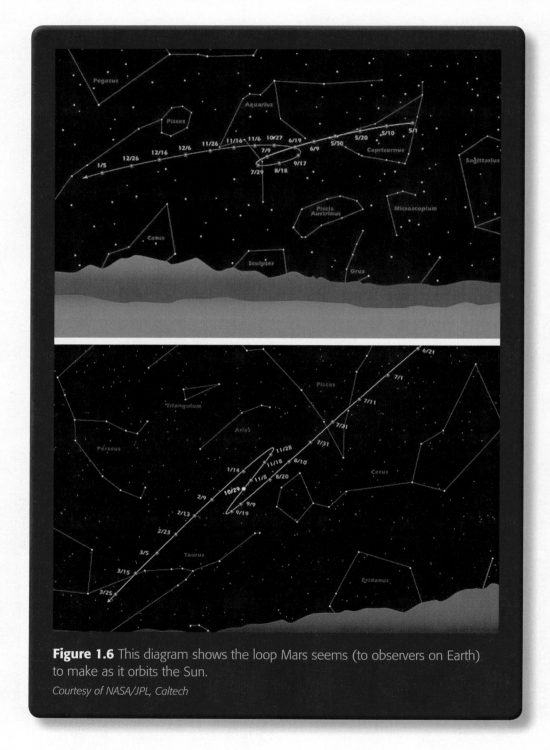

Figure 1.6 This diagram shows the loop Mars seems (to observers on Earth) to make as it orbits the Sun.

Courtesy of NASA/JPL, Caltech

CHAPTER 1   The History of Astronomy

The Greek Earth-Centered Model

As the ancients observed the sky over the centuries, they began to develop models for explaining the relationships among heavenly bodies. Some early scientists put the Sun at the center of the universe. But the Greek models that were most influential over time put the Earth at the center.

The Greek Search for Symmetry, Order, and Unity in the Universe

Many different ancient peoples looked to the heavens for different kinds of answers for their daily lives. The Babylonians looked up to the sky for the meaning behind current affairs on Earth. They read the heavens like a horoscope. The Egyptians thought they could make crop forecasts based on clues from the sky. The Chinese believed events in the sky controlled events on Earth.

But the ancient Greeks studied astronomy out of a pure desire to understand how the universe works. They believed in, and looked for, symmetry, order, and unity in the cosmos.

The Rational Approach to Astronomy: Thales of Miletus and the Pythagoreans' Central 'Fire'

Thales of Miletus, a Greek philosopher, lived about 600 BC. He believed that rational thought could lead to an understanding of the universe. More specifically, he reasoned that the Sun and other stars were not gods but balls of fire.

Around 530 BC, another Greek philosopher and mathematician named Pythagoras (Pith-AG-oh-ras) proposed that the Earth is spherical. Some 80 years later, his followers, known as the Pythagoreans, proposed a spherical universe controlled by a central "fire" whose force controlled all motion. Their idea accounted for the Earth, the Moon, the Sun, the five known planets, and the stars. This was 2,000 years before Nicolaus Copernicus published his revolutionary model of planets revolving around the Sun.

Aristotle's Earth-Centered View of the Solar System

Like the Pythagoreans, the Greek philosopher Aristotle believed that both the Earth and the Moon were spherical. But he rejected the Pythagoreans' ideas about a Sun-centered universe. He placed the Earth at the center of things.

This seems obviously wrong today. But the evidence available at the time supported his view. If the Earth were moving, he argued, people would see changes in the relative positions of the stars in the sky. He was noticing an effect that you have probably experienced while riding down a highway: As you move, you see changes in the relative positions of nearby and distant trees and other objects.

Aristotle saw some important differences between earthly and heavenly objects. Objects on Earth have a natural tendency to fall down to the ground. Objects up in the heavens remain up in the heavens, he observed. He also noted that earthly objects tend to come to a stop. Heavenly objects keep moving. From Aristotle, the ancient Greeks developed the idea that the natural world had two sets of rules: one for things on Earth and one for celestial objects.

Parallax is *the apparent shifting of nearby objects with respect to distant ones as the position of the observer changes.* Astronomers now know that the parallax phenomenon occurs in the case of stars, too. It's just that the stars are so far from the Earth that the apparent shifts of position are quite tiny. Stellar parallax, as it is known, was not observed until 1838.

Ptolemy's Model

Aristotle was an enormously influential ancient Greek scientist. But the Greek model of the universe that won the widest acceptance was that of Claudius Ptolemy. He lived around AD 150, and for 1,300 years, his model was the conventional wisdom of the scientific world.

Ptolemy's Ideal of Circular Motion and Heavenly Perfection

Ptolemy understood that the celestial sphere was not an actual physical thing on which stars and planets hung. But, like Aristotle, he believed that heavenly bodies were heavenly, celestial, and indeed even perfect, in contrast to earthbound objects. In Ptolemy's view, the universe was based on perfect circles.

The Moon fit perfectly into Ptolemy's model: a single body orbiting another body in (he thought) a perfectly circular orbit. Yes, the Moon had some imperfections, craters, and other scars. But then again, it was at a midpoint between the obviously imperfect Earth and the perfection of the heavens. So followers of Ptolemy thought the Moon's imperfections supported his model.

The Theory of Epicycles

Once Ptolemy set his gaze beyond the Moon, though, his model got more complicated. He managed to explain the planets' motions in terms of circles, all right. But it took lots of them. Specifically, he relied on something called an epicycle.

An epicycle is a kind of small circle. Specifically, it's *the circular orbit of a planet, the center of which revolves around the Earth in another circle.* In Ptolemy's view, a planet revolved in a small orbit around a central point that was part of a perfect circle around the Earth. A planet would buzz around in its epicycle while orbiting the Earth in this larger perfect circle. Epicycles were Ptolemy's way of explaining the retrograde motion of planets (Figure 1.7). For Venus and Mercury, the two planets between the Sun and the Earth, he worked out an even more complicated theory. These two planets are never far from the Sun, and so Ptolemy suggested that the centers of their epicycles remained on a line between the Earth and the Sun (Figure 1.8).

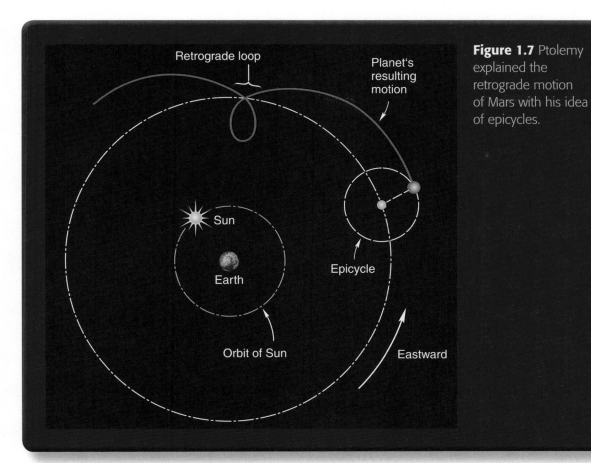

Figure 1.7 Ptolemy explained the retrograde motion of Mars with his idea of epicycles.

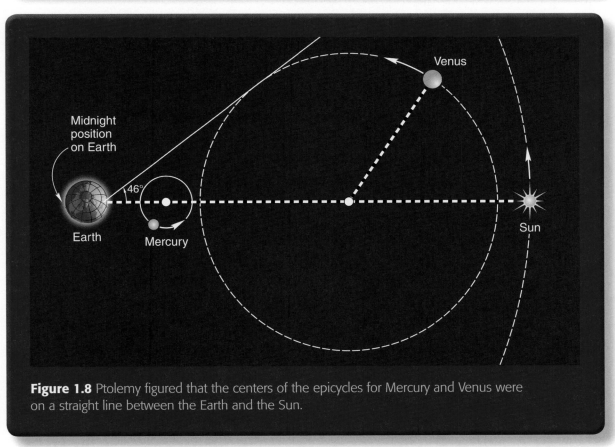

Figure 1.8 Ptolemy figured that the centers of the epicycles for Mercury and Venus were on a straight line between the Earth and the Sun.

Evaluating Ptolemy's Model

Before you leave Ptolemy, take a moment to consider how well his model holds up under modern standards of scientific method.

- Ptolemy's model fits the evidence available during his lifetime pretty well. He offered up a system of perfect circles and epicycles. The epicycles were complicated, but they convincingly explained planetary motion.

- The model includes testable predictions. It could be used to predict that Jupiter, for instance, would be at a particular place in the sky at a particular time on a given date.

- The model also assumed that the Earth was stationary. By stating his ideas this way, Ptolemy was leaving a door open to new facts or new data, in light of which he would be ready to change his idea.

- Science favors simplicity and symmetry. Ptolemy's model, with its reliance on circles and uniform motion from east to west around the Earth, largely embodies these values. But the theory of epicycles was a departure from them, especially once Ptolemy devised the special rule for Venus and Mercury. While Ptolemy's model did fit the data of his time, but in order for it to do so, it had to be continuously adjusted over hundreds of years.

In the next lesson you will read about Nicolaus Copernicus and how he arrived at his very different model of the universe.

CHECK POINTS

Lesson 1 Review

Using complete sentences, answer the following questions on a sheet of paper.

1. How did the human search for patterns lead to the naming of constellations?

2. How did the zodiac help early humans before they had calendars?

3. When do half moons appear?

4. What was it about the planets that made it hard for the ancients to make sense of them?

5. How did the ancient Greeks' interest in studying the skies differ from that of other ancient peoples?

6. What model of the universe did the Pythagoreans present?

7. What is parallax, and why did it take so long to confirm that this phenomenon occurs in the case of stars, too?

8. Why did the Moon fit so well into Ptolemy's model?

9. How did Ptolemy rely on epicycles in his model of the universe?

10. Ptolemy's model assumed the Earth was stationary. What was the significance of that?

APPLYING YOUR LEARNING

11. Why did the Greeks use circles and spheres to account for the motion of the Sun, Moon, planets, and stars?

Quick Write

_____

_____

If you were Tycho, how would you have argued for the kind of observatory you wanted from the king?

Learn About

- Copernicus and the Sun-centered model
- Kepler's laws of planetary motion

O n 21 August 1560 a young Danish nobleman named Tycho Brahe (1546–1601) saw a solar eclipse. It fascinated him—all the more so since astronomers had predicted it. His family wanted him to study law, but Tycho couldn't resist astronomy. He soon had the standard reference works in the field and began making his own observations. It wasn't long before he came to an important conclusion: that the leading theories of astronomy were built on bad data.

He became convinced that to try to account for the Sun's and stars' movements, scientists needed to observe them more systematically than they had done. It wasn't enough to monitor Mars, say, just at certain points in its orbit. Astronomers needed to follow the red planet all along its path—night after night. And these observations should continue for years to produce enough data to be useful.

Tycho came from a very wealthy and well-connected family. He traveled all over Europe to consult with leading astronomers. By the mid-1570s he was back in Denmark for a time—but had decided to settle in Basel, Switzerland.

Tycho was Denmark's leading scientist, however, and Danish King Frederick II didn't want to lose him. So the king offered to help him build an observatory. Frederick had to make a number of offers before he finally proposed something that Tycho accepted.

Frederick gave him an entire island, Ven. The project's cost was immense—roughly 5 percent of the country's entire gross domestic product, by one estimate. Uraniborg, as the observatory was known, had very large, very accurate instruments. For some 20 years, Tycho and his students studied the heavens there. They never had a telescope— that invention still lay in the future—but the data from their observations would later prove crucial to understanding how the Solar System really worked.

Copernicus and the Sun-Centered Model

Vocabulary

- heliocentric
- meridian
- ellipse

In Lesson 1, you read about Ptolemy's model of the universe. For some 1,300 years, people accepted it as the explanation for the planets' movements. But in the fifteenth century, a Polish churchman named Nicolaus Copernicus (Koh-PURR-ni-kuss) began to turn his eyes to the sky and to develop another theory.

Why Copernicus Searched for a Better Model

Copernicus seems to have had three main reasons for trying to improve on Ptolemy's model of the cosmos:

- He found that Ptolemy's predictions for the positions of celestial bodies were off from what he could observe himself—sometimes by as much as 2 degrees. This is significant in astronomical terms. To be fair to Ptolemy, his model held up pretty well for more than a thousand years. But as astronomers continued their observations over the centuries, they kept having to "reset" each planet's position on their charts. These adjustments were more than refinements of the model. They were a sign that there was something fundamentally wrong with it. Copernicus recognized this.

- His religious beliefs also pushed Copernicus to investigate a Sun-centered model of the cosmos. A heliocentric—another word for *Sun-centered*—model would provide more accurate data for the Roman Catholic Church calendar, helping ensure that holy days were observed at the right time. What's more, Copernicus also believed that a Sun-centered system would be more aesthetically pleasing. After all, he argued, the Sun was the source of light and life. The Creator would naturally put it at the center of everything.

- The third reason for exploring a heliocentric model was that the theory of epicycles didn't adequately explain the changes in Mars' brightness. While Ptolemy's model explained these changes, it didn't do so very precisely when the data were accurate.

This last point prompted Copernicus to revisit an idea proposed by the Greek scientist Aristarchus (Air-i-STAR-kuss). Observers had largely ignored it for 2,000 years.

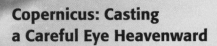

Copernicus: Casting a Careful Eye Heavenward

Nicolaus Copernicus (1473–1543) lived at a time of great cultural flowering in Europe, an exciting period of new developments and discoveries in the arts and sciences. His contemporaries included Leonardo da Vinci and Christopher Columbus.

He was well educated in many subjects. But his great contribution was as an astronomer. Educated in Italy as well as his native Poland, he spent most of his life as canon (priest) of the cathedral in the Polish city of Frauenberg (now Frombork). Like Tycho, he never had a telescope. He lived just about a century before it was invented. But his observations with the naked eye led him to redraw the map of the heavens.

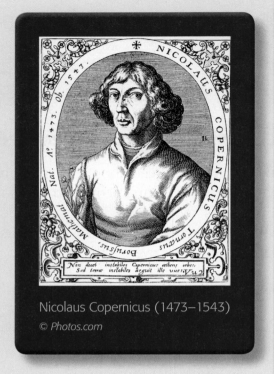

Nicolaus Copernicus (1473–1543)
© Photos.com

Star POINTS

Some 400 years before Ptolemy, Aristarchus had proposed a moving-Earth model of the universe. His calculations of the heavenly bodies' relative sizes convinced him that the Sun was much larger than the Earth, and must therefore be the central body around which the Earth and other planets revolve.

Early astronomers cared about being able to measure exactly when a celestial object crossed the meridian—*an imaginary line drawn on the sky from south to north*. To help him do this, Copernicus had a special narrow window slot built into one of the walls of his quarters. By standing in just the right spot and looking through the slot, he could be sure to tell exactly when a heavenly body crossed the meridian.

Star POINTS

Historians have often written that the Roman Catholic Church opposed scientific progress. The church had a keen interest in accurate calendar data—to ensure that believers celebrated holidays and feast days at the right time. This meant that astronomy was an important field of study for church scholars such as Copernicus. It didn't mean that the church would be happy with what Copernicus and other scholars found, however.

Copernicus's Heliocentric System

Copernicus's theory made the Earth just one of several planets revolving around the Sun. He also had the plane of the Earth's equator tilted with respect to the plane of its orbit around the Sun. This tilt causes Earth's seasons.

Like Ptolemy, Copernicus understood that the Moon revolves around the Earth. He indicated this in his drawing of the cosmos by showing the Earth encircled by the Moon's orbit (Figure 2.1).

Copernicus was able to calculate the *relative* distances of the planets from the Sun, and relatively accurately. As one example, he reckoned Mars is 1.5 times farther from the Sun than the Earth is. What he couldn't calculate were actual distances; that is, distances worked out in miles. His was a map without a scale. The early Greek scientist Aristarchus, who lived around 280 BC, had made a very rough calculation of the distance from the Sun to the Earth. But he got a crucial angle wrong. And so while he was correct in principle, his calculation was off by a factor of about 20. The Italian astronomer Cassini would be the first to measure the distance from the Sun to the Earth in 1672, after the invention of the telescope.

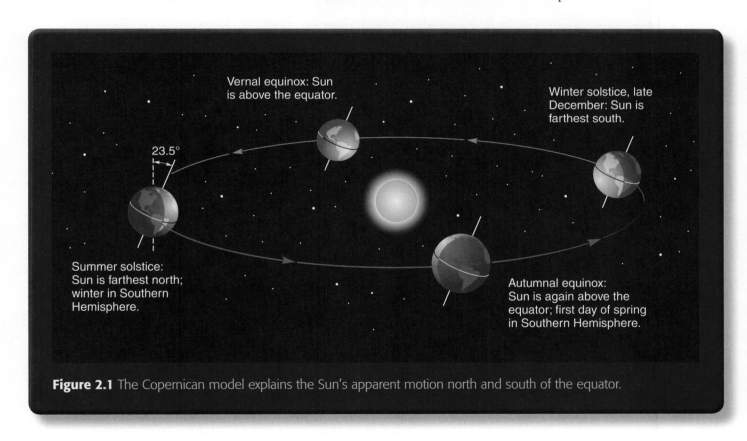

Figure 2.1 The Copernican model explains the Sun's apparent motion north and south of the equator.

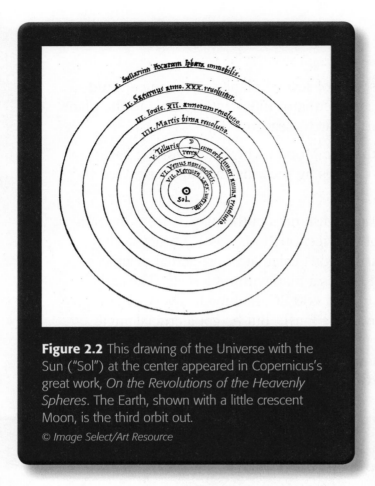

Figure 2.2 This drawing of the Universe with the Sun ("Sol") at the center appeared in Copernicus's great work, *On the Revolutions of the Heavenly Spheres*. The Earth, shown with a little crescent Moon, is the third orbit out.

© Image Select/Art Resource

One of the arguments against a heliocentric model—an argument that Ptolemy made—was that a rotating Earth would produce a great wind. Copernicus addressed this point (correctly, it turns out) by stating as an assumption that the Earth's atmosphere simply follows the Earth as it rotates (Figure 2.2).

Comparing the Copernican Model With the Ptolemaic

Scientists comparing the models of Copernicus and Ptolemy consider them in light of three different criteria. Today they accept the Copernican model. But, interestingly, Ptolemy got enough things right, and Copernicus got enough things wrong, that the competition was closer than you might think.

- *Accuracy of the data:* The heliocentric model provides a logical explanation for the observed motions of the stars, the Moon, and the planets. But the planets didn't always move quite the way Copernicus predicted. For one thing, he assumed—incorrectly—that the planets move at a constant speed. He also clung to the ideal of the circle as a planet's perfect path around the Sun. These two errors led him eventually to introduce epicycles into his own model. But the epicycles didn't really help produce more accurate predictions of the planets' motion. The two underlying errors remained. The bottom line: Using the evidence available in the 1500s, the two models score about even. Copernicus was not really better at predicting planetary motion than Ptolemy.

- *Predictive power:* Both models made testable predictions. Ptolemy predicted that stellar parallax should not exist. He held that the stars' positions relative to one another would not change because they were moving together around a stationary Earth. The Copernican model held that stellar parallax would exist. Both models pass this test. Ptolemy was actually wrong in his prediction, scientists now know. But for the purposes of this test, a prediction that is eventually disproved by new data is as acceptable as one eventually supported by new data. So again, on this criterion, the two models are in a tie.

- *Simplicity:* Here's where Copernicus pulls ahead of Ptolemy. As you read earlier, Ptolemy had to introduce epicycles to account for the planets' retrograde motion. As for the two planets between the Earth and the Sun, Venus and Mercury, he had to introduce another idea—that the centers of their epicycles remained on a line between the Earth and the Sun. But Copernicus was able to explain the motions of all the planets, in general terms at least, without epicycles and without introducing a special rule for just two planets. He used epicycles to try to make his predictions more accurate, but not to provide a basic explanation for retrograde motion.

Simplicity appealed to scientists like the devout Copernicus. They were sure that the Creator would not have made an overly complicated, disorderly universe. Simplicity still appeals to scientists today, but for a different reason: Experience has shown that in the case of two competing models, the simpler one is more likely to fit new data as they become available.

Kepler's Laws of Planetary Motion

Johannes Kepler's mission was to find a theory that fit the data. He had some ideas of his own he was eager to test out when he moved to Prague in 1600 to work as a mathematician at the imperial court. His new post would give him access to the best body of data in the world.

Johannes Kepler

When Johannes Kepler (1571–1630) was only 4 years old, he nearly died of smallpox. The disease left him with poor eyesight. That meant he was unable to observe the heavens himself. But he made great contributions to theoretical astronomy. He made sense of the data collected by others.

He was studying theology when he heard of the Copernican system. He became an advocate for it and went to work as Tycho Brahe's assistant. By that time, Tycho had left Denmark and gone to work for the Hapsburg Emperor Rudolph, whose capital was Prague. Kepler's assignment in Prague was to work out models of planetary motion.

Johannes Kepler (1571–1630)
Courtesy of National Library of Medicine

EFFIGIES TYCHONIS BRAHE O. F.
ÆDIFICII ET INSTRUMENTORUM
ASTRONOMICORUM STRUCTORIS
A° DOMINI 1587 ÆTATIS SVÆ 40

Figure 2.3 This mural shows Tycho's quadrant, designed to measure the time and elevation angle of an object crossing the meridian. The quadrant consists of the slot at upper left and the large quarter circle. Tycho is shown as the observer at the edge of the mural at the right center. The large figure pointing upward is an oversized portrait of Tycho himself. He had it painted on the wall of the observatory to remind his assistants that he was keeping an eye on them. The large quarter circle in the foreground is to the scale of the other two figures, one seated and the other standing in the front.

© Detlev van Ravenswaay/Photo Researchers, Inc.

CHAPTER 1   The History of Astronomy

Tycho Brahe's Observations of Planetary Motion

As Tycho began to study astronomy, the big question he faced was whether Ptolemy or Copernicus had the correct model of the universe. The young Dane needed more data (Figure 2.3).

Tycho's observations didn't confirm the more recent Copernican model, however. If the Earth orbited the Sun, Tycho argued, the nearby stars would show parallax. They did not, as far as sixteenth-century astronomers could tell. The model Tycho came up with put the Earth at the center of the Sun's orbit, but had other planets orbiting the Sun. You might say he came to a wrong conclusion but for a good reason.

Tycho Brahe (1546–1601)
© Photos.com

Star POINTS

You may be wondering why this text refers to Tycho by his first name. He is one of a number of artists and scientists who are known primarily by their first names. Among the others: Galileo, Leonardo (da Vinci), Michelangelo, Raphael, and Rembrandt.

Kepler's First Law: The Significance of the Ellipse

Kepler took over most of Tycho's records after he died. Kepler kept working and reworking the data, trying to devise a system of circles and epicycles that would accurately predict the positions of Mars. He had great confidence in the accuracy of Tycho's observations. After four years, he finally came up with an answer.

Both the ancient Greeks and the Christian scholars of the Renaissance had cherished the ideal of the circular orbit. But the shape that finally fit the data, Kepler found, was the ellipse. In very unscientific terms: An ellipse is like a circle but elongated, and with two centers instead of one.

Kepler's First Law of Planetary Motion holds that each planet's path around the Sun is an ellipse, with the Sun at one focus of the ellipse. The Sun is at one of the two "centers," in other words.

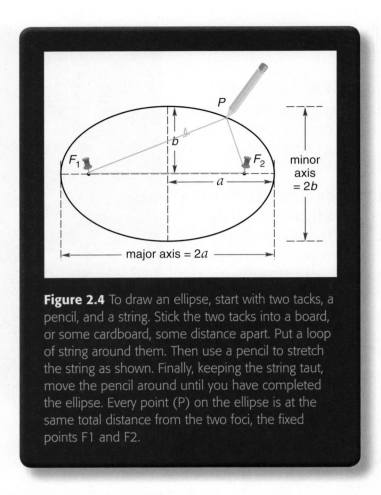

Figure 2.4 To draw an ellipse, start with two tacks, a pencil, and a string. Stick the two tacks into a board, or some cardboard, some distance apart. Put a loop of string around them. Then use a pencil to stretch the string as shown. Finally, keeping the string taut, move the pencil around until you have completed the ellipse. Every point (P) on the ellipse is at the same total distance from the two foci, the fixed points F1 and F2.

To get a little more precise: An ellipse is *a geometrical shape of which every point (P) is the same total distance from two fixed points, or foci.* ("Foci" is the plural of focus.) A circle is a set of points in a given plane at a certain distance from a given center. An ellipse is thus a somewhat more complex shape, defined with reference to two points (the foci), instead of one (Figure 2.4). (When you move the two foci closer to each other, the shape becomes more circular. When the foci lie on top of each other, the shape is a circle.) The farther apart the two foci of an ellipse are, the more "eccentric" it is said to be, and the more elongated it is. The planets' orbits are not actually all that eccentric. But they aren't perfect circles, either, and realizing this was Kepler's first major achievement.

The Second Law: The Planets' Changing Speeds

Kepler's second law tells us about a planet's speed as it moves around the Sun. He found that a planet speeds up when it is closer to the Sun. It slows down when it is farther away. He formalized this observation in his second law:

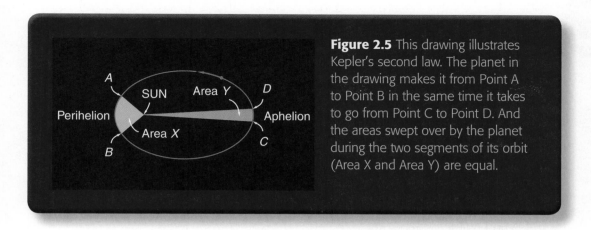

Figure 2.5 This drawing illustrates Kepler's second law. The planet in the drawing makes it from Point A to Point B in the same time it takes to go from Point C to Point D. And the areas swept over by the planet during the two segments of its orbit (Area X and Area Y) are equal.

"A planet moves along its elliptical path with a speed that changes so that a line from the planet to the Sun sweeps over equal areas in equal intervals of time (Figure 2.5)."

How Kepler's Third Law Implied the Force of Gravity

Kepler went beyond Ptolemy and Copernicus in trying to understand the planets' motions. He wanted to know *why* they moved as they did. He hypothesized that a force held the planets near the Sun—the force today known as gravity. He also thought there had to be another force that swept the planets around the Sun. Both of these forces, he believed, grew weaker over distance.

The sweeping force doesn't really exist. But the idea that there might be such a thing helped Kepler find a rule, his third law, that later led Sir Isaac Newton to discover the forces that do exist.

Kepler's third law is sometimes known as the Harmonic Law. It expresses a relationship between the time a planet takes to orbit the Sun and its distance from the Sun. The mathematical formula for this law is very concise: $P^2 = a^3$

This law means that if you know how long a planet takes to orbit the Sun, you can tell its average distance from the Sun, and vice versa. Kepler didn't know the actual distances from the planets to the Sun. But he was able to use Copernicus's estimate that Mars is about 1.5 times further from the Sun than the Earth is. That allowed him to do the necessary calculations.

So Copernicus, Tycho, and Kepler set the stage for the modern breakthroughs in astronomy and physics. You might call them the bridges from the ancient view of the universe to the modern. You'll learn more about that view in the next lesson.

Kepler's Laws in a Nutshell

The following is a greatly simplified version of this great astronomer's three laws of planetary motion.

1. Each planet's path around the Sun is an ellipse (not a circle) with the Sun at one focus.

2. A planet moves faster along its path when it is closer to the Sun and more slowly when it is farther away.

3. There is a relationship between the length of time a planet takes to orbit the Sun and its distance from the Sun. The farther away the planet is, the longer it will take to complete its orbit, the greater the distance it will travel to complete an orbit, and the slower its average speed will be.

CHECK POINTS

Lesson 2 Review

Using complete sentences, answer the following questions on a sheet of paper.

1. How did Copernicus's religious beliefs push him to investigate a Sun-centered model of the cosmos?

2. The Copernican model showed the plane of the Earth's equator tilted with respect to the plane of its orbit around the Sun. What was the significance of that?

3. Why did Ptolemy's predictions for the positions of celestial bodies lead Copernicus to seek a better model?

4. Which model of the universe did Tycho come up with?

5. What does Kepler's First Law of Planetary Motion say about the planets' orbits?

6. What does Kepler's second law say about a planet's speed as it moves around the Sun?

7. What did Kepler hypothesize about why planets orbit around the Sun?

 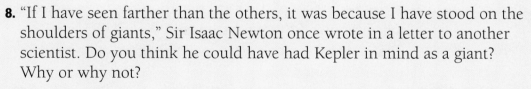 **APPLYING YOUR LEARNING**

8. "If I have seen farther than the others, it was because I have stood on the shoulders of giants," Sir Isaac Newton once wrote in a letter to another scientist. Do you think he could have had Kepler in mind as a giant? Why or why not?

Quick Write

_____

_____

Which do you think is more important for real scientific progress—curiosity and imagination, or up-to-date scientific tools?

Learn About

• Galileo and the telescope
• Newton's laws of motion and gravity
• Einstein and relativity

The German-born Dutchman Hans Lipperhey (d. 1619) often gets credit for inventing the telescope. Lipperhey was a spectacle-maker, or optician, in the Dutch town of Middelburg. Spectacles, or eyeglasses, had been developed in Italy in the late thirteenth century. Over the years, lens technology improved. By Lipperhey's day, Middelburg was a real center of the craft. (Since lenses are made of essentially the same stuff—sand—as semiconductor chips, you might call the Middelburg area a very early Silicon Valley.)

As technology improved, lenses could do more than just correct nearsightedness or farsightedness. At some point Lipperhey, and no doubt others, stumbled on the idea of combinations of lenses that would let people see objects a considerable distance away. Lipperhey applied for a patent for his telescope. But the authorities denied him, essentially on the grounds that it wasn't unique; too many others had the same idea. His patent application does go down in history, though, as the earliest record of an existing telescope.

People soon discovered they could use a telescope to get a close-up view of the local weathervane, or even read inscriptions off the tombstones in the village cemetery. The early telescopes were really just toys, albeit rather expensive ones.

Galileo and the Telescope

Galileo did not invent the telescope. But he was the first person to use a telescope to study the sky. His observations informed the debate between the heliocentric and geocentric—*Earth-centered*— theories of the universe. Some people found the whole topic so upsetting, though, that they refused to look through a telescope at the stars!

Vocabulary

- geocentric
- inertia
- accelerate
- principle of equivalence

The Significance of Galileo's Observations of Imperfections in the Sun and Moon

With his telescope, which Galileo built himself in 1609 after hearing about telescopes people were constructing in the Netherlands, he made many different kinds of observations:

- He saw mountains and valleys on the Moon
- He saw many more stars than can be observed with the naked eye
- He saw four moons of Jupiter
- He saw that Venus goes through phases like the Moon.

The first two of these didn't settle the question of Ptolemy vs. Copernicus. But they were unsettling to those who believed in the idea of the "perfection of the heavens." Galileo found the Moon full of features such as mountains and craters that seemed all too Earthlike. The (relatively) dark spots he saw on the Sun's surface looked like facial blemishes. And what about all those stars? The Bible records God creating the stars "to give light upon the earth." Did that mean stars too faint to be seen on Earth were failing in their divine mission? Or was something else altogether going on? No wonder some people didn't want to look through the telescope.

Star POINTS

The idea of "the perfection of the heavens" goes back to the ancient Greeks. But in the thirteenth century, an Italian philosopher and monk named St. Thomas Aquinas incorporated the Ptolemaic model into Christian theology.

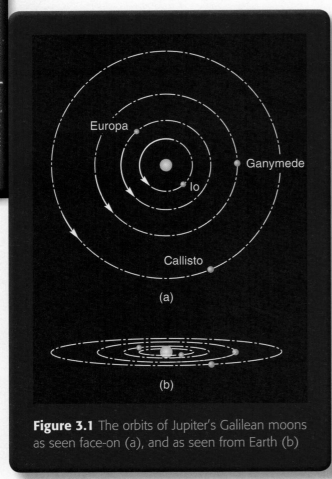

Figure 3.1 The orbits of Jupiter's Galilean moons as seen face-on (a), and as seen from Earth (b)

How Galileo's Observation of Jupiter's Four Moons Challenged Ptolemy's View

In January 1610 Galileo noticed three faint "stars" near Jupiter. He kept watching them, and eventually he saw there were four of them. And they weren't stars either, but satellites, or moons. He concluded that they revolved around Jupiter as the planets revolve around the Sun, according to the heliocentric model.

These four moons—Io, Europa, Ganymede, and Callisto, known as the Galilean moons—never appear north or south of Jupiter. That suggested to Galileo that their orbital plane aligned with that of Earth (Figure 3.1).

Ptolemy believed that everything revolved around the Earth. Galileo's observation of the four moons orbiting Jupiter suggested otherwise. Indeed, Jupiter and its moons were like a miniature solar system. That Jupiter could revolve around the Sun without losing its moons contradicted the theory that Earth must be stationary or it would lose its Moon.

Galileo's Observations of the Phases of Venus

The discovery of Jupiter's moons was important to Galileo's understanding of the Solar System. But what really convinced him that the Sun was at its center was his discovery of a complete set of phases for the planet Venus (Figure 3.2).

Seen with the naked eye, Venus is a bright dot of light. When viewed through a telescope, its phases are like those of Earth's Moon: full, gibbous, and crescent. Its full phase occurs when its entire sunlit hemisphere is visible. Its gibbous phase reveals more than half of the planet's sunlit hemisphere. In its crescent phase, less than half is visible.

In addition, Galileo noted that Venus seems to change size. A nearly full Venus is possible when Venus is on the far side of the Sun from Earth. A crescent Venus occurs when Venus on the near side. The arc of Venus's crescent is much larger than the full Venus, showing that it is significantly closer to Earth at that time. This is an important clue supporting the Sun-centered model, in which Venus orbits the Sun rather than Earth.

Galileo

Galileo Galilei was born in Pisa, Italy, in 1564—the year Shakespeare was born and Michelangelo died. The Galilei family had been wealthy and influential in the previous century, but Galileo grew up under more modest circumstances, the eldest of seven children of an accomplished musician.

He entered medical school at 17 but then took up mathematics and eventually astronomy. (In fact, in his day medical students had to study some astronomy for use in medical astrology.) He knew the Ptolemaic model. But then a book by Kepler convinced him of the rightness of the Copernican theory.

Galileo Galilei (1564–1642)
© *National Library of Medicine*

He kept quiet about his new convictions for many years, however. Then in 1613, after several years of observations with his telescope, he published his *Letters on Sunspots*. In this work he voiced his support for the heliocentric model. He stirred a great controversy—the more so since he published his work in his native Italian, rather than Latin, the language of most scientific writings.

In 1616 the Roman Catholic Church declared the Copernican doctrine "false and absurd" and barred Galileo from defending it. And it wasn't just the Catholic Church that attacked these new ideas. Protestant leaders Martin Luther and John Calvin did as well.

In 1633 an Inquisition court (a court established by the pope to denounce beliefs that conflicted with church doctrine) sentenced Galileo to house arrest for life. The court forced Galileo to abandon publicly "the false opinion" that the Sun is at the center of the universe. This meant he avoided the fate of Giordano Bruno, who had held the same view and had been burned at the stake 33 years before.

In 1992 Pope John Paul II formally acknowledged that condemning Galileo had been a mistake.

Similar things happen in societies and cultures around the world. It takes courage to defend one's ideas and be open to the possibility that you are wrong.

Ptolemy's theory would have Venus circling Earth. If that were the case, however, earthlings would never see more of Venus than a crescent (Figure 3.3). Venus would stay close to Earth, but its bright side would be the one turned to the Sun. Little of it would be visible from Earth (Figure 3.4).

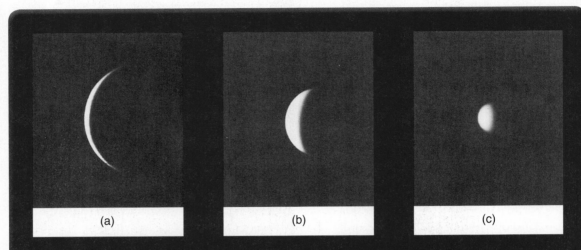

(a) (b) (c)

Figure 3.2 These views of Venus through a telescope show its full set of phases. Note how the planet's size seems to change through the cycle, as the planet is closer to, or farther from, Earth.
Courtesy of Lowell Observatory

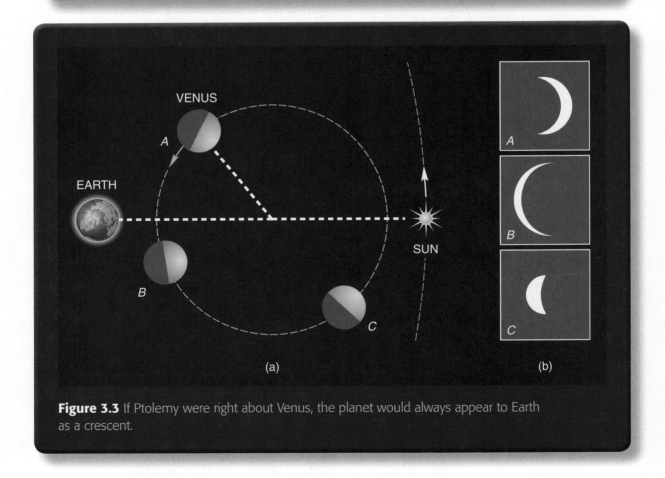

(a) (b)

Figure 3.3 If Ptolemy were right about Venus, the planet would always appear to Earth as a crescent.

CHAPTER 1   The History of Astronomy

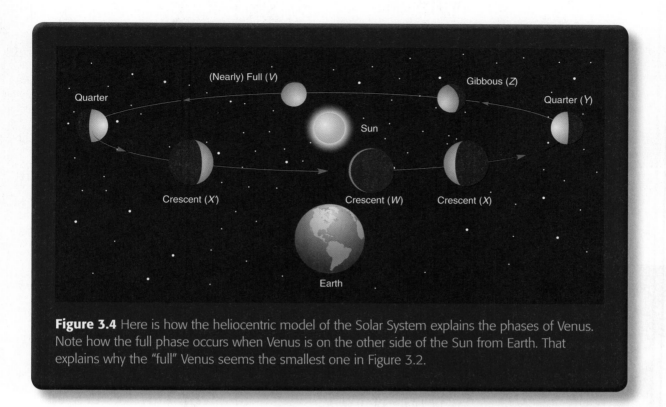

Figure 3.4 Here is how the heliocentric model of the Solar System explains the phases of Venus. Note how the full phase occurs when Venus is on the other side of the Sun from Earth. That explains why the "full" Venus seems the smallest one in Figure 3.2.

Newton's Laws of Motion and Gravity

Galileo's observations created a problem: There seemed to be no logic supporting Kepler's laws—they just worked. Building on the ideas of Galileo, Kepler, and others, Sir Issac Newton (1643–1727) was the first to create a unified model of how the universe operates. Newton summed up his conclusions about force and motion in his laws of motion and gravity.

It is difficult to exaggerate the importance of Newton's contribution to humanity's understanding of the universe. Without it, the modern science of astronomy and the exploration of space could not exist. And Newton set the stage for Einstein, who vastly deepened scientists' understanding of gravity, light, and matter.

Star POINTS

Newton understood the debt he owed to the scientists who went before him. "If I have seen further," he wrote to a rival, "it is by standing on the shoulders of giants."

Newton's Three Laws of Motion

If you talk about "inertia" in your daily life, you may use it to refer to that quality that makes it hard to get up off the couch on a Saturday afternoon. But in science, inertia has a more specific meaning: *the tendency of an object to resist a change in its motion.*

Newton stated that—in the absence of a force—an object at rest tends to stay at rest, and an object in motion tends to continue moving in a straight line at the same speed. A stationary object starts to move only when something makes it move. A moving object changes direction or stops only because something causes it to do so.

These concepts are summarized in Newton's first law of motion, sometimes stated this way: A body in motion tends to stay in motion, and a body at rest tends to stay at rest. This law is a basic observation. It cannot be directly proven or derived. But it is a powerful tool emphasizing cause and effect. It says that a force is needed to accelerate—that is, *to change the speed and/or the direction of the motion of—* an object.

Newton's second law builds on his first. It tells us how much force is necessary to produce a certain acceleration of an object: The acceleration of an object is proportional to the force exerted on it.

Look at the accompanying figures of bricks on wheels for some specific discussion on how force accelerates them (Figures 3.5 and 3.6).

Sir Isaac Newton worked in many fields. This picture shows him experimenting with a prism to investigate the nature of light.

©Photos.com

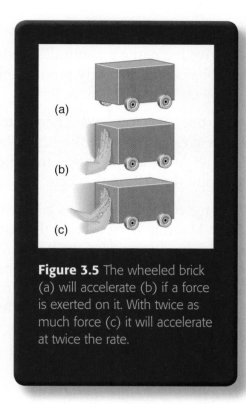

Figure 3.5 The wheeled brick (a) will accelerate (b) if a force is exerted on it. With twice as much force (c) it will accelerate at twice the rate.

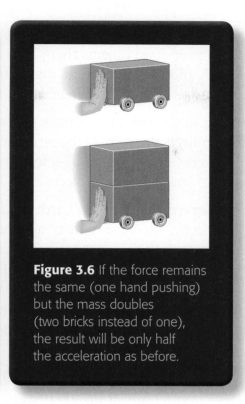

Figure 3.6 If the force remains the same (one hand pushing) but the mass doubles (two bricks instead of one), the result will be only half the acceleration as before.

As a formula, Newton's second law can be written in two ways:

Force $=$ mass $\times$ acceleration or $F = ma$

Or

Acceleration $=$ Force $\div$ mass or $a = F/m$

It can be put into words like this: A net external force applied to an object causes it to accelerate at a rate that is proportional to the force and inversely proportional to its mass.

The Law of Action and Reaction

"For every action there is an equal and opposite reaction." This is a simple statement of Newton's third law of motion.

It's a simple idea, but its implications are profound. Forces always occur in pairs. It is impossible to have a single, isolated force. When you sit down on a chair, the chair pushes back upward to you. You can call one force the action force and the other the reaction force. It doesn't matter which "comes first." Neither can exist without the other.

The Law of Gravity

You will see the importance to astronomy of Newton's three laws of motion when you combine them with another of his accomplishments: the law of universal gravitation. It's also known simply as "the law of gravity."

Here's a formal statement of it:

> Between every two objects there is an attractive force, the magnitude of which is directly proportional to the mass of each object and inversely proportional to the square of the distance between the objects' centers of mass.

Every object in the universe attracts every other one, Newton held. The greater the mass of an object, the greater the attractive force it exerts on other objects. Thus the Earth exerts a force on you—a force called *weight*—that keeps you from floating off into outer space. Nearer objects exert more pull than distant objects.

Newton observed that here on Earth the force of gravity made objects fall to the ground. He concluded that gravity also held the Moon in orbit around the Earth, and the planets in their orbits around the Sun. It's hard to underestimate the importance of the inverse square part of the law. It explains much—why planets travel at different speeds; why objects accelerate when falling; why Moon-Earth tides are greater than Sun-Earth tides, even though the Sun is thousands of times more massive than the Moon. It also greatly affects the course of spacecraft.

How Newton's Laws Confirmed Kepler's

Working from the basis of his own laws of motion and gravitation, Newton was able to prove Kepler's laws and expand on them. One of Newton's major insights was that gravity worked the same way in the heavens as on Earth. That was the final blow to Aristotle's ideas of "celestial perfection." That the laws of nature apply in a similar fashion throughout the universe is now a basic assumption of astrophysics.

Newton understood the Sun as the source of the force responsible for the motion of the planets. He used calculus to demonstrate Kepler's law about the orbits of the planets being elliptical.

Mass and Weight

You read a moment ago how *inertia* has a specific meaning in science. *Mass* is another such word—familiar from everyday conversation, but with a specific scientific definition. An object's mass is the measure of its inertia and is the same wherever the object is located in the universe. But mass is not *volume* (a brick made of clay and a brick made of plastic foam may have the same volume, but they won't have the same mass). And it's not *weight* either. Weight relates to gravity. An object's weight depends on where it is in the universe. The weight of a brick is the downward force on the brick from Earth's gravity.

Newton shows that Kepler's second law results from the fact that the gravitational force acting on an object orbiting the Sun (or other stationary body) always points toward the Sun. As a planet moves closer to the Sun, the planet speeds up. That's because the Sun pulls forward as well as sideways on the planet. And as the planet moves away from the Sun, the Sun's gravity pulls backward and sideways. The planet slows down. The sideways pull keeps the planet in orbit around the Sun.

Star POINTS

Newton, born the same year Galileo died, did much of his most important work between 1665 and 1666. The bubonic plague had struck London, and Newton, a young professor just starting his career, fled to his mother's house in the countryside. His resulting isolation led to two of the most productive years in the history of science.

Finally, Newton used his laws of motion and gravity to modify Kepler's third law. It can be applied to any two objects orbiting each other as a result of their mutual gravitational attraction.

Einstein and Relativity

Newton's work was grounded in the world of everyday observation. As scientists began to study the universe deeper and deeper in the late nineteenth and earlier twentieth centuries, they began to find situations in which Newton's laws didn't apply. This was especially true at high speeds (such as the speed of light) and in places where gravity is extremely strong. This led Einstein to develop his two theories of relativity.

The Principle of Equivalence and the General Theory of Relativity

Mass, as you read earlier, is the measure of an object's inertia. Newton proposed that mass also determines the strength of gravitational pull. Why should the same quantity be the measure of two seemingly different physical properties? Experiments have shown that these two properties—inertia and gravitational pull—are identical to at least one part in a trillion.

This seems like a coincidence, and scientists don't like coincidences. An attempt to explain this apparent coincidence is what led Albert Einstein to develop his general theory of relativity more than two centuries after Newton stated his law of gravity.

Einstein's theory begins by stating that gravity and acceleration are equivalent to each other. The principle of equivalence is *the statement that effects of acceleration are indistinguishable from gravitational effects.*

For example: Imagine a woman standing somewhere on Earth—in her backyard, say—and dropping a book. The book will land on the ground. Now imagine her on an accelerating spaceship far from Earth's gravity. If she drops her book again, it will fall to the floor. She won't be able to tell the difference between the "falling" caused by the ship's acceleration and the "falling" caused by gravity. (Note that this is an accelerating spaceship, not one in which astronauts are experiencing weightlessness.) This theory changes the way we look at gravity.

Newton's idea of gravity was that interacting objects act at a distance. One object influences the other across the empty space between them. Einstein proposed instead thinking of space as being curved by a mass so that objects move because of the curvature.

The idea of curvature of space makes you think in terms of more than three dimensions. And the way to start to get your head around this idea is to think about how you would make the transition from two dimensions to three.

Star POINTS

This illustration of multiple dimensions in space is borrowed from *Flatland,* a book written by Edwin A. Abbott in 1884.

In this world you need three dimensions to describe the position of something. You may think of them as north-south, east-west, and up-down. To get from, say, the library where you are studying this book to your bed at home, you may go half a mile north, a quarter mile east, and then 20 feet up from the sidewalk to the second floor. So far, so good.

Now imagine some tiny two-dimensional creatures on a large, expanding balloon. They have no height. Their world is two-dimensional. They understand north-south and east-west. But up-down is a blank. Yes, the balloon is really more than two-dimensional. But if it's large enough, the tiny creatures would be hard pressed to see the balloon's curvature. If one of them glimpsed the idea of a third dimension, it would be difficult to explain to the others. A third dimension is not part of their world. It would take a great stretch of the imagination.

That stretch is the one you need to make to understand the curvature of space. One way to do this is to picture the space near the Sun as being warped as the surface of a waterbed is warped by a bowling ball placed at its center (Figure 3.7).

Einstein's great insight was that matter (the bowling ball, in this case) tells space (the waterbed's surface) how to curve. The space curvature (or warping) tells other matter (say, a small ball moving across the waterbed's surface) how to move. All this adds up to the general theory of relativity.

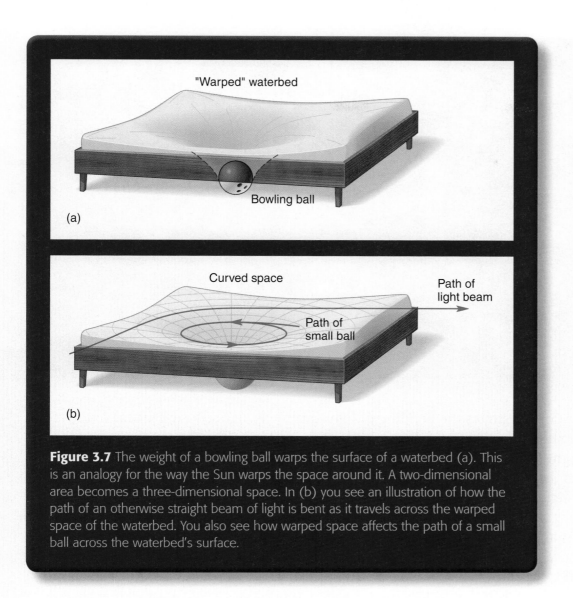

(a) "Warped" waterbed

Bowling ball

(b) Curved space

Path of light beam

Path of small ball

Figure 3.7 The weight of a bowling ball warps the surface of a waterbed (a). This is an analogy for the way the Sun warps the space around it. A two-dimensional area becomes a three-dimensional space. In (b) you see an illustration of how the path of an otherwise straight beam of light is bent as it travels across the warped space of the waterbed. You also see how warped space affects the path of a small ball across the waterbed's surface.

How Testing the Bending of Light Supported Einstein's View of Gravity

The general theory of relativity holds that if an object has enough mass, it can deflect, or bend, a beam of light away from its original straight path. The illustration with the bowling ball on the waterbed suggests as much. Einstein predicted this gravitational bending of light in 1915. Observations astronomers made during a solar eclipse in 1919 confirmed his prediction.

Newton's theory included no prediction of a gravitational effect on light, since light has no mass, and in everyday life, you cannot see light bend. It takes a very massive object to bend light with gravity in an amount you can detect. Newton's theory had never been checked in such a case.

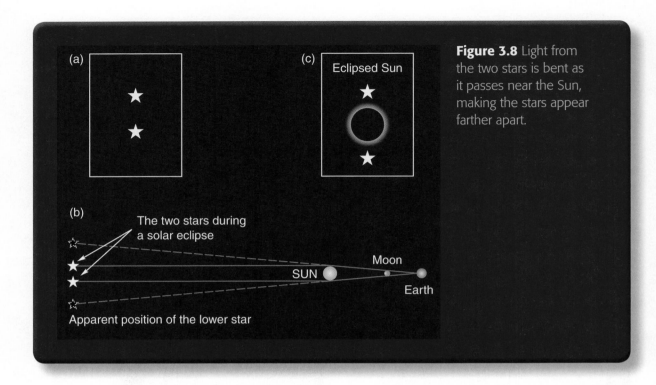

Figure 3.8 Light from the two stars is bent as it passes near the Sun, making the stars appear farther apart.

Figure 3.8 illustrates the bending of light during a solar eclipse. Part (a) of the figure shows the stars as they normally appear. During the eclipse, shown in part (b), light from the two stars bends as it passes the Sun en route to Earth. This makes the stars appear farther apart, as shown in part (c). The light is bent only very slightly—1.75 arcseconds, a tiny fraction of a single degree of a circle. But the bending was observable during the 1919 eclipse. Observations matched Einstein's theory, and they provided the first experimental support for the general theory of relativity. They showed that Einstein's theory, which predicted the effects of gravity on light, was more correct than Newton's, which did not.

How the Constant Speed of Light Proves the Special Theory of Relativity

About 10 years before he developed the general theory of relativity, Einstein proposed the special theory of relativity. It's based on two postulates, or premises:

- *All laws of physics are the same for all nonaccelerating observers, no matter what their speed.* This is another way of saying that the rules of nature are the same everywhere in the universe. There is no celestial perfection in contrast with blemished Earth, as the ancient Greeks supposed. Newton understood this, and Einstein built on it. The first postulate shows that no place in the universe is stationary, or "at rest." Everything is in motion relative to everything else. You seem to be standing or sitting still. But you aren't—the Earth on which you stand or sit is revolving and orbiting the Sun. The Sun is moving towards the star Vega. The Milky Way Galaxy you live in is rotating and moving through space. Regardless of where you are or how fast you are moving (at a constant speed), however, the laws of physics do not change.

Albert Einstein

Albert Einstein (1879–1955) can be an inspiration to anyone who doesn't fit in well in school. Born in southern Germany, he found school intimidating and boring. He dropped out before finishing high school. He studied on his own, though. His studies of geometry and science convinced him that the Bible is not literally true. That came as a shock and left him with a deep distrust of authority of any kind.

On his second try, Einstein got into the Swiss Federal Institute of Technology in Zurich. He earned a Ph. D. in 1900, but then it took him a couple of years to get a steady job—as a patent examiner in the Swiss Patent Office.

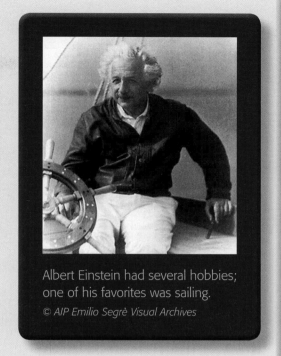

Albert Einstein had several hobbies; one of his favorites was sailing.
© AIP Emilio Segrè Visual Archives

Like Newton, Einstein did his most revolutionary work during a short period in his early 20s. In 1905 Einstein published four significant papers in an important German physics journal. He may be best known for his special theory of relativity. But he won the Nobel Prize for his work on the photoelectric effect of light. Scientists recognized his genius, but not until 1909 did he gain a full-time academic position at the University of Zurich.

Einstein's theories were slow to gain acceptance because they were so hard to confirm with experiments. But in 1919 the Royal Society of London announced that it had verified his prediction of the bending of starlight during an eclipse. After that, he was hailed as a genius. And his scientific fame gave him influence in world affairs. He happened to be visiting California when Hitler came to power in Germany. Einstein renounced his German citizenship and never returned to his home country.

- *The speed of light is the same for all nonaccelerating observers, no matter what their motion relative to the source of the light.* This prediction is so far from anything that can be observed in everyday life that it's hard to make sense of. (Remember the tiny creatures on the balloon, with no sense of "up" or "down.") You know from experience that when you catch a baseball, you see it coming at you faster if you move toward it than if you stand still. But Einstein's postulate says that light travels differently. It always travels at the same speed regardless of the observer's speed.

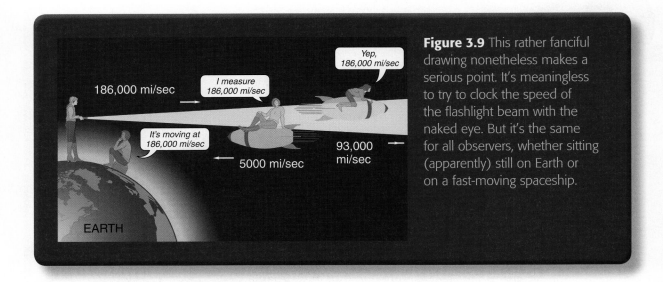

Figure 3.9 This rather fanciful drawing nonetheless makes a serious point. It's meaningless to try to clock the speed of the flashlight beam with the naked eye. But it's the same for all observers, whether sitting (apparently) still on Earth or on a fast-moving spaceship.

Scientists now consider the constancy of the speed of light a law of nature. Technological advances since Einstein's time have allowed scientists to confirm the second postulate repeatedly. He showed that Newton's laws of motion became more and more inaccurate as speed increases.

Einstein made some predictions about objects moving at very high speeds; for example, the observed passage of time becomes slower. These predictions are hard to confirm with experiments. But every time the theory has been put to the test, it has passed.

The special theory of relativity concludes that mass (m) can be transformed into energy (E) and vice versa. This conversion is expressed mathematically by the formula $E = mc^2$.

The first explosion of a nuclear bomb confirmed this equation in 1945. The energy of these bombs comes from the conversion of mass to energy. The electricity that comes from nuclear power plants is a more peaceful expression of this conversion of mass to energy.

Most people tend to view the world in terms of Newton's theories. His laws still have great practical value. At the speeds scientists can currently achieve, space navigation is still largely based on Newton's work.

For scientists studying the universe, however, Einstein's theories have replaced Newton's. Scientists' much greater understanding of the universe would not have been possible without Einstein. Yet someday another thinker may come along and—using data not available in Einstein's day—develop new theories that will replace Einstein's and more accurately describe the universe.

CHECK POINTS

Lesson 3 Review

Using complete sentences, answer the following questions on a sheet of paper.

1. What four different observations did Galileo make with his telescope?

2. How did Galileo's observation of Jupiter's moons challenge Ptolemy?

3. If Ptolemy were right about Venus, how would it appear to earthlings?

4. How does Newton's second law build on his first?

5. What is Newton's third law of motion?

6. What is a formal statement of the law of gravity?

7. How did Newton understand the Sun with regard to the planets?

8. What is the principle of equivalence?

9. What did the solar eclipse of 1919 have to do with the general theory of relativity?

10. How was the equation $E = mc^2$ confirmed in 1945?

APPLYING YOUR LEARNING

11. Suppose for a minute that someone had developed the general theory of relativity in 1800. Why might scientists of the time have rejected it?

The Moon is Earth's closest neighbor. Although very different from one another, each affects the other in important ways. The Moon's gravity helps regulate the tides on Earth. And Earth helps create a tide effect on the Moon, too.

© Photos.com

The Earth and Moon

Chapter Outline

> " There are more things in heaven
> and Earth, Horatio,
> Than are dreamt of in your philosophy. "
>
> *Shakespeare,* Hamlet

Quick Write

_____

_____

What clues does Van Allen's work give you about how scientific thinking and practice have evolved since Ptolemy first looked at the heavens?

Learn About

• Earth's interior
• Earth's atmosphere
• Earth's magnetic field

As a young man, James Van Allen became fascinated with the subject of cosmic rays. These are highly charged particles traveling through space at nearly the speed of light. By 1951 he had become head of the University of Iowa physics department. He was eager to develop ways to place scientific instruments high in the upper atmosphere to collect data. To this end, he devised something called the "rockoon." It carried instruments high above the Earth in a balloon. Then at 10 or 15 miles up, under lower air pressure, its rockets would ignite to carry the package even higher.

Authorities wouldn't let Van Allen launch his rockoons over Iowa, for fear the spent rockets would damage someone's house. So he persuaded the US Coast Guard to let him fire them from icebreakers near the north and south geomagnetic poles. In 1953 rockoons fired off Newfoundland detected the first hint that Earth is girded with immense belts of radiation.

In 1958 *Explorer 1* was launched into Earth orbit. Van Allen was on the *Explorer 1* team, running a cosmic-ray experiment. When he found fewer rays than expected, he concluded that the satellite's readings were thrown off by radiation from charged particles caught in the Earth's magnetic field. Two months later *Explorer 3* confirmed the existence of these high-radiation belts, which were later named after Van Allen.

A quarter-century later, with the help of the spacecraft *Pioneer 10* and *Pioneer 11*, Van Allen was surveying similar belts around Jupiter and Saturn. Van Allen is now recognized as one of the leading space scientists of the twentieth century.

Earth's Interior

As you continue through this book, you will be reading about other worlds beyond the Earth. But before you leave Earth, so to speak, this lesson will give you some essentials for understanding it as a physical object, made of certain materials and subject to certain forces. After all, the Earth is a planet, too.

Earth's Density

One of the first things to understand about the Earth as a physical object is how dense it is. Density is *the ratio of an object's mass to its volume.* Determining the Earth's volume is relatively easy. People have known the Earth's diameter for centuries, and so could figure its volume. Today's space program has greatly increased the accuracy of these measurements.

Calculating the mass of Earth had to wait, however, until scientists could apply Kepler's third law (as revised by Newton) to the period and radius of the Moon's orbit. Remember that this law held that the period of revolution of an orbiting object depends only on the size of the orbit and the total mass of the objects.

After making the necessary calculations scientists concluded that the average density of Earth is 5.52 grams per cubic centimeter. This compares with the density of water, 1 g/cm^3; aluminum, 2.7 g/cm^3; and iron, 7.8 g/cm^3

Knowing the density of a celestial object allows you to make a reasonable guess about its composition. Suppose you were to discover a planet with an average density 5.5 times that of water. And suppose its surface was rocky and largely covered by water. You could reasonably infer that its interior was likely to include metals such as iron and nickel. You couldn't be sure what amounts of which metal would be present. But you would have a rough idea of its composition. Such a planet would be like a snowball with a rocky core. It would also be quite like the Earth itself.

Earth's Three Layers

The Earth is made up of three layers—the crust, mantle, and core (Figure 1.1). The crust is *the outer layer of Earth.* It extends down to a depth of less than 60 miles (100 km). The crust is made up of rock and soil, and has a density of 2.5 to 3 g/cm^3

Vocabulary

- density
- crust
- mantle
- core of the Earth
- chemical differentiation
- continental drift
- rift zone
- plate tectonics
- solstice
- altitude
- equinox
- magnetic field
- aurora

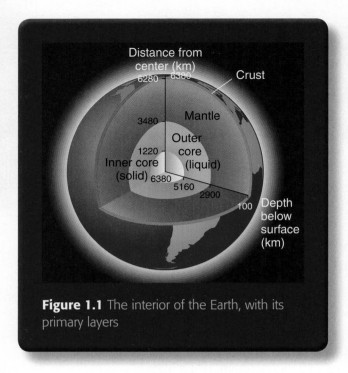

Figure 1.1 The interior of the Earth, with its primary layers

The next layer down is the mantle, *a thick, solid layer between Earth's crust and its core.* The mantle extends almost halfway to the center of the Earth, about 1,800 miles (2,900 km). This layer is solid but is able to flow under steady pressure, albeit very slowly. Under extreme pressure, it can crack and move suddenly. The mantle is denser than the crust—3 to 9 g/cm^3. The crust, therefore, floats on top of it.

The core of the Earth is *the central part of the Earth, made up of a solid inner core surrounded by a liquid outer core.* The core is even denser than the mantle. Scientists therefore think it's made up primarily of iron and nickel, the most common heavy elements.

This layering of the Earth, with the greatest density at the center, suggests to scientists a phenomenon for which they have a special term: chemical differentiation. It refers to *the sinking of denser material toward the center of planets or other objects.*

Scientists see this sinking as something that could have come about only when the Earth was in a molten state. That's when the heavier elements would have sunk through the less dense material.

Scientists understand as much as they do about the structure of the Earth because of what they have learned from studying the waves that result from earthquakes. These waves travel throughout the Earth. By analyzing their travel times, geologists can draw some conclusions about the stuff the Earth is made of.

The Motion of the Earth's Crust

You may have noticed while looking at a map of the Earth that the eastern edge of the Americas—North and South—seems to fit roughly into the western edge of Europe and Africa. It's as if they were all one land mass once upon a time but have since drifted apart.

In fact, that's pretty much what scientists believe happened. Continental drift is the scientific term for *the gradual motion of the continents relative to one another.* A German meteorologist named Alfred Wegener gets the credit for first developing the idea of continental drift in the early twentieth century. But nobody could figure out just how or why this drift occurred. And so the idea was set aside.

Star POINTS

Alfred Wegener, who developed the idea of continental drift, said, "It is just as if we were to refit the torn pieces of a newspaper by matching their edges and then check whether the lines of print run smoothly across."

Later, though, geologists discovered the rift zone, *a line near the center of the Atlantic Ocean from which lava flows upward.* These lava flows have created an extensive range of underwater mountains. All this new material oozing in, so to speak, pushed the Americas and the Euro-African landmass apart. And these forces are still at work. Scientists have studied this region, using various techniques to date the mountains and other features. They estimate Europe and North America are now moving apart about an inch (two to four centimeters) every year (Figure 1.2).

From this research has grown the present theory of plate tectonics—*the motion of sections (plates) of the Earth's crust across the underlying mantle.* Earth has about a dozen of these large plates and several smaller plates (Figure 1.3). Each extends about 30 to 60 miles (50 to 100 km) deep.

Star POINTS

Tectonics is a branch of geology concerned with the structure of the crust of a planet (as Earth) or moon and especially with the formation of folds and faults in it.

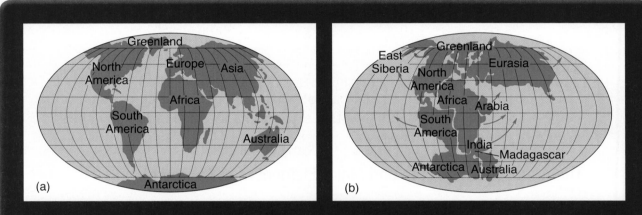

Figure 1.2 These two diagrams show the Earth as it appears today (a) and as scientists believe it was 200 million years ago (b).

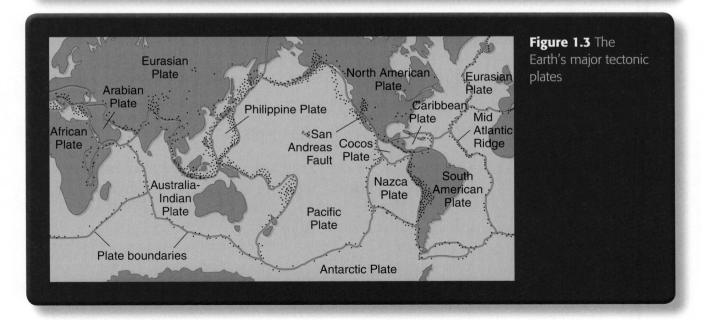

Figure 1.3 The Earth's major tectonic plates

What Makes Earth Unique

Earth is unique in our Solar System because it harbors complex intelligent life that has survived for 3.9 billion of the planet's 4.5 billion-year existence. Several factors combine to make Earth so hospitable to life:

- Liquid water near the surface
- Enough radiation from space, but not too much
- Jupiter as a neighbor protecting Earth from comets and asteroids
- A large Moon that stabilizes Earth's tilt (which helps limit seasonal extremes) and creates tides
- A stable orbit around the Sun, to which Earth is close enough but not too close
- A gaseous atmosphere and liquid water ocean
- The evolution of photosynthesis within microbial life forms, which enriched the atmosphere with oxygen.

Earth's Atmosphere

Earth's atmosphere is an essential part of what makes the planet a comfortable home for so many forms of life. In this section you'll read a little more about the atmosphere and its layers.

The Atmosphere's Depth and Major Elements

Just as the Earth's crust is a relatively thin layer of rock and soil, Earth's atmosphere, too, is a very thin layer. The higher you go from the surface of the Earth, the thinner the atmosphere becomes. It's hard to say just what its outer limit is. But by about 62 to 93 miles (100 to 150 km) up and away from Earth, the atmosphere is almost nonexistent.

Earth's atmosphere is made up mostly of nitrogen and oxygen—78 and 21 percent, respectively, by volume. That last 1 percent of the total, though, is made up of some very important constituents—water vapor, carbon dioxide, and ozone. You'll read later on about what they mean for life on earth.

The Five Layers of the Atmosphere

Star POINTS

Commercial aircraft fly at the top of the troposphere to avoid turbulence.

The *troposphere* is the lowest level of the Earth's atmosphere. It extends about seven miles, or 11 kilometers, above the surface of the Earth, and it contains about 75 percent of the atmosphere's mass. All of Earth's weather occurs in the troposphere. This layer of the atmosphere gets most of its heat from infrared radiation from the ground. So the higher up you go within the troposphere, the cooler the temperature is.

The next layer up is the *stratosphere*. In this layer of the atmosphere, the temperature rises with the altitude. This is because the stratosphere absorbs ultraviolet radiation from the Sun. At the top of the stratosphere is the *ozone layer*. Above that, from about 30 to 50 miles (50 to 85 km) up is the *mesosphere*. Once again, there's a change in the relationship of temperature and altitude. In the mesosphere, the higher up you are, the colder it gets.

The next layer is the *thermosphere*, the layer in which the space shuttle flies. Once again, the temperature rises with altitude. The outermost layer is the *exosphere*. It extends from the top of the thermosphere below to 6,200 miles (10,000 km) above the Earth. In this layer, atoms and molecules escape into space and satellites orbit the Earth (Figure 1.4).

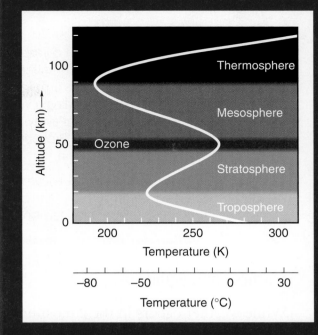

Figure 1.4 The temperature of the atmosphere varies with altitude because each layer absorbs solar energy differently. In this graphic, note the switchback path that the white line indicating temperature takes through the different layers. This drawing does not show the exosphere, which is above the troposphere.

The Importance of the Ozone Layer

The ozone layer is centered at about 30 miles (50 km) up. Ozone is a three-atom form of oxygen—written as O_3. On Earth's surface, it is a colorless, corrosive gas that is a dangerous pollutant. Thirty miles up, however, it provides important protection to life on Earth by filtering ultraviolet (UV) radiation from the Sun. UV radiation breaks apart molecules that make up living tissue. You've experienced this if you've ever been sunburned. Too much exposure to UV can cause skin cancer.

The ozone layer is under threat from modern civilization. Over much of the twentieth century, people used significant amounts of chemicals known as chlorofluorocarbons (CFCs) for purposes such as refrigeration. But these chemicals would escape into the atmosphere and damage the ozone layer. More recently, people have seen the danger and largely phased out CFCs. These chemicals remain in the atmosphere for over a century, though, and so amounts already released continue to damage the ozone layer. For example, several years ago scientists discovered that a hole in the ozone forms each year over Antarctica. Scientists believe its growth has stabilized, but the ozone is not likely to recover until about 2020 or later.

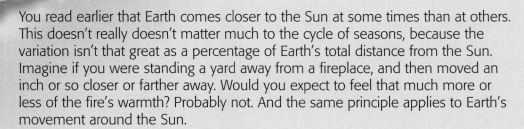

You read earlier that Earth comes closer to the Sun at some times than at others. This doesn't really doesn't matter much to the cycle of seasons, because the variation isn't that great as a percentage of Earth's total distance from the Sun. Imagine if you were standing a yard away from a fireplace, and then moved an inch or so closer or farther away. Would you expect to feel that much more or less of the fire's warmth? Probably not. And the same principle applies to Earth's movement around the Sun.

Many people mistakenly believe the seasons are caused by Earth's distance from the Sun—closer in summer and farther in winter. Actually, Earth is closer to the Sun during the Northern Hemisphere winter and farther during the summer.

Solstices and Equinoxes

While weather occurs in the atmosphere, it's Earth's tilt as it orbits the Sun that determines the cycle of seasons.

You can easily observe how the Sun behaves differently in summer than in winter. If you live in the Northern Hemisphere, you can see the Sun rising and setting farther north in the summer than in the winter (Figure 1.5). This means that the Sun has a longer journey across the sky in summer. This, in turn, means that the Sun is in the sky longer on a summer day than in winter. That hemisphere collects more solar energy during these longer days, and because nights are shorter, there isn't as much time to cool down. So overnight temperatures don't get as low during the summer.

Another difference between summer and winter is the Sun's angle toward Earth. A hemisphere gets more energy when the Sun is directly overhead. The light itself is more concentrated, and it doesn't have to cut through as much of Earth's atmosphere to reach the ground. By contrast, sunlight coming from a point closer to the horizon, as in winter, is filtered by more of the atmosphere.

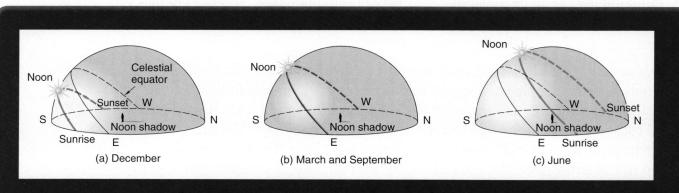

Figure 1.5 These three drawings show the Sun's path across the sky of the Northern Hemisphere at different points of the year.

CHAPTER 2   The Earth and Moon

The Sun reaches its northernmost point in the Northern Hemisphere sky every year about 21 June. Then it begins to change its (apparent) direction until about 21 December, when it reaches its southernmost point and begins to head north once again. The term for *either of the twice-yearly times when the Sun is at its greatest distance from the celestial equator* is solstice. This term comes from Latin words meaning "the Sun standing still" and refers to the Sun's apparent "stopping" as it changes direction.

In the Northern Hemisphere, 21 June is the summer solstice and 21 December is the winter solstice. In the Southern Hemisphere, these two dates are reversed. In either hemisphere, the summer solstice is the point where the Sun reaches its highest altitude—*height measured as an angle above the horizon.* The winter solstice is the point where the Sun reaches its lowest altitude (Figure 1.6).

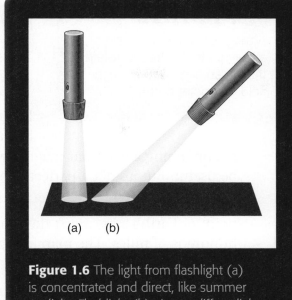

Figure 1.6 The light from flashlight (a) is concentrated and direct, like summer sunlight. Flashlight (b) gives a diffuse light, like winter sunlight.

There are a couple of other notable points along the Sun's path. The vernal and autumnal equinoxes are *the points when the Sun crosses the celestial equator.* "Equinox" comes from Latin and means "equal night." The solstices are associated with the shortest and longest days of the year. At the equinoxes, night and day are equal. That is, hours of daylight match hours of darkness. "Vernal" means "spring." The vernal equinox comes to the Northern Hemisphere around 21 March. The autumnal equinox occurs around 22 September. These dates are reversed in the Southern Hemisphere.

Earth's Magnetic Field

Another characteristic that defines Earth is its magnetic field. A magnetic field is *what exists in a region of space where magnetic forces can be detected.* You've probably done this simple experiment yourself at some point: You cover a bar magnet with a sheet of sturdy paper, such as construction paper, and sprinkle iron filings on the paper. The particles immediately respond to the magnetic force on the other side of the paper and assume a distinctive pattern (Figure 1.7).

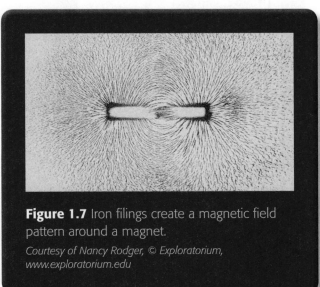

Figure 1.7 Iron filings create a magnetic field pattern around a magnet.

Courtesy of Nancy Rodger, © Exploratorium, www.exploratorium.edu

The Difference Between Magnetic North and True North

The swinging of a compass needle and the distinctive pattern of iron filings in the experiment mentioned above are two good examples of magnetic fields. In fact, the Earth's magnetic field has a shape rather like that of the iron filings. And so the idea of Earth as containing a giant bar magnet within, with two poles, north and south, may be useful. But it's not a perfect concept.

For one thing, magnetic north and the geographic North Pole (the northernmost point of Earth's axis) aren't the same. For another, magnetic north can wander hundreds of miles. The map in Figure 1.8 shows just how far it has moved over the past couple of centuries.

Another problem with the "bar magnet" comparison is that Earth's core is too hot to remain magnetic. It's not a solid magnetic region.

Changes in the Magnetic Field

Magnetic fields don't last forever, either. They decay over time. Unless something exists that will "recharge" it, Earth's current field will disappear in about 15,000 years.

Magnetic fields can also flip completely. Scientists estimate that over the past 170 million years, the Earth has undergone about 300 reversals of magnetic field. The most recent of these was 780,000 years ago. A field reversal takes place over thousands of years because of complex changes in the Earth's core.

The Van Allen Belts

The Van Allen Belts were the first scientific discovery of the space age (Figure 1.9). Early spacecraft discovered electrically charged particles swarming in a doughnut-shaped region high above Earth's surface.

These charged particles, protons and electrons—broken-off bits of atoms, you might say—come mainly from the Sun. They are captured by Earth's magnetic field. Electrons are trapped in the outer radiation belts. Protons are trapped in the inner belts. And individual particles travel through these belts on a spiral path.

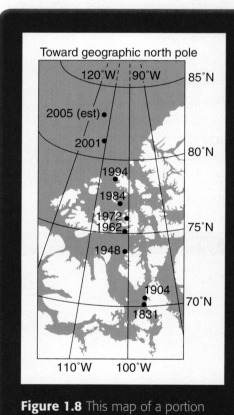

Figure 1.8 This map of a portion of the Canadian Arctic Archipelago shows how magnetic north has changed over the years.

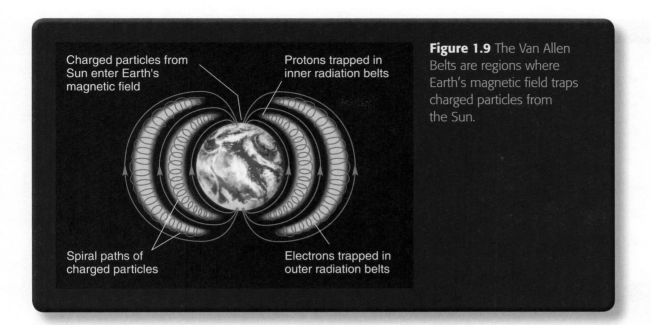

Charged particles from Sun enter Earth's magnetic field

Protons trapped in inner radiation belts

Spiral paths of charged particles

Electrons trapped in outer radiation belts

Figure 1.9 The Van Allen Belts are regions where Earth's magnetic field traps charged particles from the Sun.

James Van Allen discovered the Van Allen Belts relatively recently. But people have known for thousands of years about a spectacular phenomenon they produce on Earth: auroras. An aurora is *light radiated in the upper atmosphere because of impacts from charged particles* (Figure 1.10). These beautiful displays may be most familiar in the Arctic regions of the Northern Hemisphere as aurora borealis. (*Borealis* comes from the Latin word for northern.) They're also called the "Northern Lights." They result when charged particles penetrate Earth's magnetic field and then collide with atoms and molecules in our atmosphere. They produce "curtains" of visible light in the night skies.

The South Atlantic Anomaly

A significant feature in Earth's magnetic field is the South Atlantic Anomaly (SAA). This is a region of very dense radiation above the Atlantic Ocean off the coast of Brazil (Figure 1.11). It is the result of a sort of dip or depression in the magnetic field. It lets cosmic rays and charged particles reach down lower into the atmosphere, closer to the surface of the Earth.

Figure 1.10 An aurora results from the collisions of atoms and charged particles in the upper atmosphere.
© Roman Krochuk/ShutterStock, Inc.

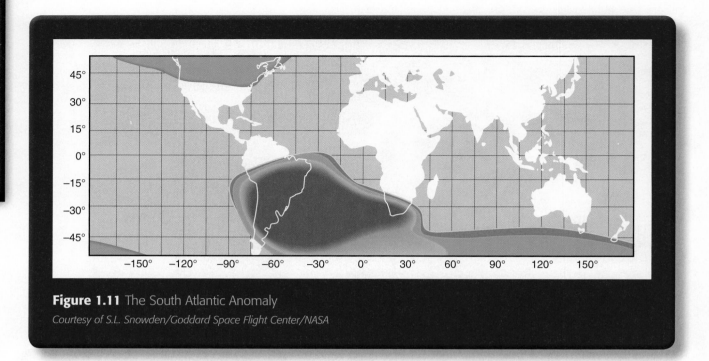

Figure 1.11 The South Atlantic Anomaly
Courtesy of S.L. Snowden/Goddard Space Flight Center/NASA

The SAA has serious effects. Its intense radiation hampers communication with satellites and aircraft. There are theories to explain the SAA—one is that the center of Earth's magnetic field and the center of Earth are not in the same place. The offset causes a weakening of the magnetic field at the SAA. Meanwhile, astronauts get higher doses of radiation when they pass through the SAA. And satellites have to be shut down, or put into "safe mode," when they pass through it. Otherwise the dense radiation could damage them.

This lesson has reviewed the most important characteristics of your home planet—characteristics that also help scientists understand other planets. Another important characteristic of any planet is how many moons it has. As everyone knows, Earth has only one. Mercury and Venus have none. The planets outside Earth's orbit have anywhere from two to dozens. The next lesson will look at Earth's Moon and the huge impact it has on the planet it orbits.

CHECK POINTS

Lesson 1 Review

Using complete sentences, answer the following questions on a sheet of paper.

1. Why is it useful to know the density of a celestial object?

2. What can Earth's mantle do under steady pressure? Under extreme pressure?

3. Why was the idea of continental drift set aside soon after it was developed?

4. What elements make up most of Earth's atmosphere? In what proportions?

5. Why does the temperature fall as you rise within the troposphere?

6. What is the importance of the ozone layer?

7. What does the Sun do every year about 21 June?

8. What is a magnetic field?

9. How long does a magnetic field reversal take, and what causes it?

10. What does the South Atlantic Anomaly do?

 APPLYING YOUR LEARNING

11. Suppose Earth's magnetic field had decayed 5,000 years ago. How might that have affected human history?

LESSON 2 | The Moon: Earth's Fellow Traveler

Quick Write

If you had been one of the Apollo 11 astronauts, what about the Moon would you have been most curious to see or experience for yourself?

Learn About

- the Moon's size and distance from the Earth
- the relationships between the Moon and the Earth
- the Moon's origin and surface

"**H**ouston, Tranquility Base here. The *Eagle* has landed." Those were American astronaut Neil Armstrong's words at 4:18 EDT on 20 July 1969. The first crewed flight to the Moon had just touched down. Within a few hours Armstrong would become the first human to set foot on the Moon.

Only a little more than eight years before, President John F. Kennedy had told a joint session of Congress, "I believe that this nation should commit itself to achieving the goal, before this decade is out, of landing a man on the Moon and returning him safely to the Earth. No single space project in this period will be more impressive to mankind, or more important for the long-range exploration of space...."

Now Armstrong and his colleague, Edwin E. (Buzz) Aldrin Jr., were there in their lunar module, the *Eagle*, on their way to fulfilling the goal that Kennedy had set, and the nation had adopted. They were about to step down onto the Moon's barren, dusty landscape. Space scientists dreamed of exploring even more distant worlds. But the Moon was the first stop.

The Moon's Size and Distance From the Earth

Since the 1960s the space program has taught scientists a great deal about the Earth and Moon. But early astronomers were able to make some important calculations about these two bodies simply on the basis of naked-eye observations.

The Distance From the Earth to the Moon

Remember what you read in Chapter 1, Lesson 1, about parallax—the apparent shift of an object when seen from different positions (Figure 2.1). An astronomer looking up at the Moon over Alexandria, Egypt, for instance, would see it in a different position—against a slightly different background of stars—from what an astronomer in Beijing would see.

The astronomer Ptolemy used his understanding of the parallax effect to determine the distance from the Earth to the Moon. He calculated the distance at 27.3 Earth diameters. Today astronomers figure the average distance to the Moon is about 30.13 Earth diameters. What's significant about Ptolemy's measurement is not only that it came so close to the actual value, but that it shows how close the ancient Greeks came to having a realistic map of the Solar System about 2,000 years ago.

As you read in Chapter 1, Lesson 2, the Greek astronomer Aristarchus came up with a heliocentric model of the universe centuries before Ptolemy proposed his geocentric theory.

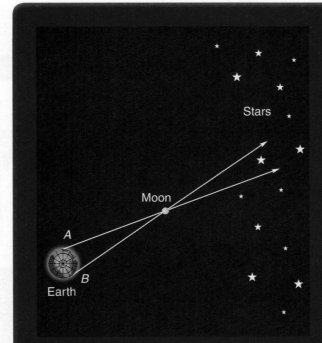

Figure 2.1 When viewed from two different spots on Earth (A and B) the Moon seems to be at two different places among the stars—the parallax effect. The drawing exaggerates the effect to help you visualize it.

Aristarchus also had correctly calculated the relative sizes and distances of the Earth, the Moon, and the Sun. He had a map, but without a scale. If his understanding of the relative positions and sizes of these bodies had been combined with Ptolemy's calculation of the distance from the Earth to the Moon, the progress of astronomy would have been advanced by about 1,300 years.

Today we know that Earth's diameter is about 8,000 miles, or 12,800 km. The Moon is about 240,000 miles (380,000 km) from Earth.

How the Moon's Size Is Estimated

Once you know how far away the Moon is from Earth, you can determine its size. Data from spacecraft can help with calculations of the Moon's diameter. But there is a more earthbound way to do the math, too.

Figure 2.2 The Moon's angular size, or angular diameter, is about one-half a degree.

One method involves calculating from the basis of the angular size of the Moon. An object's angular size is the angle between two lines that start at the observer and go to opposite sides of the object. The angular size of the Moon seen from Earth is about one-half degree. Figure 2.2 will make this clearer.

Scientists can use their knowledge of the Moon's angular size, as seen from Earth, plus the distance from the Earth to the Moon, to calculate the diameter of the Moon. They call the simple equation that they use to make this calculation the *small-angle formula*.

Using the most accurate data, scientists can calculate that the Moon's diameter is 2,160 miles (3,476 km). That's almost one-fourth the diameter of Earth.

Figure 2.3 Both characters are right. To determine size, you must know the distance. And the word "looney" does indeed come from "lunar."

Reprinted by permission of Creators Syndicate and John L. Hart FLP

One of these two photos of the Moon was taken when the Moon was closest to Earth; the other when it was farthest.

Courtesy of Galileo Project, NASA

Why the Moon Appears to Be Larger at Certain Times

Your perception of the size of a heavenly object is the result of a combination of its distance and its real size. This is the point of the cartoon in Figure 2.3.

The Moon appears to be larger at some times than others because its distance from Earth varies. Its orbit is elliptical, like those of other bodies in the Solar System. And it's a fairly eccentric ellipse—in other words, its two foci are relatively far apart. Or, to put it in more basic terms, it's not the kind of ellipse that's almost a perfect circle.

The Moon appears to be larger when it is at its perigee, or *closest distance from the Earth*, about 227,000 miles (363,000 km). It appears smaller when at its apogee, or *farthest distance from the Earth*, about 253,000 miles (405,500 km). The Moon also looks larger as it is rising and setting in the sky than it does when it is high in the sky. (So do the Sun and other objects.) This is an optical illusion. You can prove this by holding up a piece of ring-binder paper at an arm's length. The Moon will just fill the hole both when rising and when it's high in the sky.

This photo shows low tide in a bay at Campobello Island, New Brunswick, Canada. Note how much of the water simply disappears.
© Andrew J. Martinez/Photo Researchers, Inc.

The same spot a few hours later: The water has flowed back in.
© Andrew J. Martinez/Photo Researchers, Inc.

The Relationships Between the Moon and the Earth

As you read in Lesson 1, the Moon is a big reason the Earth is so hospitable to so many life forms. The Moon helps stabilize Earth in its orbit. And the gravity of both the Moon and Sun creates tides.

How the Gravitational Force of the Moon and Sun Influences the Ocean Tides

The tides are familiar to those who live near the ocean or visit there often; to those away from the coasts, less so. But, as you will read, even places far from the sea are subject to tidal forces. A tidal force is *a gravitational force that varies in strength and/ or direction over an object and causes it to deform.*

The Moon exerts a gravitational force not on the Earth as a whole, but on each individual part of the Earth. Gravity, you will recall, is the result of both mass and distance. The Moon's gravity pulls harder on those parts of Earth closer to it than on parts farther away. This pull produces a high tide on the point on Earth closest to the Moon at any given time (point A in the diagram). The Moon pulls the ocean away from Earth, in effect. It also produces a high tide on the other side of Earth (point C) by pulling the main body of Earth away from the ocean (Figure 2.4).

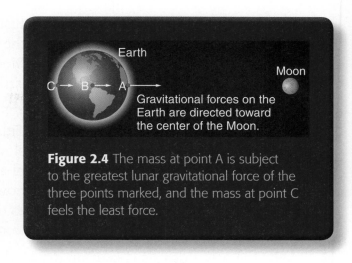

Figure 2.4 The mass at point A is subject to the greatest lunar gravitational force of the three points marked, and the mass at point C feels the least force.

The Sun is a factor in the Earth's tides, too. The Sun is vastly larger than the Moon, but the Moon is much closer to Earth. So the tidal force exerted by the Moon is 2.2 times that of the Sun. When the Sun and the Moon line up with the Earth, near the times of a full moon or new moon, the Sun's tidal forces intensify the Moon's tidal pull. Spring tides are these *exceptionally high and low tides that occur at the time of the new moon or the full moon, when the Sun, Moon, and Earth are approximately aligned.* When the Moon is in its first or third quarter, the lunar and solar tidal forces tend to partly cancel each other out. Neap tides are *the tides that occur when the difference between high and low tides is least* (Figure 2.5).

The Moon's Rotation and Revolution

The period of the Moon's rotation exactly matches that of its revolution. The Moon thus keeps the same face to Earth at all times. Scientists reckon it hasn't always been this way, though. They calculate that in times past, the Moon must have gone through a rotation period different from its revolution. Tidal friction—*the friction that results from tides on a rotating object*—has slowed the Moon down.

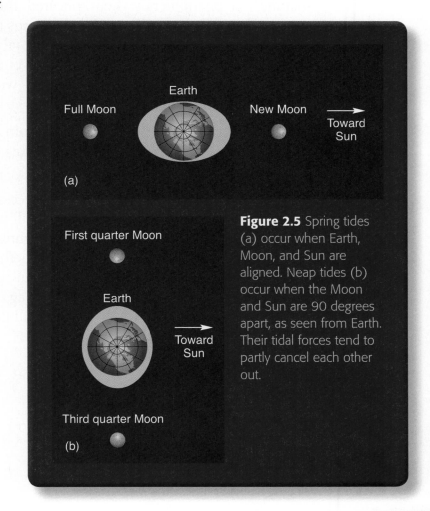

Figure 2.5 Spring tides (a) occur when Earth, Moon, and Sun are aligned. Neap tides (b) occur when the Moon and Sun are 90 degrees apart, as seen from Earth. Their tidal forces tend to partly cancel each other out.

Just as the Moon and Sun cause tides on Earth, the Earth and the Sun cause tides on the Moon. It has no vast oceans. But the Moon's surface may move as much as four inches (10 centimeters) over a month.

Tidal friction is at work on the Earth as well, slowing the planet down by about 25 billionths of a second every day. Billions of years from now, the Earth's rotation time will increase to the point that the Earth always keeps the same face toward the Moon.

By the way, tides also occur on dry land on Earth. Rocks and dirt actually stretch under the force of gravity. Parts of a continent may move as much as 20 inches (half a meter) in a day. So even if you live in Iowa, you're subject to tidal forces!

How the Moon's Location Influences the Earth's "Wobble"

If you've ever spun a child's top on a smooth table, you know that the axis of the top doesn't usually stay completely upright. It wobbles as it spins. The math behind this phenomenon is complicated. But the gist of it is that the top has a tendency to fall over, and its rotation keeps this from happening. The top wobbles around, keeping the same angle with the table's surface until friction slows it down. Whenever a force acts on a spinning object to change the orientation of its axis, the object will wobble. Scientists call this wobble precession and define it as *the conical shifting of the axis of a rotating object.*

The force acting on Earth to change its spin is gravity—the gravity of the Moon and the Sun. It's a slow process. Earth takes about 26,000 years to complete a precession cycle.

The precession of the Earth means that eventually there will be a new "North Star." Polaris, currently the brightest star close to the north celestial pole, now fills this role—in fact, its name means "pole star." But scientists estimate that in about 12,000 years, Vega, the brightest star in the constellation Lyra, will be at the celestial North Pole.

The Moon's Origin and Surface

The Moon is a very near neighbor. But in contrast with Earth's rich diversity of environments and life forms, the Moon is a very different place. Cold and lifeless, it has little interior heat and no plate tectonics. Still, there is much to learn from studying the Moon.

The Moon's Lowlands (Maria) and Craters

The moonscape has two principal features: the maria or "seas"—*lunar lowlands that resemble seas when viewed from Earth*—and the cratered mountainous regions. (The singular form of "maria" is "mare." It is the Latin word for "sea.") Craters almost completely cover the Moon's far side. Until the mid-twentieth century, scientists assumed volcanoes formed the Moon's craters. After all, most craters on Earth arise from volcanic activity. Today, however, scientists have determined that the craters are the result of the impact of meteorites—*interplanetary chunks of stone or matter that have crashed into a planet or moon from space.*

Volcanic eruptions, however, did produce the maria in the Moon's past. Impacts from large asteroids formed the craters to begin with. Later, dark lava flooded the basins of the craters to form the maria.

The Features of the Moon's Crust

The Moon's crust ranges from about 35 miles to 60 miles (about 60 km to 100 km) deep (Figure 2.6). It's thinner on the side facing the Earth. The molten lava released when the maria formed flowed toward the side with the thinner crust. This is natural, since lava has a greater density than the rocks of the highland areas. As you read earlier in this lesson, tidal forces slowed the Moon's rotation. They also acted on the Moon's uneven distribution of mass to make the denser side face the Earth.

The Moon's mountains formed differently from Earth's. Earth's came about by the shifts of tectonic plates and the explosion of volcanoes. Lunar mountains, by contrast, are the result of millions of ancient craters piled up on top of one another.

Star POINTS

The Apollo astronauts left sensors on the Moon to measure moonquakes. Some of these were artificially produced by striking the Moon at various places. But about 3,000 natural moonquakes occur each year. Tidal interaction between the Moon and the Earth causes these quakes.

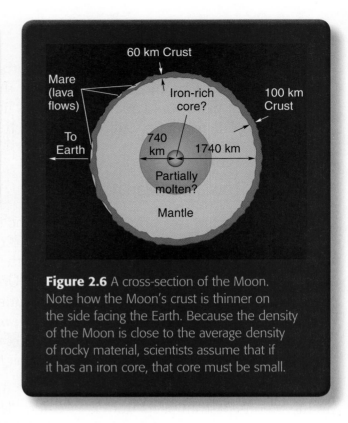

Figure 2.6 A cross-section of the Moon. Note how the Moon's crust is thinner on the side facing the Earth. Because the density of the Moon is close to the average density of rocky material, scientists assume that if it has an iron core, that core must be small.

Four Theories of the Moon's Origin

Since the early nineteenth century, scientists have advanced three main theories of the Moon's origin.

The double planet theory holds that *the Moon was formed at the same time as the Earth*. This theory goes back to the early nineteenth century. The idea is that as the Earth formed from a spinning disk of material, some leftover material that wasn't absorbed into the planet formed an orbiting Moon instead. The problem with this theory is that it doesn't account for the fact that the Moon is much less dense than Earth: 3.35 grams per cubic centimeter (3.35g/cm^3) for the Moon, compared with Earth's 5.52 g/cm^3

In 1878 Sir George Howard Darwin (Charles Darwin's son) proposed the fission theory. It held that *the Moon formed from material spun off from the Earth*. Darwin proposed that the combination of a fast rotation of Earth and the force of solar tides caused a large glob of stuff to spin off from the area where the Pacific Ocean is now. This theory has two main drawbacks. No one has offered a satisfactory explanation of just how this spinoff might have occurred. And if the Moon had been ejected from Earth, it would presumably orbit in the plane of Earth's equator. It does not.

CHAPTER 2 The Earth and Moon

In the early twentieth century the capture theory held that *the Moon is made up of Solar System debris captured by Earth*. The problem with this theory is that colliding celestial objects don't easily "capture" one another. If one actually collides with another, capture is possible. Alternatively, if one object actually came very near another, it might slow the second one down so that a third object could capture it. But such a three-way near miss seems highly unlikely.

In the 1970s A. G. W. Cameron and William Ward of Harvard University proposed the large impact theory of the Moon's origin. It holds that *the Moon formed as the result of an impact between a large (Mars-sized) object and the Earth* (Figure 2.7). The metal cores of the two bodies combined to form the massive core of Earth. The lighter material became the Moon. Since the mid-1980s a scientific consensus has been building for this theory as the one that best fits the data. It explains both the geologic similarities and differences between the Earth and the Moon. It also fits data from calculations about other possible rates of rotation for Earth.

Understanding the Moon and the Earth allows scientists to better understand the Solar System and humanity's place in it. Helped by what they learned from the *Apollo* missions, scientists are able to explain the features on the Moon's surface and the mystery of the Moon's origin.

The next chapter will take a look at the various members of the Solar System, starting with its central player—the Sun.

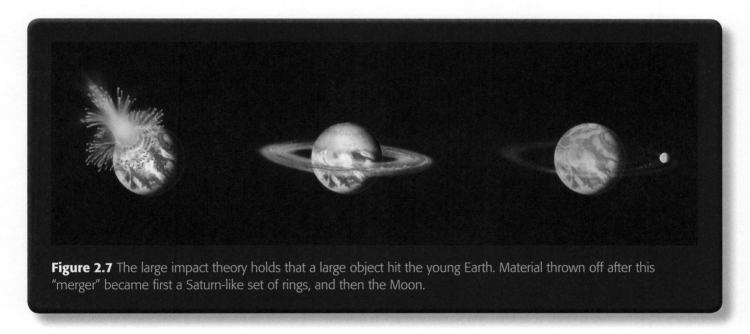

Figure 2.7 The large impact theory holds that a large object hit the young Earth. Material thrown off after this "merger" became first a Saturn-like set of rings, and then the Moon.

Reading the History of the Moon

Scientists can "read" the Moon's craters and determine which ones are relatively newer than others. If one crater overlaps another, the overlapping crater is surely newer than the one overlapped. Scientists thus know that the crater Tycho, named for the Danish astronomer you read about in Chapter 1, Lesson 2, is relatively young. Lunar rays provide another way to "date" features of the lunar landscape. These rays streak out from craters like the "rays" in a child's drawing of the Sun. Lunar rays darken over time, and so brighter ones are likely to be newer.

Knowing the order in which things happened isn't the same as knowing when they happened, though. Not until the *Apollo* astronauts returned with 840 pounds of Moon rocks could scientists work out a reliable timescale for the Moon's history. Now they know the Moon formed about 4.6 billion years ago. The oldest Moon rocks the astronauts brought back were about 4.42 billion years old. The Moon's surface was molten for millions of years. As the Moon solidified and cooled, impacts with meteoroids marked its surface with craters. Most craters formed between 4.2 billion and 3.9 billion years ago. Giant impacts toward the end of this period, followed by lava flows into crater basins, led to the creation of the maria we see today. After the cratering period came a period of considerable volcanic activity. This ended about 3.1 billion years ago.

Cratering continues today, but at a slower pace. Our part of the Solar System has been largely swept clear of most large chunks of matter, and so the Moon isn't taking as many hits nowadays.

CHECK POINTS

Lesson 2 Review

Using complete sentences, answer the following questions on a sheet of paper.

1. What was significant about Ptolemy's measurement of the distance from the Earth to the Moon?

2. Scientists can calculate the Moon's diameter by knowing the angular size of the Moon, plus what other measurement? What is the simple equation they use to make this calculation called?

3. Why does the Moon appear to be larger at some times than others?

4. What are neap tides and why do they occur?

5. What effect do tides have on the Moon?

6. What big change in the northern skies do scientists expect to see in about 12,000 years?

7. What assumption about the Moon's craters did scientists make until the middle of the twentieth century? What assumption do they make today?

8. How were the lunar mountains formed?

9. What are the two main drawbacks of the fission theory of the Moon's formation?

APPLYING YOUR LEARNING

10. Suppose you lived 2,000 years ago. What observations could you make that would lead you to conclude that the Moon plays a role in the Earth's tides?

Every planet in the Solar System, except Earth, takes its name from a Roman or Greek god. But it's the Sun in particular that some ancient people worshipped. This fiery globe provides the light, heat, and energy that make life on Earth possible.

The Sun and the Solar System

Chapter Outline

> In the center of all rests the Sun. For who would place this lamp of a very beautiful temple in another or better place than this wherefrom it can illuminate everything at the same tme? As a matter of fact, not unhappily do some call it the lantern; others, the mind and still others, the pilot of the world.... And so the Sun, as if resting on a kingly throne, governs the family of stars which wheel around.

Nicolaus Copernicus, De Revolutionibus

LESSON 1 The Sun and Its Domain

Quick Write

_____

_____

What are some of the differences in the way you regard the Sun, and the way an ancient Egyptian teenager would have regarded it?

Learn About

- the Sun's energy
- the Sun's core, atmosphere, and sunspots
- the Solar System's structure

For most of human history the Sun has been an object of mystery, of reverence and sometimes, of fear. Early civilizations learned about seasonal changes in the sky very early on. When the Sun is high up in the sky at high noon, it is summer and it is warm. When it is low in the sky at noon, it is winter and it is cold. They also learned that the equinoxes meant a change over to spring or fall would soon occur.

Historically, people thought that humans were the center of everything, and the acts they performed made them responsible for keeping everything going in this eternal cycle—through the seasons and through the years. Beyond that, most civilizations had some sort of mythology about the Sun. They either thought it was a deity of some sort or a supernatural force that demanded respect. In some grisly situations, there were sacrifices—not always pigs, dogs, and other animals, but human sacrifices as well.

It's only been fairly recently, in the last few thousand years, that people learned that eclipses of the Sun resulted from the motion of the Moon. They came up with primitive ways to predict when eclipses would happen. Just as the Sun's presence gives Earth light, the Sun's absence in broad daylight or during the middle of the day has terrified humans down through the ages. People who learned to predict when eclipses would happen became very powerful—they became shamans or early scientists.

Once scientists developed telescopes and the ability to observe the Sun as an object—up close and in detail—they learned even more. After Galileo began using the telescope in 1609, the Sun was a popular item to study. It no longer was a mythical object in the sky, but became for the first time a physical object scientists could investigate.

The Sun's Energy

The Sun is the center of the Solar System in many ways. The Sun provides the light, heat, and energy that make life on Earth possible.

People have wondered for thousands of years where the Sun's energy comes from. Theories abounded on how to explain the source of the Sun's energy. For example, some proposed that chemical reactions—such as the burning of a fuel—are the source of the Sun's energy. We now know that this cannot be the case, simply because if the Sun were made of a fuel such as coal or oil, it would burn out in a few centuries at the rate that it is releasing energy.

Vocabulary

- protons
- neutrons
- nuclear fusion
- equilibrium
- hydrostatic equilibrium
- conduction
- convection
- radiation
- photosphere
- chromosphere
- corona
- solar wind
- sunspots
- solar flares
- conservation of angular momentum

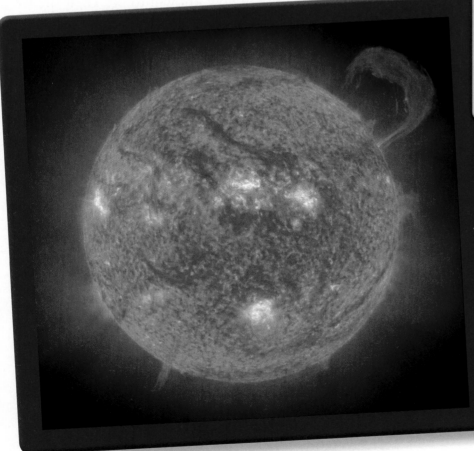

This photograph, taken from space, gives a hint of the tremendous energy within the Sun.

Courtesy of SOHO-EIT Consortium/ESA/NASA

It wasn't until the mid-twentieth century that scientists were able to measure the amount of energy the Sun has produced for billions of years. The solar energy the Earth's surface receives—assuming you could collect and harness it efficiently—is enough to cover the entire world population's energy needs 10,000 times over.

The Source of the Sun's Energy

In the mid-19th century, Hermann von Helmholtz and Lord Kelvin proposed that the source of the Sun's energy was a very slow gravitational contraction—shrinking caused by the force of gravity. They believed this compressed the gases inside the Sun, raising their temperature. It's similar to the air in a bicycle tire getting warmer when you compress it with a pump. The theory said that when the Sun's gases became hot enough, they began radiating energy out into space.

Helmholtz and Kelvin's calculations showed that gravitational contraction could have produced the Sun's energy output with a reduction in the Sun's diameter of only a few tens of meters per year. That's a negligible amount throughout the course of recorded history.

They assumed that the Sun formed from a large diffuse cloud. They figured that gravitational contraction could not have started more than a few hundred million years ago. This period seemed long enough in the nineteenth century, as scientists thought the Earth to be much younger than they now know it to be. Their theory seemed to be a good one. It fit the available data.

Then, in the twentieth century, geologists discovered the Earth is not a few hundred million years old, but rather a few *billion* years—10 times older. Scientists abandoned the contraction theory. The source of the Sun's energy was again an open question.

Star POINTS

Bethe was awarded the Nobel Prize in 1967 for his theory that stars, including the Sun, get their energy from nuclear reactions in which hydrogen is converted into helium.

In the early twentieth century Einstein proposed his special theory of relativity, and scientists began considering that mass can be converted to energy, and energy into mass. In the late 1920s, scientists theorized that this process could be the source of the Sun's energy. Then during the 1930s, Hans Bethe at Cornell University worked out a theory that explains how the Sun has produced its tremendous power for the past 4 billion to 5 billion years, and how it will continue this production for another similar period.

The Nuclear Fusion Power of the Sun

Bethe's theory requires an understanding of atoms. You may already know that an atom is made up of a nucleus surrounded by orbiting electrons. That nucleus is what you should consider when thinking about the Sun's nuclear power. An atom's nucleus, which is most of the atom's mass, is made up of two kinds of particles: protons, *positively charged particles*; and neutrons, *particles with no electrical charge.* (Note that most hydrogen nuclei have no neutrons.)

The Sun's core is too hot to allow for complete atoms to exist. Instead, nuclei and electrons are separated from one another and bounce around at great speeds. In nuclear fusion, *two nuclei combine to form a larger nucleus.* They "fuse." The primary source of the Sun's energy, then, is a series of nuclear fusion reactions. Four hydrogen nuclei fuse to form one helium nucleus. In the process a small amount of the mass of the nuclei is changed into energy. This is where Einstein's theory comes into play (Figure 1.1).

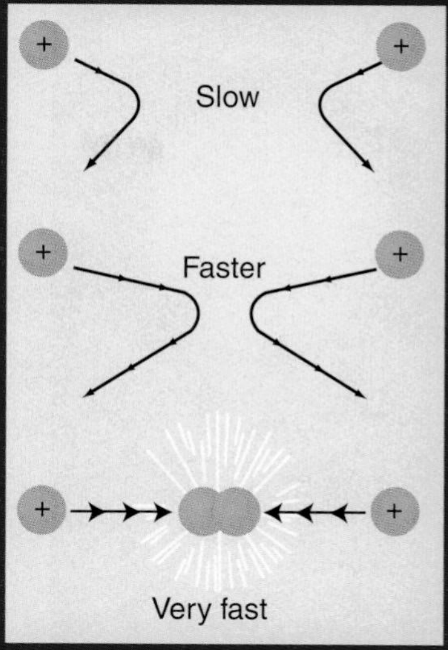

In order for the Sun to produce its enormous power, a large number of fusions of hydrogen nuclei must take place every second. As a result, the density of the matter in the region of the Sun where fusion occurs is very high.

Figure 1.1 Nuclei moving toward one another at too slow a speed will repel each other because of their positive charges. If they are moving fast enough, however, the repelling force won't be strong enough to prevent them from colliding and fusing.

Star POINTS

Nuclei is the plural of the Latin word *nucleus.*

The Sun's Core, Atmosphere, and Sunspots

The same factors determine what the Sun's core is like. Gravity holds the Sun together, just as the force of gravity holds the atmosphere of the Earth near its surface. The Earth's atmosphere is denser near the surface, not simply because the force of gravity is greater there than higher up, but also because the pressure exerted by the gas above compacts the lower layers. Gases lower in the atmosphere have to support the gases above.

The Sun's Equilibrium

The same logic applies to the Sun. At any particular depth below the Sun's surface, the pressure of the gas at that point must be enough to support the gas above. Thus, it is convenient to think of the Sun as having layers like the various layers of an onion, as shown in Figure 1.2. Keep in mind, however, that in the Sun there are no distinct boundaries between layers but rather a continuous change moving toward or away from the center. The Sun is in a state of equilibrium, or *balance, neither noticeably contracting nor expanding*, so the pressure downward on any thin layer must be equal to the upward pressure on that layer.

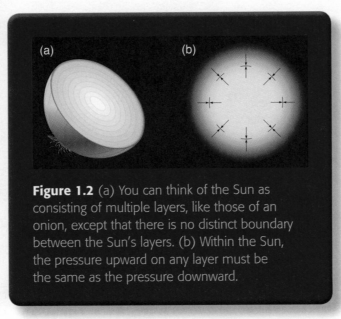

Figure 1.2 (a) You can think of the Sun as consisting of multiple layers, like those of an onion, except that there is no distinct boundary between the Sun's layers. (b) Within the Sun, the pressure upward on any layer must be the same as the pressure downward.

The equilibrium conditions in the Sun are known as hydrostatic equilibrium. "Hydro" refers to the fluid state. This is basically just a more complex case of the situation with the inflated bike tire. In that case, the stretched rubber holds the air inside in a compressed state. As long as the outward pressure of the compressed air inside is enough to support the inward pressure of the rubber, equilibrium is maintained.

The gas at the center of the Sun is supporting the weight of the gas all the way out to the surface, so you should expect great pressures at the center.

This tremendous pressure pushes protons close enough together that hydrogen fusion can take place. Only near the center of the Sun are the temperature and density of hydrogen great enough to support fusion. The solar core, where fusion takes place, extends out to perhaps 25 percent of the Sun's radius.

The fusion reactions in the core provide a heat source that scientists must take into account when calculating conditions within the Sun. When a gas is heated, it tends to expand. A balloon expands when its temperature rises, but then it stabilizes at a (different) equilibrium condition. Likewise, the Sun exists in a state of equilibrium, with the force of gravity balanced by forces tending to expand the gas.

The Three Modes of Solar Energy Transfer

The ways in which the energy produced at the Sun's core are transferred to the surface are the same ways heat transfers here on Earth. Indeed, the same process occurs everywhere in the universe.

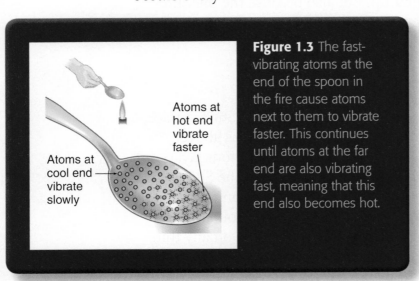

Atoms at hot end vibrate faster

Atoms at cool end vibrate slowly

Figure 1.3 The fast-vibrating atoms at the end of the spoon in the fire cause atoms next to them to vibrate faster. This continues until atoms at the far end are also vibrating fast, meaning that this end also becomes hot.

If you put one end of a spoon on the burner (or in the flame) of a kitchen stove and hold the other end, you'll feel your end of the spoon gradually getting warmer until it is too hot to hold. Energy is transferred by vibration through the metallic crystal structure of the spoon. This first method of transfer is called conduction, *the transfer of energy in a solid by collisions between atoms and/or molecules.*

Imagine the atoms near the end of the spoon on the fire. As the spoon gets hotter, the atoms vibrate at a faster speed (Figure 1.3). These atoms exert forces on adjacent atoms of the metal, and cause those atoms to start moving faster. Gradually, the increased vibration spreads up the spoon until the atoms at the other end are also vibrating faster. In transferring energy by conduction, atoms do not move from one region to another. Instead energy from vibration—thermal energy—is transferred.

In order for conduction to take place, the particles of the substance must be in close contact. The same is true for the atoms in a solid. This is rarely the case in stars, so conduction is not a significant factor in transporting energy from within the Sun.

The second method of energy transfer, convection *occurs when the atoms of a warm liquid or gas move from one place to another*. You may have heard the expression "Heat rises." Put your hand about a foot above a hot stove burner. You will feel hot air rising from the burner. This happens because hot air is not as dense as cold air, and therefore, the hot, less-dense air rises.

On Earth, convection currents let hang gliders travel long distances, carried along with the rising air. In a star, convection between neighboring layers is significant only when the temperature difference is great compared with the pressure difference. In the case of the Sun, this condition is met only in the region within about 125,000 miles (200,000 kilometers) of the surface (Figure 1.4). In this region, convection constantly mixes the solar material as hot gas rises and cooler gas descends.

The third method of energy transfer, radiation is *the transfer of energy by electromagnetic waves*. If you hold the palm of your hand exposed to the stove burner, you can tell that your hand is being heated by something else besides the rising hot air. To emphasize this, hold your hand off to the side, where it is not in the stream of hot air, and you will still feel the heat.

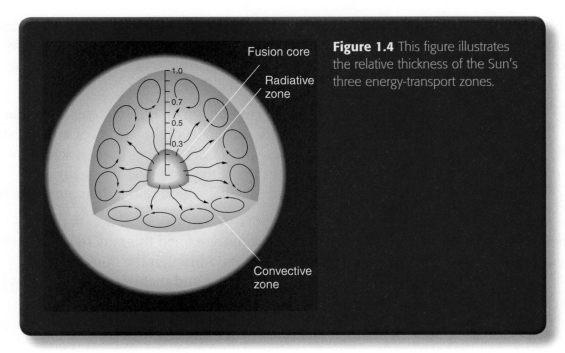

Fusion core

Radiative zone

1.0
0.7
0.5
0.3

Convective zone

Figure 1.4 This figure illustrates the relative thickness of the Sun's three energy-transport zones.

Radiation of energy occurs in all portions of the electromagnetic spectrum. Its effect on another object depends only on whether the receiving object absorbs the particular wavelengths of radiation the radiator emits. The air of your kitchen, for example, is transparent to most electromagnetic radiation produced by the stove burner and so is not heated by it directly. Your hand, however, absorbs the radiation and is heated by it.

Star POINTS

The speed of light is 186,000 miles per second.

Radiation is the principal means of energy transfer inside the Sun. If the Sun were transparent, the electro-magnetic radiation produced in the core would travel outward at the speed of light and reach the surface in about 2 seconds. In reality, energy may take hundreds of thousands of years to reach the Sun's surface.

The Three Layers of the Sun's Atmosphere

It might seem odd to say that the Sun has an atmosphere. But it does. It's made up of material beyond the Sun's surface that you cannot see with your eyes.

Star POINTS

Warning: To prevent severe damage to your eyes, *never* look directly at the Sun.

The Sun's atmosphere has three levels—the photosphere, the chromosphere, and the corona. The first is the photosphere, *the visible part of the Sun and that part of the solar atmosphere that emits light.* The photosphere is the level you see when you look at pictures of the Sun.

The photosphere is about 250 miles (400 kilometers) thick. The variation in brightness is a result of hot areas within the photosphere rising and falling within the region—variations in temperature produce variations in brightness.

The next layer within the Sun's atmosphere is the chromosphere, *the region between the photosphere and the corona* (Figure 1.5). This is where solar flares and eruptions come from. This region is only visible from Earth during a solar eclipse, when the Moon covers the photosphere, creating a bright red flash.

The final layer of the Sun's atmosphere is the corona, which is *the outermost portion of the Sun's atmosphere.* This region extends for hundreds of thousands of miles from the Sun and can be seen during total solar eclipses. Centuries ago, when scientists were first exploring the Solar System, they believed the corona was actually an optical illusion caused by the eclipse instead of a real region in the solar atmosphere.

As you read about the various layers and regions of the Sun, keep this in mind—the boundaries between them are artificial, and were named this way to distinguish between them based on certain selected properties. Although the boundaries between various regions appear sharp in the graphics and pictures, in reality, they are not as well defined. This is especially true of the outer limits of the corona, where the coronal material becomes the solar wind.

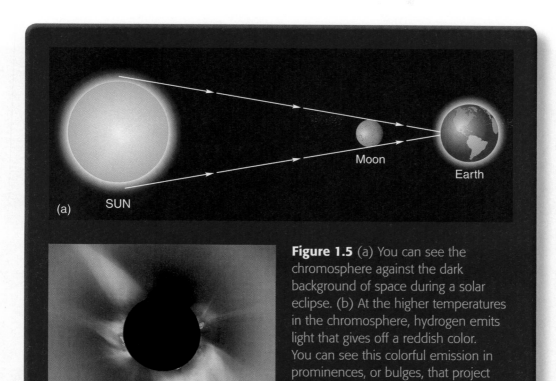

Figure 1.5 (a) You can see the chromosphere against the dark background of space during a solar eclipse. (b) At the higher temperatures in the chromosphere, hydrogen emits light that gives off a reddish color. You can see this colorful emission in prominences, or bulges, that project above the Sun during total solar eclipses. *Courtesy of UCAR/NCAR/High Altitude Observatory/NASA*

This is a composite photo of the Sun and its corona taken in March 1988. The surface of the Sun is a combination of X-ray and visible-light images.

The photo of the corona was taken during the March 1988 eclipse. Note that corona has an irregular shape and streams outward to form the solar wind.
1988 © University Corporation for Atmospheric Research

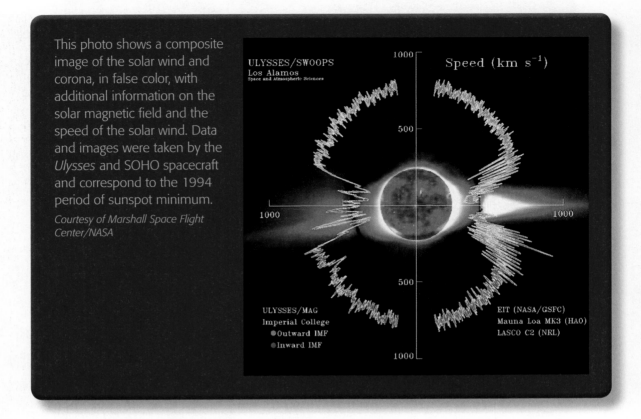

This photo shows a composite image of the solar wind and corona, in false color, with additional information on the solar magnetic field and the speed of the solar wind. Data and images were taken by the *Ulysses* and SOHO spacecraft and correspond to the 1994 period of sunspot minimum.

Courtesy of Marshall Space Flight Center/NASA

Star POINTS

Space weather is the conditions and phenomena in space—especially in the environment near Earth—that may affect space assets or space operations. Space weather may affect both spacecraft and ground-based systems. Solar flares and solar wind affect space weather significantly.

The Solar Wind

This solar wind is *a continuous outflow of charged particles from the Sun*, mostly in the form of protons and electrons. Moving at about 249 miles (400 kilometers) per second near Earth, it plays a significant role in space weather activity. One of its effects is to cause comet tails to point away from the Sun as its particles sweep comet material along with them. Another dramatic effect is the auroras you can see near the Earth's poles. You read about these in Chapter 2, Lesson 1.

The Impact of Sunspots and Solar Flares on Earth's Climate

Another important feature of the Sun is the dark spots people have observed for thousands of years. Sunspots are *the dark spots appearing periodically in groups on the Sun's surface*. The Chinese reported seeing sunspots as long ago as the fifth century BC. Europeans, though, did not report seeing sunspots until Galileo observed them using his telescope.

Alexander Wilson proposed in the late eighteenth century that sunspots were actually windows of a sort, showing glimpses of a cooler surface beneath the hot outer layer of the Sun. William Herschel, who discovered the planet Uranus, even suggested that the interior of the Sun might be cool enough to support life.

Modern science has determined that sunspots are about 1,227 degrees C cooler than the surrounding photosphere, but are still far too hot for life to exist. They appear to be dark because they emit roughly three times less radiation than the surrounding area.

While sunspots put out less radiation than the surrounding area, they also have a much stronger magnetic field. (See the pictures below.) Scientists in the early part of the twentieth century determined that the magnetic field in a sunspot is about 1,000 times stronger than the surrounding photosphere. Also, sunspots often appear in pairs, with each pair having opposite polarity, one north and one south, like the magnets you may use in science classes.

Sunspots are also only temporary. They can last a few hours or even a few months. Sometimes many sunspots are visible, and sometimes there are none. German chemist and amateur astronomer Heinrich Schwabe discovered in 1851 that there is a fairly regular cycle in the number of sunspots. Single sunspots don't last very long, but Schwabe found the sunspot cycle lasts about 11 years.

Sunspots are important to the Earth and your daily life. Between 1645 and 1715, scientists saw very few sunspots. At the same time, the earth went through a "little ice age" when global temperatures were cooler than normal. There is also evidence that similar instances of fewer sunspots and cooler temperatures on Earth happened further back in history. These incidents suggest a clear connection between changes in solar activity and the Earth's climate.

The Sun's turbulent magnetic field also causes the colossal flare-ups called solar flares. A solar flare is *an explosion near or at the Sun's surface, seen as an increase in activity such as prominences (bulges)*. In just a few seconds, flares can heat solar material to tens of millions of degrees. They also eject solar particles at very high speeds.

In the largest flares, these particles reach Earth in less than an hour. They not only cause spectacular auroras, but also disrupt radio transmissions on Earth.

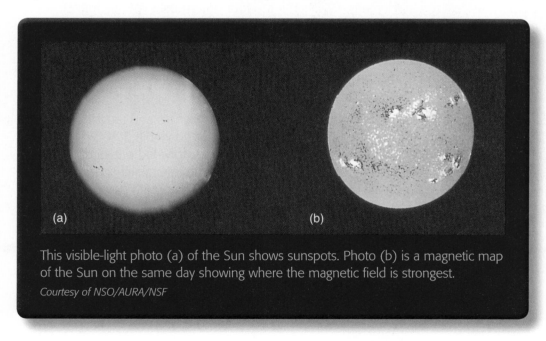

(a)   (b)

This visible-light photo (a) of the Sun shows sunspots. Photo (b) is a magnetic map of the Sun on the same day showing where the magnetic field is strongest.
Courtesy of NSO/AURA/NSF

LESSON 1 ■ The Sun and Its Domain

The Solar System's Structure

The Sun is obviously the dominant object in the Solar System. It makes up more than 99 percent of the Solar System's mass. It is about 10 times larger in diameter than Jupiter, the largest planet. The Sun's diameter is about 862,000 miles (1,390,000 kilometers), compared to Earth's diameter of about 8,000 miles (13,000 kilometers). In other words, the Sun's diameter is about 110 times that of Earth.

To picture this better, think of the Sun as an object the size of a basketball, a sphere 9.4 inches in diameter. On this scale the Earth would be about the size of the head of a pin, one-tenth of an inch in diameter (Figure 1.6).

Jupiter is the largest planet in the Solar System, with a diameter about 11 times that of the Earth. On the scale of the basketball, Jupiter would have a diameter of about an inch. The dwarf planet Pluto has a diameter about one-fifth that of Earth. In your scale model it would be a grain of sand, about $^1/_{64}$ inch across.

Now consider the distances between the planets. To continue the model in which the Sun is a basketball, put the basketball at one end of a tennis court. A pin at the opposite end of the tennis court would be the Earth. A one-inch ball one and a half football fields away would be Jupiter. Pluto would be a grain of sand a kilometer (six-tenths of a mile) away. Between these objects you put nothing— or almost nothing. The only thing between them is the other planets, all smaller than Jupiter, as well as other smaller objects—asteroids and comets—that you'll read about in later lessons.

The *Voyager* Spacecraft

When scientists realized in the early 1970s that the four outermost planets—Jupiter, Saturn, Uranus and Neptune—would all be lined up close together facing the Sun for the first time in more than 170 years, they decided to send out space probes to explore these planets. Since 1977 NASA scientists have been learning more and more about these outer planets and the farthest reaches of the Solar System from two separate spacecraft, *Voyager 1* and *Voyager 2*.

These space probes gave scientists an up-close view of the planets and taught them about previously unknown moons and rings. *Voyager 2* traveled within 3,100 miles of Neptune in August 1989. Since then both space probes have traveled outward, farther from the Sun and the Earth. By 2015, *Voyager 1* will pass into interstellar space, outside the Solar System. *Voyager 2* will follow five years later. There is enough power on the two probes to allow them to continue to operate and send back information to NASA until at least 2020.

Figure 1.6 This shows the Sun, the planets, Pluto, and a few of the large moons drawn to scale.
Jupiter and Saturn: Courtesy of NASA

How the Solar System Formed

So how did the Solar System come to be? The story begins more than 5 billion years ago, when the atoms and molecules that now make up the planets— and your body—were spread out in a gigantic cloud of dust and gas.

A study of the Solar System's beginnings is interesting as an example of the way science in general (and astronomy in particular) makes progress, because the theory is still in its early development, and many gaps remain.

The search for answers here resembles a mystery story in which there are many clues. New ones appear all the time, and some of the clues seem to contradict others. There are two main categories of competing theories to explain the origin of the Solar System: evolutionary theories and catastrophe theories. This section examines the evidence for each and shows why one is gaining favor among astronomers.

Any theory of the Solar System's formation must explain the following data:

1. All the planets revolve around the Sun in the same direction and their orbits are nearly a circle.
2. All of the planets lie in nearly the same plane of revolution.
3. All, except Venus and Uranus, rotate in the same direction as they circle the Sun.

4. Most moons revolve around their parent planet in the same direction as the planets rotate and revolve around the Sun.

5. The spacing of the planets forms a pattern as one moves out from the Sun. The outer planets are less dense than the inner planets.

6. All the planets and moons that have a solid surface show evidence of craters similar to those on Earth's Moon.

7. All the outer (Jovian) planets have rings.

8. Asteroids, comets, and meteoroids exist in the Solar System along with the planets. Each category has its own pattern of motion and location in the Solar System.

9. Planetary systems in various stages of development exist around other stars.

Evolutionary Theories

There is no single evolutionary theory for the Solar System's origin. But there are several theories that share common ideas about how the Solar System came to be as part of a natural sequence of events.

These theories began with one proposed by René Descartes in 1644. He suggested that the Solar System formed out of a gigantic whirlpool, or vortex, in some type of universal fluid. He suggested that the planets formed out of small eddies in the fluid. This theory was rather elementary and didn't explain the nature of the universal fluid. It did, however, explain the observation that the planets all revolve in the same plane, the plane of the vortex.

During the next 350 years various scientists and astronomers debated and analyzed new discoveries and conclusions regarding the formation of the Solar System. But they refuted many of these ideas because the theories violated longstanding laws of astronomy and science. For example, Issac Newton showed that Descartes's theory would not obey the rules of Newtonian physics. So by the mid-twentieth century, evolutionary theories lost favor to what became known as the catastrophe theory.

Catastrophe Theory

In contrast to what the name may imply, catastrophe theory does not refer to a disaster. Rather, it refers to an unusual event—in this case, the formation of the Solar System by an unusual incident. In 1745, Georges Louis de Buffon proposed such an event: the passage of a comet close to the Sun. Buffon suggested that the comet pulled material out of the Sun to form the planets. Astronomers now know this was not possible, but it seemed plausible at the time.

More recently, some thought that perhaps the Sun was once part of a triple-star system—three stars revolving around one another. Such star systems are common, so this is not a far-fetched idea. This particular catastrophe theory holds that the configuration was unstable and that one of the stars came close enough to cause

CHAPTER 3 The Sun and the Solar System

a tidal disruption of the Sun, producing the planets. The close approach of this star also caused the Sun to be flung away from the other two stars.

Starting around the 1930s, however, astronomers began to find major problems with catastrophe theories. First, calculations showed that material pulled from the Sun would be so hot that it would scatter rather than condense to form planets.

Finally, as discussed later, other nearby stars have planetary systems around them. A catastrophe theory would predict that such systems are rare, since they are produced by unusual events. If planetary systems are found elsewhere, there is probably some common process that forms them.

Modern Evolutionary Theories

As these problems have become apparent, catastrophe theories have been nearly abandoned in favor of modern evolutionary theories. These hold that the formation of the Solar System resulted from gravitational forces and pressures on heat and mass.

In the 1940s, German scientist Carl von Weizsaecker showed that different areas of a gas disk rotating around the Sun would not all rotate at the same speed. The inner portion would move faster than the outer. Eddies would form as a result (Figure 1.7).

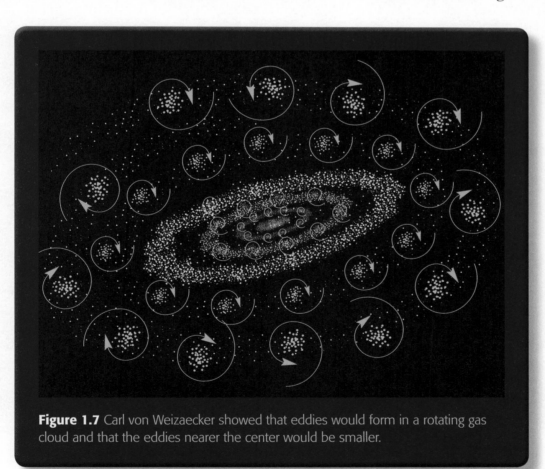

Figure 1.7 Carl von Weizsaecker showed that eddies would form in a rotating gas cloud and that the eddies nearer the center would be smaller.

The farther from the Sun, the larger these eddies would be. These eddies are the beginnings of planet formation (Figure 1.8).

Scientists now believe that stars such as the Sun form from vast clouds of interstellar gas and dust. When the cloud collapses, any rotation that it had at the beginning greatly increased in the central portion. Newton's laws predict this increase, which results from the law of conservation of angular momentum. This law says *an object will spin more slowly as resistance increases and spin faster as resistance decreases*. Over millions of years, the material in the central portion becomes a star (Figure 1.9).

As the gases in the disk around the forming star cool, they condense. These form small chunks of matter that attract more matter to their surfaces. Their orbits are elliptical, and they collide with one another, forming even larger chunks. These then develop orbits that are closer to a circle.

Figure 1.8 At the stage of development shown here, planetesimals have formed in the inner Solar System, and large eddies of gas and dust remain at greater distances from the forming Sun.
Courtesy of D. Berry/STSCI AUL

CHAPTER 3 The Sun and the Solar System

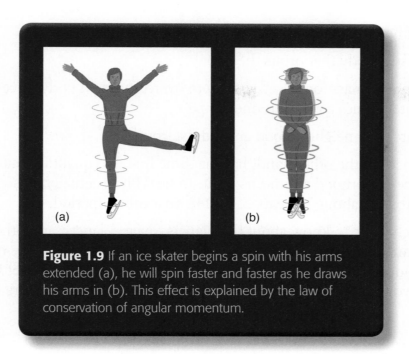

Figure 1.9 If an ice skater begins a spin with his arms extended (a), he will spin faster and faster as he draws his arms in (b). This effect is explained by the law of conservation of angular momentum.

These chunks develop gradually into miniature planets called *planetesimals*, which eventually become planets. Some of them have dust and gas orbiting them—material that eventually condenses into their moons. Today's asteroids are leftover planetesimals that Jupiter's large mass and gravity prevented from pulling together into a planet.

Classifying Objects in the Solar System

Scientists today classify the planets in the Solar System into three types. They call the four innermost planets—Mercury, Venus, Earth, and Mars—the *terrestrial planets*, because of their similarity to Earth. The next four planets—Jupiter, Saturn, Uranus, and Neptune— are called the *Jovian planets* because of their similarity to Jupiter.

The terrestrial planets are smaller than the Jovian planets. Only Earth and Mars have moons, whereas the Jovian planets have many moons each. The Jovian planets also have rings— orbiting debris ranging in size from mere dust to objects several feet long.

Until recently, people considered that there were nine planets in the Solar System. A few years ago, however, the International Astronomical Union reclassified the ninth planet, Pluto. New data and discoveries of other objects in the Solar System caused scientists to reconsider their view of Pluto. Now scientists count only eight planets. Pluto is now considered a *dwarf planet* under the new rules.

Star **POINTS**

The word *terrestrial* comes from the Latin word *terra*, which means "earth."

Star **POINTS**

Jupiter was named after the most powerful of the Roman gods. *Jove* is another form of the name Jupiter, and *Jovian* is an adjective form of the name.

What's the difference between a planet and a dwarf planet? A planet:

- is a celestial body that orbits the Sun

- has sufficient mass for its self-gravity to overcome rigid body forces so that it assumes a nearly round shape

- has cleared the neighborhood around its orbit.

A dwarf planet, on the other hand, has the same first two qualities, but has not cleared the neighborhood around its orbit. In fact, Pluto actually crosses inside Neptune's orbit for about 20 years of its 248-year orbital period.

There is much more to know about the planets, dwarf planets, and other objects that circle Earth's Sun. Scientists have gained much of this knowledge only in the last few decades. This means it's an exciting time to learn about space! In the next few lessons you'll read about the residents of the Solar System in more detail.

✔ CHECK POINTS

Lesson 1 Review

Using complete sentences, answer the following questions on a sheet of paper.

1. How old is the Sun?

2. What is the primary source of the Sun's energy?

3. Which forces are balanced so that the Sun can be in a state of equilibrium?

4. What is the principle means of energy transfer inside the Sun?

5. Where do solar flares and eruptions come from?

6. What causes a comet's tail to point away from the Sun?

7. How long do sunspots last?

8. Six of the eight planets rotate in the same direction as they revolve around the Sun. Which two planets rotate in an opposite direction?

9. What are the three classifications of objects in the Solar System?

APPLYING YOUR LEARNING

10. If scientists want to understand more about distant stars and planets, why is it important to understand as much as possible about the Sun and the Solar System?

Quick Write

_____

_____

What drives people to explore? If you got the chance to visit space, where would you want to go and why?

Learn About

• Mercury
• Venus
• Mars

The Vikings sailed to Newfoundland, in what is now Canada, around AD 1000. Christopher Columbus and other Europeans first set foot on North America about 500 years later. But who was the first to discover Mercury, Venus, and Mars? The answer is, no one knows.

The Chinese, the Greeks, the Romans, and no doubt countless others, noticed the bright lights in the sky that were different from the other stars thousands of years ago. The early astronomers gave them names based on their observations, and they stuck. The color of Mars, for example, is red and resembles blood, so the Greeks named it after the god of war.

Ancient astronomers didn't have telescopes or satellites. Instead they noticed that some lights in the sky moved against the background of the stars. The ones that "wandered" they called planets. Up until the sixteenth century almost all astronomers believed Earth was the center of the universe and the planets revolved around it, and not the Sun. Roughly 100 years later with the invention of the telescope, astronomers began to study the known planets up close—and to discover new planets.

In the twentieth century spacecraft from the United States and the Soviet Union began to explore the far reaches of the Solar System and the surfaces of its planets. This exploration continues today, although (so far) no human has set foot on any other planet. But many unmanned probes have done so, and the information they have sent back has greatly changed scientists' understanding of Earth's neighbors.

Mercury

The terrestrial planets are the four closest to the Sun—Mercury, Venus, Earth, and Mars. Since you already read about Earth in Chapter 2, Lesson 1, this lesson will focus on the other three terrestrial planets. Your trip begins with the planet closest to the Sun: Mercury.

The View of Mercury From Earth

It should come as no surprise that the smallest planet is also one of the hardest to see from Earth with the naked eye. Because Mercury is closest to the Sun, you can only see it shortly before dawn or after dusk when the Sun is below the horizon (Figure 2.1).

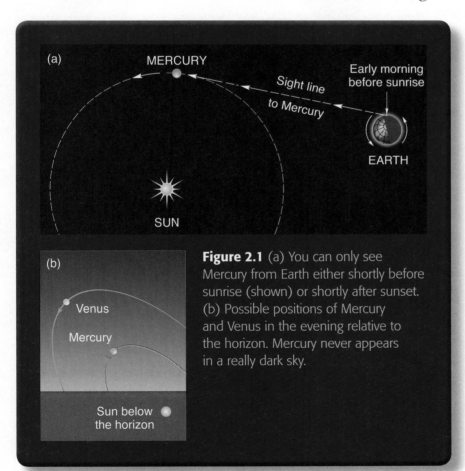

Figure 2.1 (a) You can only see Mercury from Earth either shortly before sunrise (shown) or shortly after sunset. (b) Possible positions of Mercury and Venus in the evening relative to the horizon. Mercury never appears in a really dark sky.

Mercury was named for the Roman god of commerce and travel because this tiny planet is the fastest in the Solar System. When you look at it through a telescope you can see some bright and dark areas. It was hard for astronomers to get a clear picture of what the surface was like until space probes traveled close to the planet and took thousands of pictures. The probes sent these pictures back to Earth, and scientists continue to examine them today.

Mercury's surface is very similar at first glance to the Moon's. Craters cover both of them. They are not exactly alike, however. The walls of Mercury's craters are less steep than those of the Moon. This makes Mercury's craters less prominent than the Moon's. Since Mercury's surface gravity is about twice that of the Moon, this isn't surprising. Because it has stronger gravity, materials don't stack as high on Mercury.

Another difference is that the pattern of materials ejected from the craters is less extensive on Mercury than on the Moon. Mercury's gravity kept the material ejected by meteorite impacts from flying as far as it did on the Moon. Mercury also lacks the large maria seen on the Moon.

Mariner 10 and *MESSENGER*

The United States has sent two space probes to explore Mercury. *Mariner 10*, launched in 1973, was the first.

Mariner 10 was also the first space probe to use the gravitational pull of one planet (Venus) to explore another (Mercury). Its main goal was to explore Mercury's surface. The spacecraft took more than 4,000 pictures, giving scientists the first clear views of what the planet's surface is like.

This photograph, a mosaic of pictures taken from *Mariner 10*, shows Mercury's crater-filled surface.

Courtesy of NSSDC/NASA

But the mission was able to view less than half of Mercury's surface area. NASA shut down *Mariner 10* in 1975, and the probe is now in a perpetual orbit around the Sun.

In 2008 and 2009 *MESSENGER*—the MErcury Surface, Space ENvironment, GEochemistry, and Ranging mission—took more pictures of Mercury's surface. Now more than 98 percent of the surface area has been photographed. In 2011 *MESSENGER* will enter Mercury's orbit and will examine even more of the planet.

MERCURY
4879 km

CALLISTO (JUPITER)
4821 km

PLUTO
2390 km

GANYMEDE (JUPITER)
5262 km

TITAN (SATURN)
5150 km

MOON (EARTH)
3474 km

Figure 2.2 The diameter of Mercury compared to some other objects in the Solar System. In the case of moons, the parent planet is identified in parentheses. Pluto is smaller than Mercury, but one of Saturn's moons and one of Jupiter's are bigger.

Courtesy of NASA/Stephen P. Meszaros

Mercury's Density and Gravitational Pull

Not only is Mercury the smallest of the eight planets—two moons in the Solar System are bigger. Its diameter is 3,033 miles (4,880 km). To give you an idea of how small this is, Mercury's surface area is only a little bigger than that of the Atlantic Ocean (Figure 2.2).

Mercury's mass—calculated from its gravitational pull on *Mariner 10*—is only 0.055 of Earth's mass. It has a density only a little less than Earth's: 5.43 times the density of water versus Earth's 5.52 times.

Since Mercury experiences less gravitational pull than does Earth, its materials are not as compressed as those on Earth. However, the fact that Mercury's density is only slightly less than Earth's density means that Mercury has a higher concentration of heavy elements than Earth does. While *Mariner 10* discovered that rocks similar to those on Earth also make up Mercury's surface, scientists have further concluded that Mercury's higher-density materials are below its surface. Its core must be very large, made up of mostly iron and some nickel, and accounts for about 65 percent to 70 percent of the planet's mass.

Besides exploring Mercury's surface composition, *Mariner 10* detected a magnetic field on Mercury, although it is not very strong—about 1 percent as strong as the Earth's field. The field appears to be shaped like the Earth's and has magnetic poles nearly aligned with the planet's spin axis.

The magnetic field's presence has puzzled scientists since they first identified it in 1974. First, astronomers hypothesized that Mercury's metallic core is molten because the dynamo effect—*the generation of magnetic fields due to circulating electric charges, such as in an electric generator*—requires molten magnetic material within the planet.

More recent measurements indicate, however, that Mercury does not release more heat than it receives from the Sun. Therefore, the planet's interior must no longer retain heat from its formation. This makes it difficult to imagine a molten core. Many astronomers still think that some form of the dynamo effect is responsible for the magnetic field, but the question remains open. It could be that the core is partly molten because of impurities such as sulfur, which would lower the temperature at which the metal would solidify. Or the answer could lie in tidal interactions between Mercury and the Sun.

Star POINTS

Mercury moves faster in its orbit than any other planet, at an astonishing rate of speed: 48 km per second, or about 107,000 miles per hour.

Mercury's Rotation and Orbit Around the Sun

The closest planet to the Sun, Mercury circles the Sun more quickly (88 days) and moves faster in its orbit than any other planet. Mercury's orbit is the most eccentric of the eight planets: Its distance from the Sun varies from 29 million to 44 million miles (46 million to 70 million km).

Because Mercury is located so close to the Sun and makes an elongated orbit, it exhibits an interesting effect: One solar *day* on Mercury lasts two Mercurian *years*—176 Earth days. To explain this, think about the Moon. You know because of tidal effects, the Moon keeps the same face toward the Earth. Astronomers once thought that Mercury also always faced the Sun the same way. But radar observations indicate this is not the case (Figure 2.3).

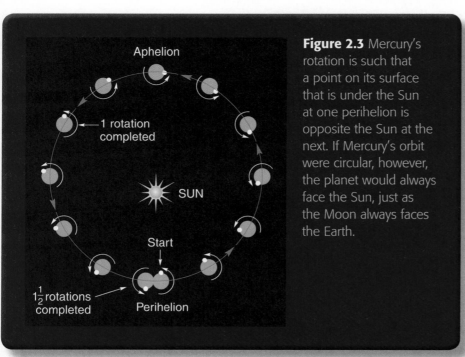

Figure 2.3 Mercury's rotation is such that a point on its surface that is under the Sun at one perihelion is opposite the Sun at the next. If Mercury's orbit were circular, however, the planet would always face the Sun, just as the Moon always faces the Earth.

They show that Mercury rotates on its axis once nearly every 59 Earth days—precisely two-thirds of its orbital period of nearly 88 days. This means that the planet rotates one and a half times for every time it goes around the Sun.

This happens because Mercury is not perfectly balanced; one side is more massive than the other. As a result, the Sun exerts a torque, a twisting force, on the planet, especially when Mercury is in the perihelion stage—*when a planet is closest to the Sun.* (An aphelion, by contrast, is *the point in a planet's orbit when it is farthest from the Sun.*)

Because of this, the planet's rotational period is aligned with its orbital period. Astronomers believe an object that hit Mercury's surface and created its large Caloris Basin must have been made of dense material that is still under the planet's surface. This causes Mercury to be lopsided and to have periods of rotation and revolution coupled as they are.

Venus

Venus is the second closest planet to the Sun. And except for the Sun and Moon, it's the brightest object in the sky. The Romans thought it such a beautiful sight that they named it after the goddess of beauty and love. However, you'll see that as a possible home for a human colony, it is not very hospitable.

The View of Venus From Earth

This bright planet is never seen farther than 46 degrees from the Sun. So, like Mercury, it is visible only in the evening sky after sunset or in the morning sky before sunrise. Stargazers often refer to it as the Morning Star and the Evening Star.

Seeing such a bright object above the horizon has fooled many people into thinking they were seeing something else. During World War II, fighter pilots sometimes shot at Venus, thinking that it was an enemy plane. Today, reports of UFOs often increase when Venus is at its brightest in the sky.

Venus's Density and Magnetic Fields

Venus is often called Earth's sister planet. It is similar to Earth in many ways: Its diameter is 95 percent of Earth's; its mass is 82 percent of Earth's; and of all the planets, its orbit is located closest to Earth's. From the values for its diameter and mass, you can calculate that Venus has a density 5.24 times that of water—not much different from Earth's 5.52.

Soviet spacecraft that landed on Venus determined that its surface rocks have a similar composition to basaltic volcanic rocks on Earth. Thus, to be as dense as it is, Venus must have a very dense interior with a metallic core.

Scientists have not found a magnetic field on Venus. If one does exist, its strength is less than 1/10,000 of Earth's. Because Venus has a slow rotation (about 243 days), scientists wouldn't expect it to have a strong magnetic field. However, according to present theories for the origins of planetary magnetic fields, Venus's field should be strong enough to detect. Astronomers know that Earth's field reversed its direction in the past. Perhaps Venus's magnetic field is now in the process of reversing as well, which would explain why it is so weak.

Venus's Rotation and Orbit Around the Sun

Venus orbits the Sun in a more-circular orbit than any other planet. Its period is about 225 days. Its rotation is very slow, making a solar day on Venus about 117 Earth days. Dense clouds mask Venus's surface, so it is not visible from Earth. Radar waves, however, can penetrate this cloud cover. Since 1961 scientists have been bouncing radar signals off its surface. That's how they first discovered its rotation rate and surface features.

The radar studies taught scientists that Venus rotates backward compared with most other objects in the Solar System. Remember that revolving is how the planets circle the Sun. Rotation is how they spin on their axis. While all eight planets *revolve* around the Sun in a counterclockwise fashion, Venus and Uranus are the only two that *rotate* in a clockwise fashion.

Venus's Surface and Atmosphere

Venus may have many similarities to Earth. But just as its solar days are far longer than Earth's, its surface and atmosphere are quite different, too.

Star POINTS

The Soviet *Venera* probes visited Venus between 1961 and 1984. NASA's *Pioneer Venus 1* started orbiting Venus in 1978 and sent data until 1992. The US *Magellan* orbiter sent data from 1990 to 1994.

The Surface

It took a lot of technology to penetrate Venus's cloud cover. As noted earlier, radar mapped out the planet's surface in extensive detail. In the latter part of the twentieth century Soviet and American missions to the planet produced extensive images. The Soviets landed on Venus nearly a dozen times. The pictures developed from these trips showed rocks scattered near the landing sites. Some were older and weathered smooth, while others were jagged and not yet affected by the planet's fairly calm winds.

Rolling hills cover about two-thirds of Venus's surface, with craters here and there. Highlands occupy less than 10 percent of the surface, while lower-lying areas make up the rest. All of Venus, including the rolling plains that make up most of its surface, is drier than the driest desert on Earth.

Venus has about a thousand craters that are more than two miles in diameter. Earth has only a few craters this size, while the Moon has many more. In fact, Venus has far fewer craters than both the Moon and Mars. Scientists know that there was more cratering early in the Solar System's history than occurs now because there was more interplanetary debris. The cratering rate decreased as this debris was swept away.

Since Venus has no craters older than about 500 million years, scientists believe that the average age of the planet's surface is no older than that. This is about twice as old as Earth's surface, but much younger than the Moon's. Also, as a surface becomes smoother due to weathering, it becomes a poorer reflector of radio waves. This allows scientists to compare the age of lava flows over the planet. Some *Magellan* radar data suggest that Venus has regions of very young lava flows— less than 10 million years old.

There's more evidence for recent volcanic activity on Venus—the amount of sulfur dioxide in the atmosphere has decreased, and scientists have detected lightning bursts. This decrease in sulfur dioxide suggests that a major eruption may have occurred in the 1970s.

The presence of volcanoes suggests that Venus has a molten interior. The evidence suggests that without plate tectonics as on Earth, instabilities build in Venus's interior until eruptions occur on the planet's entire surface— erasing signs of past activities and older craters. Most of Venus's surface is covered with lava rock.

The thick atmosphere of Venus shields the surface from small interplanetary debris, but impacts of larger objects result in craters such as these.

The foreground crater (named Howe) is 23 miles (37 km) wide. Most of the crater floors are flat because lava has flooded them.

Courtesy of NASA/JPL-Caltech

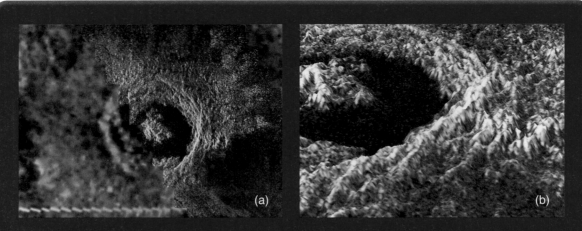

(a) The Venusian crater Golubkina is shown as a mosaic of two images. The image on the left side was produced by the Soviet *Venera 15/16* craft that discovered the crater; the one on the right was produced from *Magellan* data. The crater, named after Anna Golubkina (1864–1927), a Soviet sculptor, is about 20 miles (32 km) across. Its features, including the central peak, terraced inner walls, and surrounding ejected material, are characteristic of impact craters on the Earth, Moon, and Mars. (b) This *Magellan* three-dimensional image enhances the height of structural features of Golubkina Crater.

Courtesy of NASA/Soviet Academy of Sciences, Venera 15/16

The lava near Maat Mons on Venus is the youngest crust material yet found on the planet.

This mountain is five miles (eight km) high—exaggerated in this image made from *Magellan* radar data.

Courtesy of NASA/JPL-Caltech

The Atmosphere

The Soviet *Venera* landers showed scientists that Venus's atmosphere is also quite different from Earth's.

Earth's atmosphere consists of nearly 80 percent nitrogen and 20 percent oxygen, with small amounts of water, carbon dioxide, and ozone. Venus, on the other hand, has an atmosphere made up of about 96 percent carbon dioxide, 3.5 percent nitrogen, and small amounts of water and sulfuric acid—the same acid used in car batteries. Venus's clouds are in large part made of sulfuric acid droplets. Chemical reactions between the sulfuric acid in the atmosphere and fluorides and chlorides in surface rocks result in very corrosive substances that can dissolve even lead. Venus is indeed inhospitable.

As the space probes descended through Venus's atmosphere, they encountered hazards from the acidic atmosphere, the great winds found at the level of its cloud tops, and tremendous pressures. The atmospheric pressure on Venus's surface is about 90 times that on the Earth's surface because of the weight of the gases in Venus's atmosphere. That's about the same as the pressure at a depth of nearly a mile in Earth's oceans, and the heaviest atmosphere of any planet. As if the acidic atmosphere and high pressure were not enough, Venus is also inhospitable because of its high temperatures: about 867 degrees F (464 degrees C) near the surface.

The European Space Agency's *Venus Express* probe, which arrived on the planet in 2006, will help answer many more questions about the planet's atmosphere and climate.

Mars

Like Venus and Mercury, Mars gets its name from mythology. The ancient Greeks called the fourth planet from the Sun "Ares," the name of their mythical god of war, perhaps because of the association of its red color with blood. In Roman mythology the god of war is called Mars.

The View of Mars From Earth

Each time you look at Mars through a telescope, you will not necessarily see the same thing. First, it can look different because of big dust storms moving around the planet. But Mars can also look different because it's not always the same distance from the Earth.

The best time to look at Mars is when it is in opposition. Opposition is *when a planet is directly opposite the Sun in the sky*. In other words, the objects are aligned as Sun-Earth-planet. Mars is in opposition about every 2.2 years. Because of its orbit, which is closer at some points to the Sun than at others, Mars can be as close to the Earth as 28 million miles (55 million km) when it is in opposition, or as far as 62 million miles (100 million km).

The bright, red, cloudy region in the middle of this October 2005 color composite HST image is a large regional dust storm. Bluish water-ice clouds are at the top (the north, winter, polar region), whereas the south polar ice cap has mostly turned to gas or vapor as summer approaches.

Courtesy of NASA/ESA/The Hubble Heritage Team, STSc1/AURA/J. Bell, Cornell University/M. Wolff, Space Science Institute

Scientists have been recording their views of the Mars' surface since the early seventeenth century. Based on these observations, they have determined how fast Mars rotates. Scientists have also seen ice caps on Mars' north and south poles similar to those in the Arctic and Antarctica. Like Earth, Mars tilts so you can see each ice cap growing larger or smaller as it tilts closer to or farther from the Sun. The same thing happens on Earth—Winter is in the Northern Hemisphere when the North Pole tilts away from the Sun, and summer when the North Pole tilts toward it.

Scientists have noticed other seasonal changes on Mars. Different parts of the planet have changed from a light color to a darker color and back, which led some people to believe Mars had plant life. If plants grew on Mars, people wondered, did animals roam there, too? Could there be intelligent life, like humans? Only in the 1970s, when the first Earth probes landed on the red planet, could scientists give a negative answer to these questions.

Mars' Density and Magnetic Field

Mars has a diameter about half of Earth's. Its mass is only one-tenth that of Earth. While Mercury and Venus have densities fairly close to Earth's, Mars does not. Using Mars' diameter and mass, scientists calculate that Mars' density is 3.93 times that of water, about 0.7 of Earth's density. Furthermore, the red planet has only about a quarter as much surface area as Earth and one-eighth its volume— although Mars' surface area equals that of Earth's land mass.

The rocks on Mars are very rich in iron and silicon. Mars rotates almost as quickly as Earth, with one day on Mars being only 40 minutes longer than a day here. But Mars doesn't have a magnetic field as Earth does. From these data and the planet's low density, scientists conclude that the iron on Mars is spread throughout the planet, instead of being concentrated in its core.

Mars' surface has only the remains of its once active magnetic field, scattered around its surface in areas of highly magnetized rock. Observations suggest that Mars probably lost its magnetic field about 4 billion years ago. Since then, the planet's atmosphere has been feeling the full force of the Sun's solar wind. This erodes Mars' atmosphere and contributes to its loss of water.

Mars' Rotation and Orbit Around the Sun

On average, Mars orbits the Sun from about 143 million miles (230 million km) away. Its orbit, however, is fairly eccentric, and its distance from the Sun varies from about 130 million to 155 million miles (210 million to 250 million km). The red planet's day is almost the same as Earth's—24 hours and 40 minutes.

Mars' equator tilts 25.2 degrees with respect to its orbital plane. This is very similar to Earth's 23.4 degrees. The tilt of a planet's axis causes opposite seasons in the planet's two hemispheres, and this is indeed the case on Mars as on Earth.

But there is another feature of Mars' motion that affects its seasons. The eccentricity of Mars' orbit causes it to be much closer to the Sun during parts of its year than at others. Mars is 19 percent closer to the Sun during the northern hemisphere's winter than during its summer. Being closer to the Sun in winter and farther away in summer means that the temperatures in the northern hemisphere aren't all that different between seasons.

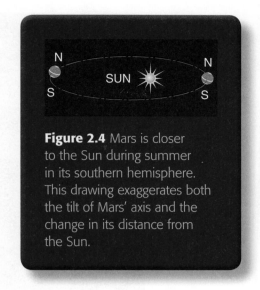

Figure 2.4 Mars is closer to the Sun during summer in its southern hemisphere. This drawing exaggerates both the tilt of Mars' axis and the change in its distance from the Sun.

In the southern hemisphere, however, the reverse is true. Mars is closer to the Sun during the southern hemisphere's summer and farther from the Sun in winter (Figure 2.4). Thus, the southern hemisphere experiences greater seasonal temperature shifts than the northern hemisphere. The same effect occurs on Earth, which is closest to the Sun in January (winter in the northern hemisphere). But the Earth's orbit is so nearly circular that it is only about 3 percent closer to the Sun at one time than the other (rather than Mars' 19 percent).

Mars' Surface and Atmosphere

When thinking about the surface and atmosphere of Mars, it's helpful to put it in the context of a world geography class.

If you measure from sea level, the tallest mountain on Earth is Mount Everest. It's more than five miles high. Still, airplanes can and do fly above it. This mountain is little more than a molehill in comparison with Olympus Mons on Mars (Figure 2.5). Olympus Mons is the largest known mountain in the Solar System—at 15 miles high! By comparison, the highest that planes usually fly is around eight or nine miles high.

Like the High Cascade Mountains in the US Pacific Northwest, Olympus Mons is a volcano. With such an astonishing height, it's not surprising it has an expansive base. If you placed its base in the United States, it would cover much of Oregon and Washington state. The base of Olympus Mons has a diameter of 400 miles and the collapsed depression at its top is 50 miles across and two miles deep.

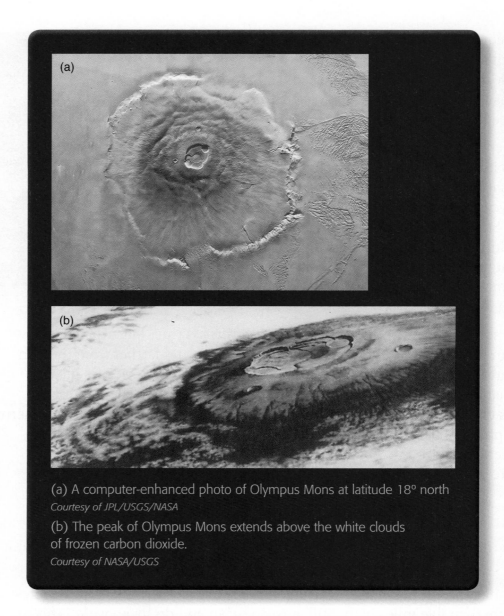

(a) A computer-enhanced photo of Olympus Mons at latitude 18° north
Courtesy of JPL/USGS/NASA

(b) The peak of Olympus Mons extends above the white clouds of frozen carbon dioxide.
Courtesy of NASA/USGS

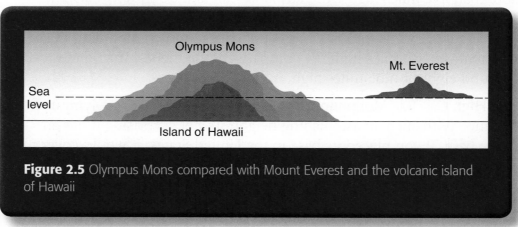

Figure 2.5 Olympus Mons compared with Mount Everest and the volcanic island of Hawaii

CHAPTER 3 The Sun and the Solar System

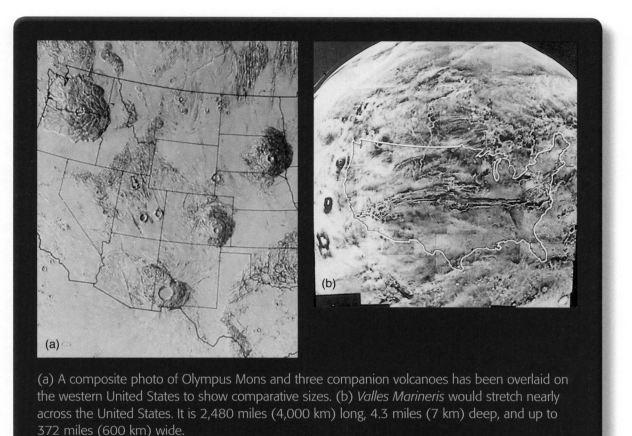

(a) A composite photo of Olympus Mons and three companion volcanoes has been overlaid on the western United States to show comparative sizes. (b) *Valles Marineris* would stretch nearly across the United States. It is 2,480 miles (4,000 km) long, 4.3 miles (7 km) deep, and up to 372 miles (600 km) wide.

Courtesy of NASA/Stephen P. Meszaros

The Surface

Stretching away from the area of Olympus Mons is a canyon that more than matches the volcano. If placed on Earth, *Valles Marineris* (named for the *Mariner* spacecraft) would stretch all the way from Charleston, South Carolina, to Los Angeles, California (Figure 2.6).

In 1976, the seventh anniversary of the first human landing on the Moon, *Viking 1* landed on a cratered plain of Mars and took the first close-up photographs of the surface. They showed that the red color you see from Earth is real.

Another spacecraft called *Pathfinder* landed on Mars on 4 July 1997. It sent back photos of a rock-strewn area.

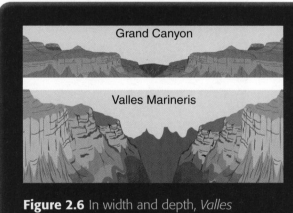

Figure 2.6 In width and depth, *Valles Marineris* dwarfs the Grand Canyon, which is 279 miles (450 km) long, just more than a mile (1.6 km) deep, and up to 15 miles (29 km) wide. It is the product of stresses in the Martian crust, not of erosion by flowing waters, as is a canyon.

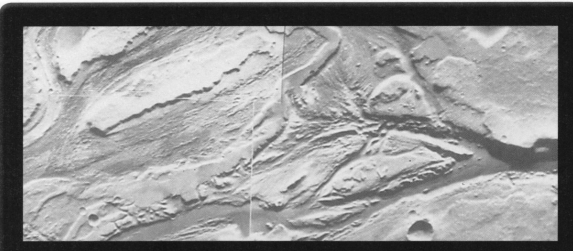

What appear to be dry riverbeds on Mars lead scientists to believe that water once flowed on the surface.

Courtesy of NASA/JPL-Caltech

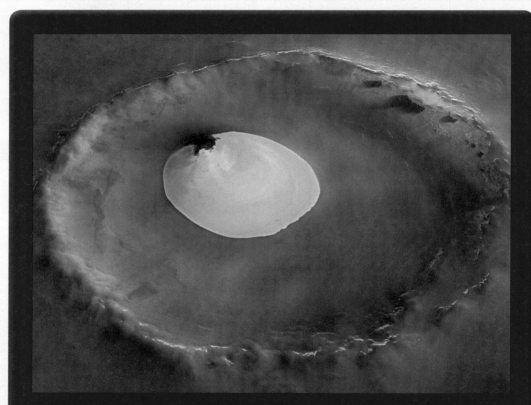

This *Mars Express* image (in almost natural colors) shows water ice on the floor of a crater near the north pole of Mars. Faint traces of water ice are also visible around the rim and on the walls. The crater is 21 miles wide and has a maximum depth of about 1.25 miles from the rim. The image was taken during late summer in the Martian northern hemisphere.

Courtesy of ESA/DLR/FU Berlin (G. Neukum)

Pathfinder's roving robot, *Sojourner*, began a slow journey to investigate the rocks' composition. Scientists learned that Mars is more like Earth than scientists had thought. The first rock investigated was rich in silica, the quartz material found in sand. A volcanic action or a meteorite impact is a possible explanation of how the rock got there in the first place.

The *Viking* orbiting modules and *Mariner 9* found evidence to support the hypothesis that running water was once common on Mars. The channels, gullies, and layers of sedimentary rock discovered on Mars clearly suggest that the planet was very different 3.5 billion years ago.

Liquid water seems to have played an important role in shaping the planet's surface, but it is likely that it was not the only force at play. Other factors such as volcanoes, tectonic shifts, ice, and strong winds almost certainly contributed to the features seen today.

But Mars' atmosphere today is so thin that rainfall is not possible—there is only a small amount of water vapor present. The amount of water vapor in the atmosphere would barely fill a small pond. The water volume in the ice on Mars' north polar cap, however, is about 4 percent of the Earth's south polar ice sheet.

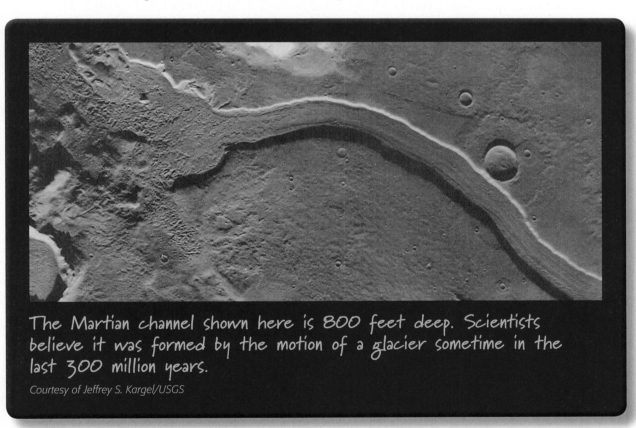

The Martian channel shown here is 800 feet deep. Scientists believe it was formed by the motion of a glacier sometime in the last 300 million years.

Courtesy of Jeffrey S. Kargel/USGS

The Atmosphere

Mars' climate is one of extreme changes. Near the Martian equator, the temperature around noontime can reach as high as 86 degrees F (30 degrees C), a comfortable temperature for humans. At night, however, the temperature might drop to minus 210 degrees F (minus135 degrees C). By comparison, the coldest temperature ever recorded on Earth is minus 128.6 degrees F (minus 89.2 degrees C).

Mars' thin atmosphere is the reason for this extreme difference in temperature. A planet's atmosphere works like a blanket. It protects the planet from the Sun during the day and provides warmth at night by capturing the day's heat and reflecting it back to the surface.

The atmosphere's size and composition determine the amount of shielding and reflection. Mars' atmosphere is 95 percent carbon dioxide. You might suppose that this would make Mars hot as it does Venus. But the low atmospheric pressure at Mars' surface means that there is simply too little atmosphere of any type to significantly moderate the temperature. When vast dust storms sweep across the planet, however, the sunlight-absorbing dust heats the dry atmosphere. A 2001 storm increased Mars' temperature 86 degrees F (30 degrees C) (Figure 2.7).

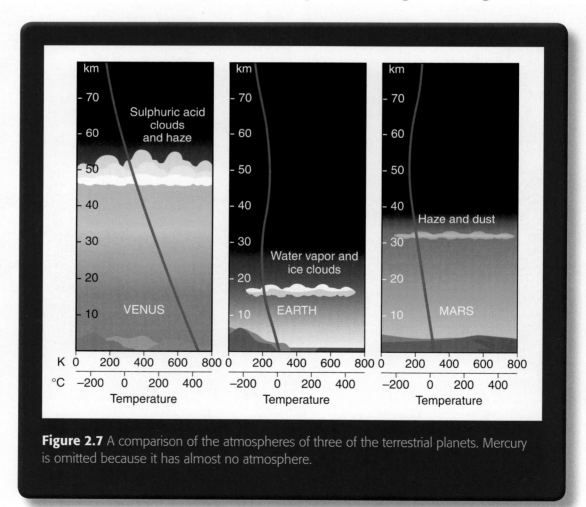

Figure 2.7 A comparison of the atmospheres of three of the terrestrial planets. Mercury is omitted because it has almost no atmosphere.

CHAPTER 3 The Sun and the Solar System

Mars' Moons

Mars has two natural satellites, Phobos ("fear" in Greek) and Deimos ("terror"). Both are small and irregularly shaped. Phobos, for example, is about 17 miles (27 km) across its longest dimension, but 14 miles (22 km) and 11 miles (18 km) across its other dimensions. Deimos is even smaller and is similarly shaped. In other words, they have a form like potatoes.

Since these moons are so similar, scientists believe they are captured asteroids. A capture like this could have taken place in three ways: The asteroid was either slowed by friction with the Martian atmosphere (which was much denser in the past); by collision with a smaller asteroid; or by gravitational pull from another asteroid. This might seem unlikely, but considering the number of asteroids that have passed by Mars over billions of years, it is certainly a possibility. The moons could also be the result of a meteorite impact with the red planet.

The masses of Phobos and Deimos are so small that you would weigh almost nothing standing on their surfaces. A 120-pound person on Earth would weigh about one ounce on the lowest point of Deimos. The Martian moons are not only small; they orbit very close to the planet and have short periods of revolution—about 0.3 days for Phobos and 1.3 days for Deimos.

Both revolve in the same direction as most other Solar System objects, counter-clockwise as seen from a point above the Solar System, north of Earth. But given the speed at which Phobos circles Mars, it would appear from the Martian surface to rise in the west and move across the sky to the east.

N ↗
5 km

Mars Express took this color image of Phobos from a distance of less than 124 miles (200 km). The scale bar is valid only for the center of the image.

Courtesy of ESA/DLR/FU Berlin (G. Neukum)

Scientists are now learning even more about Mars from several unmanned NASA missions to the red planet, including the exploration rovers *Spirit* and *Opportunity*. Data received from these missions ensure that by the time you read this book, scientists will have made new discoveries. If humans ever do set foot on another planet, it will probably be Mars, as you'll learn in a later lesson.

As you have seen, the terrestrial planets—although each quite different—share several similarities with Earth. The outer planets, however, are very different from their little brothers and sisters closer to the Sun. You've learned a bit about them already. The next lesson will explore the Jovian planets in more detail.

CHECK POINTS

Lesson 2 Review

Using complete sentences, answer the following questions on a sheet of paper.

1. When is Mercury visible to the naked eye on Earth?

2. Why do scientists believe Mercury doesn't have a molten core?

3. Why does Mercury rotate one and a half times for every time it goes around the Sun?

4. Why did fighter pilots in World War II shoot at Venus?

5. Why is Venus called the Earth's sister planet?

6. How does Venus's rotation differ from that of most other planets?

7. Why was there more cratering on the surface of the planets early in the Solar System's history and not so much now?

8. When is the best time to see Mars?

9. Why does Mars feel the full force of the solar wind?

10. In addition to the tilt of its axis, what affects Mars' seasons?

11. Why is rainfall not possible on Mars, even though there is water on the planet?

APPLYING YOUR LEARNING

12. Venus and Mars are very different, yet they are often considered the planets most similar to Earth. Why is Mars more likely to serve as the target for the first human visit than Venus? List as many reasons as you can.

LESSON 3 The Outer Planets

Quick Write

Can you think of an experience in your own life, or in someone else's, where something that you observed turned out to be very different from what you thought it was? What does that experience suggest about the mindset a scientist must have to deal with new data and observations?

Learn About

- Jupiter
- Saturn
- Uranus
- Neptune

Saturn is the planet with rings: That's something everyone knows today—even people who don't know much about space science at all. (In fact, people today who *do* know something about space know that all four of the outer planets have ring systems. You'll read more about that later in this lesson.) But there was a time when no one knew about Saturn's rings.

Galileo was the first to observe them. When he looked at them through his 20-power telescope in 1610, he thought the rings were "handles" or maybe large moons on either side of the planet. "I have observed the highest planet [Saturn] to be tripled-bodied," he said. "This is to say that to my very great amazement Saturn was seen to me to be not a single star, but three together, which almost touch each other."

Two years later, he looked again for them, and they were gone. He was astounded. At that time, he wrote, "I do not know what to say in a case so surprising, so unlooked for, and so novel." The rings hadn't really disappeared, however. It's just that as Saturn and Earth's relative positions shifted, Galileo was seeing the ring edge-on, or from the side. With only a 20-power telescope, that meant he saw nothing at all.

But a few years later, in 1616, he observed the rings as two half-ellipses. He wrote, "The two companions are no longer two small perfectly round globes … but are present much larger and no longer round … that is, two half ellipses with two little dark triangles in the middle of the figure and contiguous to the middle globe of Saturn, which is seen, as always, perfectly round."

In 1655 the Dutch astronomer Christian Huygens advanced a better explanation for what astronomers had seen: He proposed that a solid ring, "a thin, flat ring, nowhere touching, and inclined to the ecliptic," surrounded Saturn. Then in 1659 he published a book in which he explained that every 14 to 15 years the Earth passes through the plane of Saturn's ring. That is, most of the time Earth looks either up to Saturn's rings from slightly beneath, or down on them from slightly above. The transition from one position to the other is known as a Saturn ring plane crossing. Galileo unintentionally became the first person to see one.

Vocabulary

- differential rotation
- oblate
- protoplanet
- occultation

Voyager 1 took this photo of Jupiter when the probe was about 25 million miles (40 million kilometers) from the planet.

Courtesy of Voyager 1, NASA

Jupiter

You read in Lesson 1 that the outer, or Jovian planets, are different in many ways from the terrestrial planets. One of those differences is size. Jupiter, named for the most powerful ancient Roman god, is the Solar System's largest planet. It's the closest to the Sun of the Jovian planets.

Even without a telescope, Copernicus figured out that Jupiter was larger than Venus. Venus at its brightest is brighter than Jupiter. But Copernicus knew that Jupiter was much farther away. And, since he knew Jupiter shines only by the Sun's reflected light, he concluded that Jupiter must be very large to appear as bright as it does.

Jupiter as Seen From Earth and From Space

Scientists have calculated Jupiter's mass by observing its moon's periods of revolution and the radii of their orbits. And they have concluded that Jupiter is big indeed. It has more than twice the combined mass of all the other planets, their moons, and the asteroids. Jupiter has 318 times the mass of Earth.

Table 3.1 Distances of the Terrestrial and Jovian Planets From the Sun

Planet	Distance
Mercury	0.4 AU
Venus	0.7 AU
Earth	1.0 AU
Mars	1.5 AU
Jupiter	5.0 AU
Saturn	10.0 AU
Uranus	19.0 AU
Neptune	30.0 AU
Note: 1 Astronomical Unit (AU) = 93 million miles (150 million km)	

Jupiter's volume exceeds Earth's by an even greater ratio. Jupiter's diameter is about 11 times that of Earth. And so its volume is 1,400 times that of Earth.

Jupiter isn't quite the lumbering, slow-moving giant you might picture, though. True, it takes Jupiter nearly 12 Earth years to cycle around the Sun. But it spins around on its axis once every nine hours, 55 minutes. Fast rotations like this are the norm in this part of the Solar System.

The *Pioneer* and *Voyager* missions during the 1970s taught scientists a lot about Jupiter. Scientists not only got thousands of photographs out of the missions, but the spacecraft instruments picked up data on charged particles, radiation from the planet, and Jupiter's magnetic field. Between 1995 and 2003 another spacecraft dubbed *Galileo* orbited Jupiter and its moons and then finally plunged into the huge planet's atmosphere. Before that, *Galileo* dropped a probe into Jupiter's atmosphere that gathered information on the planet's weather patterns.

Jupiter's Rotation

Jupiter's rotation differs from Earth's in more than just its speed. Even through a small telescope, you can see that Jupiter has light- and dark-colored bands parallel to its equator. These bands rotate at slightly different speeds. The bands near the equator move slightly faster than those around the poles, completing their circuit in nine hours, 50 minutes. Bands closest to the poles complete a rotation in nine hours, 56 minutes.

The phenomenon of different parts of a planet having different periods of rotation is known as differential rotation. It indicates that Jupiter's visible surface isn't solid. It must be at least partially fluid. When you stir a pot of stew, the liquid moves easily, but the chunks of meat may not. Something similar takes place on Jupiter.

Another thing to notice about the planet: It's somewhat oblate, or *flattened at the poles*. This is an effect of Jupiter's swift rotation. Mars and Earth are also oblate, but not to the same degree. Jupiter's equatorial diameter is 6.5 percent greater than its diameter measured over the poles.

You can easily see the bands around Jupiter in this photo taken from Earth. The circle has been drawn to show how oblate Jupiter is.

Courtesy of California Institute of Technology

The Composition of Jupiter's Atmosphere

The terrestrial planets are hard and rocky. But Jupiter and the other Jovian planets are different. They're more like the Sun. The *Galileo* spacecraft found Jupiter's atmosphere to be about 90 percent hydrogen and 10 percent helium, with small amounts of methane, ammonia, and water vapor. Scientists believe the original solar nebula had a similar makeup.

Galileo also found small amounts of certain heavier elements—carbon, nitrogen, and sulfur. These were present in Jupiter's atmosphere at three times the level at which they are found in the Sun. That suggests that meteorites and other small objects have made the planet what it is. *Galileo's* instruments came across few signs of complex organic compounds. Thus, scientists are unlikely to find life on Jupiter as we know it on Earth.

But the *Galileo* probe did *not* detect the thick, dense clouds scientists expected. The probe instead revealed that Jupiter has the same helium content as the Sun's outer layers, but only about one-tenth as much neon. The Sun's outer layers lose helium. And so the *Galileo* findings suggest that Jupiter has some mechanism that removes helium and neon from its upper atmosphere. Jupiter may have helium rain "showers" in its upper atmosphere, with neon dissolving in helium raindrops under certain conditions.

Galileo made yet another discovery, one that challenges current theory on planet formation. The spacecraft's probe found that three so-called "noble" gases—argon, krypton, and xenon—are two to three times as prevalent on Jupiter as in the Sun. The only way for Jupiter to get such quantities of these gases would be to trap them by condensation or freezing. But that requires very cold temperatures, colder than Pluto, in fact.

As you read in Lesson 1, current theory holds that material from the original solar nebula were the building blocks for Jupiter and the other giants. Planetesimals clumped together into a protoplanet, according to this theory. A protoplanet is *a hypothetical whirling gaseous mass within a giant cloud of gas and dust that rotates around a sun and becomes a planet.* After it got big enough, a protoplanet would start sweeping up gas directly from the nebula.

The new data, however, suggest that the material that makes up Jupiter must have come from a much colder place than Jupiter's current location. This may indicate that the original solar nebula was much colder than we think. Or it may be that planetesimals began to form even before the solar nebula formed. Or Jupiter may have formed farther away from the Sun and drifted toward it over time. This last explanation is particularly interesting because scientists have recently discovered many planetary systems where large planets are very close to their stars.

Jupiter's Weather and Its Great Red Spot

Imagine a hurricane with wind speeds of up to 300 miles (500 km) per hour—about twice the speed needed for classification as a Category 5 storm. Imagine that this storm covers an area as big as the entire Earth. Then imagine that this storm lasts for centuries.

Now you have a handle on Jupiter's Great Red Spot. Humans have been studying it since about the middle of the seventeenth century—since the advent of the telescope. This Jupiter-size hurricane has continued to rage ever since—rising and falling in intensity, but never going away.

The Great Red Spot follows the same basic laws of physics as the much smaller storms your hometown weather forecaster tells you about. It's a system of rising high-pressure gas whose cloud tops are colder and higher (about five miles higher) than the surrounding regions. It's in Jupiter's southern hemisphere, and so it rotates counterclockwise, completing a circuit about once every six days.

Jupiter's weather patterns may be larger scale than Earth's. But they're simpler, in part because they unfold far above whatever solid surface the planet may have. That's probably why a hurricane can last for three centuries.

Besides the Great Red Spot, Jupiter's most significant visible features are the bands mentioned earlier. The light-colored bands are called *zones* and the dark-colored bands are named *belts*. Both are made up of clouds. The Great Red Spot is in between a zone and a belt.

Scientists expect to learn a lot about weather on Earth from studying Jupiter's weather. The practice of studying a simple system to figure out the workings of a complicated one is common in the scientific world.

This photo shows Jupiter's Great Red Spot in the upper right corner, with a photo of Earth superimposed at the same scale. The Great Red Spot is an immense hurricane that has been raging at least since the middle of the seventeenth century.

Courtesy of NASA/Stephen P. Meszaros

Jupiter's Three Groups of Moons

As of 2009 the family of Jupiter's known moons numbers 63. Scientists put these moons into three groups. The first four orbit very close to Jupiter. They are called *fragmented moonlets*.

The second group is the four Galilean satellites: Io, Europa, Ganymede, and Callisto. They are about the same size as Earth's Moon. Their orbits are nearly perfect circles. The smallest of the "big four" that Galileo discovered, Europa, is more massive than the largest of the non-Galilean moons by a factor of 7,000.

The Galilean moons present some amazing features. Io is covered with sulfur volcanoes and geysers that constantly renew its surface. Europa has an iron core covered by a frozen and liquid ocean of water 60 miles (100 km) deep. Its magnetic field reverses every 5½ hours. Ganymede is the largest moon in the Solar System— bigger than Mercury. Photos show a surface of ice, but also craters. This indicates that its surface is less active than Europa's. It appears to have a small iron or iron/sulfur core surrounded by a rocky mantle and a shell of ice perhaps 497 miles (800 km) thick. Finally, Callisto has the most cratered surface and the largest impact crater of any observed object in the Solar System. This is typical of a very inactive surface. Callisto appears to have a relatively uniform mixture of ice (40 percent) and rock (60 percent), with an increasing percentage of rock toward the center.

The third group consists of the remaining 55 moons. Many of these orbit clockwise, unlike most objects in the Solar System, and their orbits are fairly eccentric. Astronomers speculate that these moons are captured asteroids; that is, that they used to orbit the Sun but veered close enough to Jupiter to be captured by its gravity. Scientists also suggest that long ago, Jupiter had a more extensive atmosphere. This might have slowed down the asteroids so the planet could capture them.

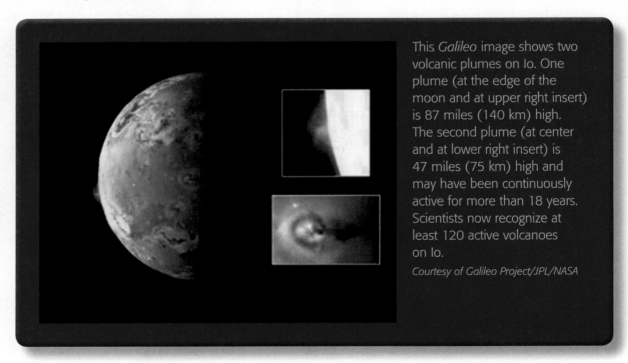

This *Galileo* image shows two volcanic plumes on Io. One plume (at the edge of the moon and at upper right insert) is 87 miles (140 km) high. The second plume (at center and at lower right insert) is 47 miles (75 km) high and may have been continuously active for more than 18 years. Scientists now recognize at least 120 active volcanoes on Io.

Courtesy of Galileo Project/JPL/NASA

Jupiter's Rings

One of the surprise discoveries of *Voyager I* was that, like Saturn, Jupiter has rings. Not until the explorations of *Galileo*, though, did scientists get good images of Jupiter's rings. The rings could only be seen looking back toward the Sun. The light that reached the camera from the rings was therefore scattered by the ring material. Because large chunks of matter cannot scatter light in this manner, scientists know that the rings are made up of tiny particles, as fine as cigarette smoke, and they orbit fairly close to the planet. Scientists calculate that the particles making up the rings have not been there since the Solar System's beginning. Radiation pressure from the Sun, plus Jupiter's own strong magnetic field, would send particles either into Jupiter's surface or off into space. So scientists think the rings are continually replenished, probably because of meteoroid impacts on nearby moonlets.

The *Galileo* spacecraft caught this image of Jupiter's rings on 9 November 1996.
Courtesy of Galileo Project?JPL/NASA

Saturn

Like Jupiter, the planet Saturn has a Roman god for its namesake. In Roman mythology, the god Saturn was Jupiter's father. The planet Saturn is probably the most impressive object visible with a small telescope. The astronomer Galileo first observed Saturn in 1610. He didn't know what to make of the "bumps" he saw on either side of Saturn. Some 50 years later the Dutch physicist and astronomer Christian Huygens recognized that the "ears" on Saturn were really rings.

Saturn's Size, Mass, and Density

Saturn is not much smaller in diameter than Jupiter. But it's only half as dense. It has only 0.7 the density of water. Saturn likely has a less dense core and less liquid metallic hydrogen than Jupiter does. Saturn's atmosphere is like that of Jupiter and the Sun: about 96 percent hydrogen, 3 percent helium, and 1 percent heavier materials.

Star POINTS

Christian Huygens (1629–1695) was a Dutch physicist and astronomer. He made major advances in the field of optics and discovered Saturn's moon Titan. The probe dropped onto Titan from the *Cassini* spacecraft was named for him.

Saturn

These 1996 images from the Hubble Space Telescope show changing views of Saturn as it moves from fall to winter in its northern hemisphere.

Courtesy of NASA and the Hubble Heritage Team, STScI/AURA/Acknowledgement: R.G. French (Wellesley College), J. Cuzzi (NASA/Ames), L. Dones (SWRI), J. Lissauer (NASA/Ames)

Saturn's Speed of Rotation and Solar Orbit

Saturn takes 29.5 Earth years to orbit the Sun. Like Jupiter, it has a short "day"—it rotates on its axis in 10 hours, 39 minutes. Also like Jupiter, Saturn experiences differential rotation. The figure just given for its day refers to its rate of rotation at the equator. Saturn is also oblate, more so than any other planet. This is because it doesn't have enough gravitational force to keep a more-spherical shape.

At various points in its orbit, Saturn will appear very different to someone observing the planet from Earth. Saturn's rings are in the plane of its equator, and the planet tilts 27 degrees with respect to its orbital plane. So as both Earth and Saturn orbit the Sun, earthlings will sometimes see the edge of Saturn's rings, and at other points during Saturn's orbit they will observe the rings' "top" and "underside."

Titan, Saturn's Largest Moon

Titan, the largest of Saturn's more than 60 moons, may turn out to be the most interesting moon in the Solar System. It's the second largest, after Ganymede, Jupiter's largest moon. It has an atmosphere that extends 10 times as far out into space as Earth's does today. And scientists think Titan's atmosphere is like Earth's long ago. It is mostly nitrogen, plus some methane and argon, along with traces of ethane and carbon monoxide. As ultraviolet light from the Sun breaks down the methane in Titan's atmosphere, it forms organic molecules that slowly drift down to the surface. It's like smog over a big city.

That methane is still present on Titan suggests that the gas is probably locked with water ice in a crust above an ocean of ammonia and liquid water. Methane escapes into the atmosphere as part of what are known as "outgassing" events.

Titan is only slightly larger than Mercury, which has no appreciable atmosphere. Titan, however, does have an atmosphere—much denser and more massive than Earth's, in fact. Titan's low temperatures have let it hang on to its atmosphere. Surface temperatures of minus 180 degrees slow gas molecules down to prevent their escape. However, data from the Huygens probe suggest that at one point, Titan had an atmosphere five times as dense as it is today. That implies that Titan is losing material from the top of its atmosphere into space.

The Particles That Form Saturn's Rings

Saturn's rings are extremely thin. You might think of them as a compact disc. The rings are made up of chunks of water ice and smaller bits of rock and organic matter. There's a lot of empty space between the chunks, too. Each individual rock or snowball making up the rings is, in a sense, a separate satellite of Saturn.

As seen from Earth, Saturn has three rings, labeled (from outer to inner) A, B, and C. Photographs taken in space show that there are actually more than three rings, and that the spaces between rings have different causes. One cause of the spaces is gravitational tugs from small moons orbiting nearby. The moon Mimas, for example, tugs regularly at particles caught in the break between rings A and B and tends to pull them out of it. The Italian astronomer G. D. Cassini discovered this gap, now known as Cassini's division.

The Origin of Saturn's Rings

Scientists don't know for sure what caused Saturn's rings. The most likely scenario is that an icy moon once orbited near Saturn but then was struck by a passing asteroid or comet. The impact would have shattered the moon. Another theory is that an object came too close to Saturn and the planet's gravity shattered it. In either case particles would have ultimately arranged themselves in flat rings like those orbiting Saturn today.

The Hubble Space Telescope took this photo of Uranus in 2003 using filters sensitive to red, green, and blue light. This is how the planet would appear if you could see it at this size through a telescope.

Courtesy of NASA/Erich Karkoschka, University of Arizona

Uranus

The third planet of the Jovian quartet, Uranus, is visible to the naked eye—just barely. So the ancient Greeks surely saw it. They just didn't recognize it as a planet.

What Led Herschel to Discover the Planet Uranus

By 1690, when telescopes had been in use for nearly a century, Uranus was showing up on star charts. But astronomers didn't call it a planet because, with their still fairly rudimentary telescopes, Uranus appeared only as a speck of light, not as a disk. It took so long to orbit the Sun that astronomers found Uranus hard to track.

Then in 1781 the English astronomer William Herschel, who had some of the best telescopes available at the time, noticed that this object did not appear as a point of light, as stars do. Rather, it appeared as a disk when magnified. Over the period of a few nights, it moved. That suggested a different kind of celestial object, one evidently much closer to Earth than anyone had realized.

Herschel, the Musician/Astronomer

Friederich Wilhelm Herschel (1738–1822) was a German-born astronomer known as William once he settled in England at age 19. He continued a tradition of early astronomers who excelled in more than one major field. A musician by day, he spent his evenings observing the heavens. He built his own telescopes, too, and their quality was an immense advantage to him. He could see things others couldn't. This meant, though, that others hesitated to welcome him into scientific circles. Eventually, though, his great discovery of Uranus won his critics over. King George III gave him a pension that allowed him to devote himself full time to astronomy.

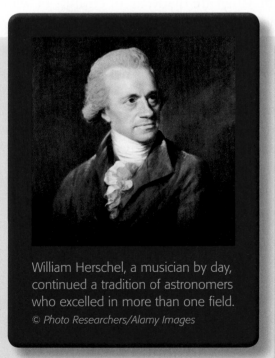

William Herschel, a musician by day, continued a tradition of astronomers who excelled in more than one field.

© Photo Researchers/Alamy Images

At first Herschel thought it was a comet. But some calculations showed that the object's orbit was nearly circular. Herschel realized he had discovered a new planet. It was the first discovery of a new planet in recorded history.

The Motion of Uranus and the Tilt of Its Equatorial Plane

Uranus takes 84 Earth years to orbit the Sun. No wonder the ancients didn't recognize it as a planet!

Uranus has another claim to fame, too. All the planets you have read about so far rotate about axes that are tilted less than 30 degrees from their plane of revolution around the Sun. That's true of Saturn, whose relatively sharp angle gives people on Earth such a widely varying view of the ringed planet over time. It's true of Venus, too—even though Venus rotates clockwise, unlike the other planets. What's unique about Uranus: Its axis is tilted nearly 90 degrees to its orbital plane (Figure 3.1). In other words, it spins on its side. Sometimes its north pole points right at the Sun; at other times, it points away.

You might expect this to affect the weather on Uranus. After all, you know that the tilt of Earth's axis of rotation creates the seasons here. Earth's poles alternate with six months of sunlight and six of darkness. Because of its tilt, the poles on Uranus alternate between 42 years of sunlight and 42 years of darkness. But on Uranus temperatures are fairly uniform: uniformly cold, that is—about minus 319 degrees F (minus 195 degrees C) over the planet's entire surface (Figure 3.2).

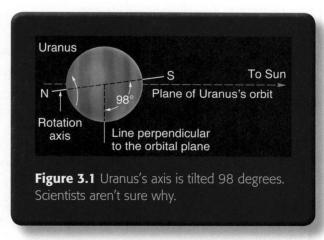

Figure 3.1 Uranus's axis is tilted 98 degrees. Scientists aren't sure why.

Scientists generally think Uranus's odd tilt might have been the result of a glancing blow from another object. But the planet's 27 moons still orbit in its equatorial plane, which doesn't quite square with the glancing-blow theory. A new hypothesis speculates that an impact may have occurred during a very early stage of the moons' formation. Computer simulations show that Uranus's tilt could have arisen naturally as a result of strong gravitational interactions among the bodies of the Solar System during its early years.

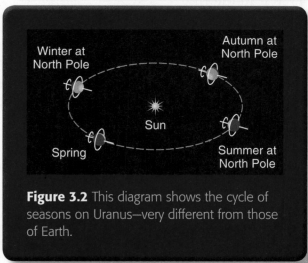

Figure 3.2 This diagram shows the cycle of seasons on Uranus—very different from those of Earth.

How Uranus's Moons Act as Shepherds for the Particles of Its Rings

It wasn't until 1977—nearly two centuries after Hershel's discovery of Uranus as a planet—that scientists made the first reliable determination of its diameter. They used an occultation—*the passing of one astronomical object in front of another*—to do this. They knew how fast Uranus travels and how far from the Sun its orbit was. They waited for Uranus to pass in front of a star. This way they could measure how long the planet blocked the star's light and from that calculate Uranus's diameter. (*Occult* has many meanings, but as a verb in astronomy, *to occult* means to hide, cover, or conceal.)

This occultation in 1977 not only helped measure Uranus. It revealed that the planet has five rings. These rings are narrow, with well-defined edges. These discoveries soon prompted astronomers Peter Goldreich and Scott Tremaine to offer an explanation as to why the rings stay in these bands. Their hypothesis was that a narrow planetary ring will have a pair of "shepherd moons"—one orbiting just inside and the other orbiting just outside the ring.

According to Kepler's law, the inner moon will orbit faster than the ring's particles. As it passes individual particles—like one car overtaking another on the highway— it pulls them along, giving them an energy boost. This tends to push them away from the planet, making them orbit it slightly farther away. Similarly, particles passing the outer moon are slowed somewhat while doing so. This tends to make them move closer to the planet.

Goldreich and Tremaine formed their hypothesis in response to discoveries about Uranus. They thought they would have to wait some time before spacecraft could provide photo confirmation. But they got their answer sooner than expected. *Voyager 2* photographed Saturn's rings and found a pair of tiny moons on either side of the F ring. As predicted, they act as shepherds for the flock of particles in the ring.

Neptune

If the story of Uranus was one of recognizing a planet that's visible, the story of Neptune was one of astronomers knowing there had to be something there— and looking until they found it.

What Made Scientists Search for Neptune

Once Herschel identified Uranus as a planet in 1781, astronomers went back to earlier charts to track its orbit. (The data points in these older charts were still valid, even if scientists had misclassified the object as a star.) Soon it became clear that Uranus wasn't behaving quite the way Newton's laws predicted. Most astronomers thought another planet was somehow disturbing Uranus's orbit.

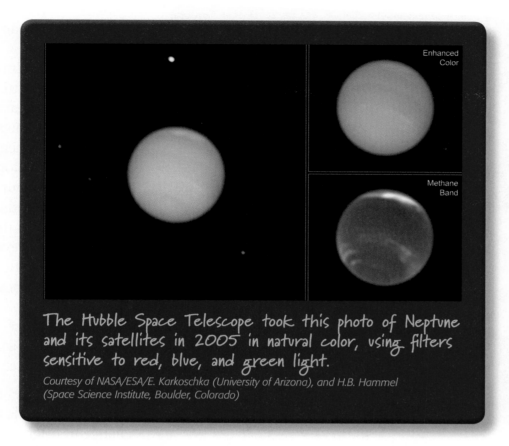

The Hubble Space Telescope took this photo of Neptune and its satellites in 2005 in natural color, using filters sensitive to red, blue, and green light.

Courtesy of NASA/ESA/E. Karkoschka (University of Arizona), and H.B. Hammel (Space Science Institute, Boulder, Colorado)

But an influential minority thought that maybe the law of gravity applied a little differently that far from the Sun.

A young British mathematician named John C. Adams set about calculating the position of the unknown planet he was sure was out there. When his work was finally complete, he could not interest Sir George Airy, the British Astronomer Royal, in his ideas or get permission to use telescope time to look for the new planet. Meanwhile, in France, the mathematician Urbain Le Verrier was making similar calculations—and likewise having trouble getting telescope time to search for the planet.

Finally Le Verrier wrote to Johann Galle at the Berlin Observatory to ask him for help. Galle convinced his boss of the worthiness of the request the day Le Verrier's letter arrived. Galle located the planet that evening, 23 September 1846—and almost exactly where Le Verrier said it would be.

Neptune's Wind Speeds and Differential Rotation

Scientists have known of Neptune's existence for well more than a century and a half. But most of what they really know about the planet has come from *Voyager 2*, which flew past Neptune in 1989.

Star POINTS

Neptune's discovery is especially interesting because it highlights some very human aspects of science: the challenge young scientists have in getting the attention of the scientific establishment; the competition for telescope time; and the controversies over who found what first.

The Great Dark Spot of Neptune has clouds made up of methane crystals.

Courtesy of NASA/JPL-Caltech

Like the other Jovian planets, Neptune experiences differential rotation—to an extreme degree, in fact. Near the equator, it rotates in 18 hours, but near the poles, its rotation takes about 12 hours. This variation holds only at the top layer of the atmosphere, however. Neptune's magnetic field rotates every 16 hours, seven minutes, so scientists consider this period the basic "day" on Neptune.

Neptune's surface has parallel bands that change in brightness over time. *Voyager 2* images revealed a Great Dark Spot on Neptune's surface. Like Jupiter's Great Red Spot, it was as big as Earth, was essentially a storm system, and had extremely high-speed winds. In fact, the Great Dark Spot's winds reached up to 1,500 miles (2,400 kilometers) an hour—nearly five times the wind speed of Jupiter's Great Red Spot. The Dark Spot had vanished when the Hubble Space Telescope viewed Neptune in 1994, but then another similar storm appeared.

Star POINTS

By the late twentieth century, scientists believed that Neptune had rings—but incomplete ones that didn't circle the entire planet. *Voyager* photos have since shown that Neptune's rings are indeed complete—but they're a bit "lumpy." This may be because of undiscovered moons within them. In any case, scientists now know that each of the Jovian planets has a ring system.

Neptune shows signs of much stronger weather patterns than Uranus. And, remarkably for a planet so far from the Sun, Hubble's observations of strong winds and storm systems suggest that Neptune has an annual cycle of seasons. But like Uranus—which doesn't seem to have seasons—Neptune has notably uniform temperatures at the poles and the equator.

The Unusual Orbits of Neptune's Two Major Moons

Astronomers have long known Neptune has two moons—Triton and Nereid. Since Voyager visited, the tally is up to 13. But the two longest-known are the most interesting. Triton, the largest moon, and the seventh-largest in the Solar System, revolves clockwise around its planet. As you read earlier, some of Jupiter's smaller moons revolve this way—but Triton is the only major moon to do so.

Triton is as unusual as its orbit. The area around the south pole features very irregular terrain. The rest of the moon is much smoother. The surface consists primarily of water ice, with some nitrogen and methane frost. The surface is very young, and has active volcanoes that shoot nitrogen ice six miles (10 km) above the moon's surface. Triton may give scientists a glimpse of what the surface of Pluto looks like, although important differences exist between the two.

Nereid, the other major moon of Neptune, revolves in the "correct" direction—counterclockwise. But it has the most eccentric orbit of any known moon in the Solar System. This means that its distance from Neptune varies widely over the course of its orbit. Its apogee—farthest point from Neptune—is five times the distance of its perigee—its closest point. Nereid takes 360 days to orbit Neptune. Its "month" lasts nearly as long as an Earth year, in other words.

The southern part of Triton has been exposed to sunlight for 30 years, and Voyager photographed erupting volcanoes (not obvious here).

Courtesy of NASA/JPL-Caltech

Scientists speculate that both Triton and Nereid are objects that Neptune "captured" after the Solar System first formed. Much remains to be learned, however, before this thesis can be completely accepted.

While the Sun, planets, and moons are the largest objects in the Solar System, there are many other things whirling around the Sun. Dwarf planets, asteroids, comets, and other objects are also members of the neighborhood. Scientists have known about some of these for hundreds of years. They learned about others only well into the twentieth century. The next lesson will take a look at some of what they have learned and what it reveals about the Solar System.

CHAPTER 3   The Sun and the Solar System

CHECK POINTS

Lesson 3 Review

Using complete sentences, answer the following questions on a sheet of paper.

1. How big is Jupiter compared with the rest of the planets in the Solar System?

2. What is Jupiter's differential rotation and what does it indicate?

3. What did Galileo find out about Jupiter's atmosphere?

4. What do astronomers speculate about Jupiter's 55 outermost moons?

5. Saturn's atmosphere is like those of which two other bodies in the Solar System? How so?

6. Why does Saturn appear so different from Earth at different points in its orbit?

7. Why do scientists believe that Titan is losing material from the top of its atmosphere into space?

8. What makes up Saturn's rings?

9. The astronomer Herschel's calculations of a celestial object he was studying in 1781 showed that its orbit was nearly circular. What was the significance of that?

10. What's unique about Uranus's rotation?

11. What feature of Uranus did scientists discover in 1977 while measuring the planet? What are the key characteristics of this feature?

12. What did most astronomers think when it became clear that Uranus wasn't behaving quite the way Newton's laws predicted?

13. How was Neptune's Great Dark Spot like Jupiter's Great Red Spot?

14. What is unusual about the orbit of Triton? Of Nereid?

APPLYING YOUR LEARNING

15. What does the discovery of noble gases on Jupiter tell you about current theories on the Solar System's origins?

LESSON 4 | Dwarf Planets, Comets, Asteroids, and Kuiper Belt Objects

Quick Write

Halley's experience calculating the orbits of comets tells you something about the great timeframes astronomers must deal with. Describe an instance in which you or someone you know of had to think in terms of a long time frame— perhaps by making plans well in advance, or starting in on a task without being sure what the outcome would be?

Sir Edmund Halley did not discover the comet that bears his name. But he did something even more significant: He demonstrated that comets, like planets, orbit the Sun. He realized that the comet he had observed in 1682 was the same comet that had also appeared in 1531 and 1607. He applied Kepler's ideas about the elliptical paths of orbiting objects to comets as well as planets. By calculating the comet's orbit, he provided a direct confirmation of the work of his friend, Sir Isaac Newton.

In 1705 he predicted that the comet would return in 1758 or 1759. He knew he wouldn't live to see it. But he expressed hope that "candid posterity will not refuse to acknowledge that this was first discovered by an Englishman."

Comet Halley most recently passed near Earth in 1986. Several Soviet, Japanese, and European satellites passed by and photographed the comet.

Learn About

- Pluto
- asteroids
- comets
- the Oort cloud and Kuiper Belt

Pluto

As you read in the last lesson, Uranus was the planet that the ancient Greeks undoubtedly saw, but failed to recognize as a planet. Neptune was the planet that scientists concluded had to be there, somewhere in the heavens. They could tell that something was tugging on Uranus, affecting its motion in ways not explained by Newton's laws or the presence of Jupiter and Saturn. But after the discovery of Neptune, astronomers found that it didn't explain all of Uranus's movements. They determined that something else was out there pulling on Uranus. So they went looking again. What they found was Pluto.

The Discovery of Pluto

Percival Lowell, a businessman and astronomer, predicted the existence of an additional planet in 1905. He searched for "Planet X" until his death in 1916, but with no success. On 1 April 1929, however, Clyde W. Tombaugh started a new search. Instead of scanning the skies for a small, faint disk, he looked for signs of planetary motion.

Tombaugh used a device called a blink comparator. It allows an observer to look alternately at two different photographs of the sky. An astronomer could take two pictures of the same part of the sky a few days apart. If a moving object, such as a planet or a comet, is in that part of the sky, it will appear in different positions in the two pictures. The quick blink from one image to another makes a moving object appear to jump from one spot to another.

It was unbelievably tedious work, made even more difficult because the place in the sky that Lowell's data pointed to was a region with many faint stars—a very crowded neighborhood in space. But on 18 February 1930, more than 10 months into his search, Tombaugh spotted the "jump" of a tiny dot on a photograph. After confirming his discovery by comparing even more photos, he announced his discovery on 13 March 1930. It would have been Lowell's 75th birthday. Tombaugh named his discovery, hailed at the time as the ninth planet, Pluto, for the Greek god of the underworld. The abbreviation for Pluto is PL—Percival Lowell's initials.

Vocabulary

- astronomical unit
- asteroid belt
- nucleus
- coma
- tail of a comet
- Oort cloud
- Kuiper Belt
- meteor
- fireball

Star POINTS

After other astronomers confirmed his discovery of Pluto, young Clyde Tombaugh set off for college, enrolling as a freshman at the University of Kansas.

These are the two photos that Clyde Tombaugh compared to identify Pluto, marked with arrows in each image.

Courtesy of Lowell Observatory

Pluto's Orbit and Atmosphere

Pluto is a true eccentric in the Solar System. Its orbit is more eccentric than that of any of the eight planets. That is, all of them are more nearly circular in their orbits. This eccentricity means that Pluto varies widely in its distance from the Sun. It averages 40 astronomical units from the Sun, but it ranges from 30 units to 50 units. An astronomical unit, or AU, is *the mean distance between the Earth and Sun, about 93 million miles (150 million km)*. This means that it sometimes swings inside Neptune's orbit, as it did from 1979 to 1999. Pluto's orbit also tilts 17 degrees to the ecliptic. None of the planets' orbits tilt more than 7 degrees.

Scientists have determined that Pluto has an atmosphere, but not year-round. When the planet is at perihelion—its closest distance from the Sun—it "warms up," relatively speaking. At this point it has an atmosphere, including nitrogen and traces of methane and carbon monoxide. But when Pluto is at aphelion—the point of its greatest distance from the Sun—it receives so little solar energy that its surface is below the temperature at which methane freezes. This chilling causes the gas to condense on Pluto's surface.

Star POINTS

In August 2006 the International Astronomical Union decided to downgrade Pluto's status to that of a "dwarf planet."

Pluto may have lost some status when scientists downgraded it to a dwarf planet in 2006. But they are finding it a more complex system than they used to think. Pluto's atmosphere is undergoing a global cooling, but its surface is getting warmer. Its surface temperatures, moreover, are not uniform. In 1956 scientists discovered that Pluto's brightness changes slightly every 6.4 days. From this they have concluded that it has a dark area on its surface and that its period of rotation is 6.4 days.

The Synchronous Orbit of Pluto's Moons

In 1978 James W. Christy of the US Naval Observatory noticed that Pluto seemed to have a bump on one side. Once continued observation showed that the bump moved from one side to another, scientists concluded that Pluto had a moon. They named it Charon, for the mythical boatman who ferried souls to the underworld, where Pluto reigned. Charon's orbit tilts 61 degrees relative to Pluto's orbit around the Sun (Figure 4.1). That's an even greater angle than Pluto's is to the ecliptic. Charon and Pluto are locked in synchronous orbit. Each keeps the same "face" to the other constantly, and the two bodies circle around their common center of mass once every 6.4 days—the same period as Pluto's "day." This arrangement means that each of these objects always hovers in the sky of the other. There is no moonrise or "planet-set," unlike on Earth and its moon.

Since discovering Charon, scientists have learned a lot about Pluto, but there's still plenty they don't know. Between 1985 and 1990 Charon occulted (concealed from view) Pluto during each of its revolutions. By measuring the time each occultation took, scientists got a better idea of how big each object is. Still, they don't know much about Pluto's atmosphere, and that makes it hard to know just what its diameter is.

Pluto's mass is about eight times that of Charon, but only about one-fifth of the mass of Earth's Moon's. Pluto's density suggests an interior that's probably 30 percent ice and 70 percent rock. Scientists actually know more about Charon's diameter (about 750 miles or 1,212 kilometers) and density. This density, slightly less than Pluto's, suggests a composition of rock and ice, with a surface mostly of water ice. In 2005 the Hubble Space Telescope discovered two more moons around Pluto, Nix and Hydra, each less than 100 miles (160 kilometers) in diameter.

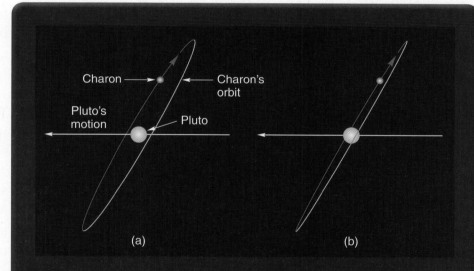

Figure 4.1 These two images illustrate how the view from Earth of Pluto and Charon has shifted since Charon was discovered in 1978. (a) This represents the view from Earth when Charon was discovered. (b) A decade later, Pluto had moved such that Charon passed in front of and behind Pluto. The two bodies are not drawn to scale.

Asteroids

Besides planets and dwarf planets, such as Pluto, astronomers estimate there are millions, even billions of other objects floating around in the Solar System. On 1 January 1801, Father Giuseppe Piazzi discovered the first asteroid, Ceres, and named it for the patron goddess of his native Sicily. Recall that asteroids are, at their most basic, rocks that range in size from about 1 mile (1.6 km) in diameter to several hundred miles wide and that travel through space.

Ceres is the Solar System's largest asteroid, with a diameter of 580 miles (930 km). Scientists now classify Ceres as a dwarf planet. About 130 other asteroids also have impressive diameters, ranging between 60 miles and 310 miles (100 km to 500 km). Some millions of others are about two-thirds of a mile wide (about 1 km) or less.

This time-exposure photo shows how asteroids appear as streaks against a background of stars. Note the two arrows identifying the asteroids.

Courtesy of Yerkes Observatory

The spacecraft *Galileo* took this picture of the asteroid Gaspra while en route to Jupiter. Gaspra measures about 12 miles (19 km) from end to end.

Courtesy of Galileo Project/NASA/JPL

Most asteroids go nameless. Astronomers don't bother naming them until they have calculated their orbits. And calculating an asteroid's orbit can take a lot of work without yielding much astronomical knowledge. Scientists have two main ways of classifying asteroids: by position and by chemical composition.

The Location of the Asteroid Belt

The asteroid belt, as its name suggests, is *the region between Mars and Jupiter where most asteroids orbit*. Most asteroids orbit the Sun at distances from 2.2 astronomical units (AU) to 3.3 AU. This corresponds, according to Kepler's third law, to orbital periods ("years") of 3.3 years to 6 years. Most asteroids have fairly circular orbits, but these orbits are still more elliptical than the planets.

CHAPTER 3 The Sun and the Solar System

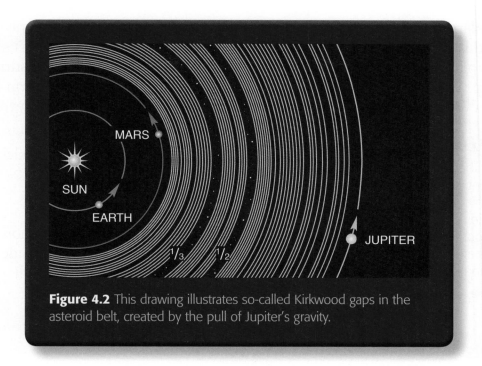

Figure 4.2 This drawing illustrates so-called Kirkwood gaps in the asteroid belt, created by the pull of Jupiter's gravity.

The Orbits of Asteroids Around the Sun

About 100,000 asteroids are big enough to show up in photographs. This may suggest that the asteroid belt is a sort of rocky traffic jam in orbit around the Sun. In fact, though, the asteroid belt is a mostly empty zone. Even if all the biggest 100,000 asteroids were in exactly the same plane, spread evenly across the asteroid belt, there would be more than a million miles (2 million km) separating them. And there's even more space than that because the asteroids aren't all in the same plane.

They aren't spread evenly either. In Chapter 3, Lesson 3, you read about how the moons of the Jovian planets exert a gravitational pull on the rings of those planets to create gaps in them, such as the Cassini division. The same phenomenon occurs at a larger scale in the asteroid belt. Jupiter's gravitational pull has created gaps in the asteroid belt at 2.50 AU and 3.28 AU. These gaps, first explained by the American astronomer Daniel Kirkwood in 1866, aren't completely empty. But they contain fewer asteroids than neighboring areas of the asteroid belt (Figure 4.2).

The Origins of Asteroids

Astronomers once thought that asteroids were the remains of a planet that had exploded. Nowadays they've abandoned this theory for two compelling reasons. For one, scientists have no idea how a planet could explode. They don't understand how that could be physically possible. The other problem is that the asteroids don't really add up to much. That is, if all the asteroids were combined into one object, it would be much smaller than Earth's Moon. Such a small object doesn't fit the pattern of planetary sizes.

The explanation that seems more plausible today is that asteroids are simply primordial material that never formed into a planet. The reason they didn't seems to be Jupiter. Its gravity keeps pulling on objects in the asteroid belt and stirring them up. Without Jupiter, these objects would tend to pull together to form larger objects. But Jupiter's gravity keeps that from happening.

And although there isn't a model for a planet exploding, there is evidence that Jupiter causes collisions between asteroids that result in their fragmentation. Some asteroids that orbit outside the main belt may be the result of such collisions.

Star POINTS

In 2006, for the first time, scientists linked a breakup event in the asteroid belt with a large quantity of interplanetary dust deposited on Earth. Samples of the Earth's core taken from the ocean floor led them to conclude that 8.2 million years ago, an asteroid named Veritas blew apart and dusted Earth with its remains.

Comets

Also circling the Sun, but from much farther out than asteroids, are comets. A comet is one of the most spectacular astronomical sights available to the naked eye. But if you know comets only from photographs, you may expect them to be even more spectacular than they are. They don't flash across the sky like Fourth of July fireworks. Rather, they move across the sky like one of the planets, so that you could see their motion over a period of days. But many do have distinctive streaking tails.

Comets' Predictable Orbits Around the Sun

Sir Isaac Newton first proposed that comets, too, orbited the Sun. It's just that comets have very elongated orbits.

His friend Edmund Halley used Newton's methods, his own observations, and descriptions of previous comet sightings to calculate the orbits of several comets. Halley noted that the comets of 1531, 1607, and 1682 held very similar orbits, but the time between appearances wasn't absolutely regular. Then he realized that one of the planets, especially Jupiter or Saturn, might affect the orbit of the comet through gravitational pull.

With this in mind, he concluded that the three appearances were all of the same comet. In 1705 he boldly predicted the comet's return in 1758. He didn't live to see it, having died some 16 years before. But on Christmas night of 1758, Comet Halley, as it's now known, appeared in the sky. Drawing on reports of comets in literature, scientists have traced Halley sightings back to 239 BC.

A Comet's Three Main Parts

A comet consists of a head and a tail. The head is in turn made up of a nucleus—*the solid core of a comet*, the rock at the heart of the snowball, so to speak—plus the coma (Figure 4.3). The coma is *the part of a comet's head made up of diffuse gas and dust*. (This has nothing to do with "coma" as a medical term.) The tail of a comet is *the gas and/or dust swept away from the comet's head*.

Harvard astronomer Fred L. Whipple proposed a model of a comet in 1950. With some modifications, that model remains the one scientists use to explain comets. As Whipple described it, a comet's nucleus is essentially a dirty snowball that's only a few miles in diameter. It's made up of water ice, frozen carbon dioxide ("dry ice") and some other frozen substances, plus some grains of "dirt." Scientists now believe there's a crusty layer on the surface of the nucleus, with the ices and the dirt inside. When the comet is far from the Sun, it is frozen solid, and the nucleus is all there is to it. Approaching the Sun, the comet warms. The ices inside melt and vaporize. These materials break through the crust and form the coma.

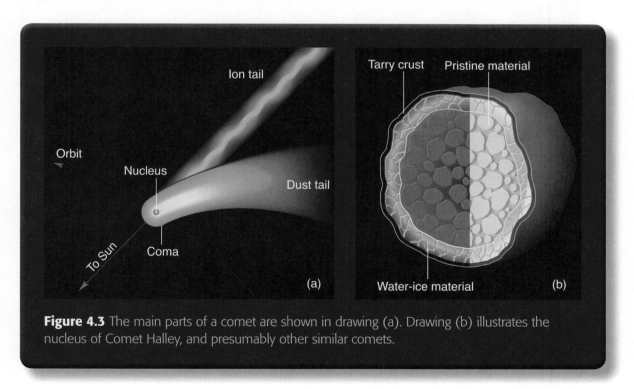

Figure 4.3 The main parts of a comet are shown in drawing (a). Drawing (b) illustrates the nucleus of Comet Halley, and presumably other similar comets.

The Tails of Comets

Comets usually have two tails, although one or both may be very small and they may change greatly as time passes. Solar wind—the stream of protons coming from the Sun—sweeps away one of the tails composed of ions, or charged molecules. These molecules move away from the comet at a high speed and form a straight tail pointing away from the Sun.

Weak solar radiation pushes away grains of dust in the coma to form the second tail. This tail curves and points back along the orbit path, no matter what the motion of the nucleus. Comet tails can be as long as an astronomical unit—the distance from the Earth to the Sun.

This photo, made 29 March 1997, clearly shows the two tails of Comet Hale-Bopp: the straight blue ion tail and the curved white dust tail.

© Datacraft/Age fotostock

The Oort Cloud and the Kuiper Belt

Comet Halley's 76-year orbit means that, for most people, seeing it is a once-in-a-lifetime experience. But by the standards of comets, Halley is a near and familiar neighbor. Comet Hale-Bopp was visible in 1997 but won't return again for more than 2,300 years (Figure 4.4). In this section you will read about celestial objects on the very far fringes of the Solar System.

The Shell of Comets Surrounding the Solar System

In 1950 the Dutch astronomer Jan Oort revived an idea that astronomers had considered before: that in a space far beyond Neptune's orbit, a great number of comets orbit the Sun. This space is the Oort cloud. It is *a theoretical sphere, between 10,000 AU and 100,000 AU from the Sun, containing billions of comet nuclei.*

Three observations prompted Oort's hypothesis:

- No comet has ever been observed that was not gravitationally bound to the Sun

- The orbits of long-period comets (those greater than 200 years) tend to have aphelia—maximum distances from the Sun—around 50,000 AU

- Long-period comets seem to come into the inner Solar System from all over the sky.

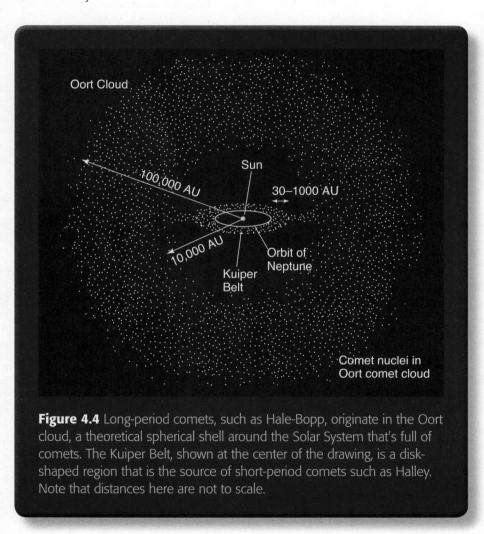

Figure 4.4 Long-period comets, such as Hale-Bopp, originate in the Oort cloud, a theoretical spherical shell around the Solar System that's full of comets. The Kuiper Belt, shown at the center of the drawing, is a disk-shaped region that is the source of short-period comets such as Halley. Note that distances here are not to scale.

Comet Shoemaker-Levy 9 was broken by Jupiter's gravitational forces into more than 20 fragments (bottom photo), all following the same path toward a collision with Jupiter.

The top photo of Jupiter shows the impact spots left by some of the fragments.

Top: *Courtesy of Hubble Space Telescope Comet Team/NASA* Bottom: *Courtesy of STSci/NASA*

CHAPTER 3   The Sun and the Solar System

The Small Band of Comets That Makes Up the Kuiper Belt

The Oort cloud does not explain all comets, however. In 1951, the year after Oort proposed his cloud, another Dutch astronomer, Gerard Kuiper, proposed a second, smaller band of comets within the Oort cloud. The Kuiper Belt is *a disk-shaped region beyond Neptune's orbit, 30 AU to 1,000 AU from the Sun and the presumed source of short-period comets.* Short-period comets have orbits that take less than 200 years.

The first object in the Kuiper Belt was discovered in 1992. Since then, scientists have discovered about 1,000 objects. Most of these have been close to the ecliptic. They range from 60 miles to 600 miles (100 km to 1,000 km) in diameter. Their surfaces are mostly dust and ice. Scientists estimate that there are a billion short-period comets in the Kuiper Belt, and about 100,000 icy little worlds with a diameter of at least 60 miles (100 kilometers). This compares with the 200 asteroids in the asteroid belt known to be that large. Scientists also estimate that all the Kuiper Belt objects put together would amount to about one-tenth of Earth's mass.

In 2005 scientists at the California Institute of Technology discovered a Kuiper Belt object larger than Pluto. Informally known as Xena, and now formally named Eris or UB313, it is now the largest known dwarf planet and the largest object found in orbit around the Sun since 1846. Eris is three times more distant from the Sun than Pluto—some 10 billion miles.

Meteors vs. Meteorites

Almost everyone has seen the flash of light in the sky often called a *falling star* or *shooting star*. Such "stars" are essentially rocks falling from the heavens and burning up as they pass through Earth's atmosphere. Ancient writings of the Greeks, Romans, Hebrews, and Chinese reflect an understanding of this concept, although some people found it hard to accept.

A better and more exact term for a shooting star is a meteor—*a streak of light in the sky caused when a rock particle falling to Earth is so heated by friction with the atmosphere that it emits light.* Most meteors are very dim. A fireball, however, is *an extremely bright meteor.* The object that causes the meteor is a *meteoroid*, which as you read in Chapter 2, Lesson 2, is an interplanetary chunk of matter smaller than an asteroid. Remember, too, that a *meteorite* is an interplanetary chunk of matter that has struck a planet or a moon.

Under ideal viewing conditions, a person can see an average of five to eight meteors an hour. Scientists calculate that over the entire Earth, there must be some 25 million meteors every day that are visible to the naked eye. Very small meteoroids—ranging in size from a grain of sand to a marble—produce most of these meteors. But estimates are that 1,000 tons of meteoritic materials—rocks falling out of the sky, in other words—land on Earth every day.

The Solar System is a collection of diverse objects. As space probes venture out to study it in more detail, astronomers are continually surprised by the diversity they find. While great differences exist among planets, asteroids, comets, meteors, and moons, there are also significant similarities.

Since the objects in the Solar System formed at about the same time, these similarities and differences can tell scientists much about the conditions that existed then. Scientists are just beginning to understand why these objects are different and why they are the same, but there's much still to learn. That's what makes studying the Solar System so exciting.

But the Solar System is only a minuscule part of its home galaxy—the Milky Way. The next lesson will venture out into the galaxy and explore it properties.

CHAPTER 3 The Sun and the Solar System

CHECK POINTS

Lesson 4 Review

Using complete sentences, answer the following questions on a sheet of paper.

1. What was it about Percival Lowell's data that made Clyde Tombaugh's work with the blink comparator even more difficult?

2. In 1956 scientists discovered that Pluto's brightness changes slightly every 6.4 days. What conclusions did they draw from this?

3. What does it mean for Charon and Pluto that they are locked in synchronous orbit?

4. What orbital periods does Kepler's third law predict for most asteroids?

5. What effect does Jupiter's gravitational pull have on the asteroid belt?

6. What two compelling reasons led scientists to abandon the idea that an exploding planet created asteroids?

7. What was Edmund Halley's great achievement?

8. What are the three main parts of a comet?

9. What idea did Jan Oort revive in 1950?

10. What idea did Gerard Kuiper propose in 1951?

11. What is a better and more exact term for a shooting star?

APPLYING YOUR LEARNING

12. Why do comet tails always point away from the Sun instead of trailing behind the comet as it streaks through space?

CHAPTER 4

The Coma Cluster is a collection of galaxies located more than 300 million light-years from Earth. It is one of the densest groupings of galaxies in the universe. The fuzzy gold-brown galaxies in this image contain old stars.

Courtesy of NASA/ESA/Hubble Heritage Team/STScI/AURA

Deep Space

Chapter Outline

" Torrent of light and river of air,
Along whose bed the glimmering stars
  are seen,
Like gold and silver sands in some ravine
Where mountain streams have left
  their channels bare! "

H. W. Longfellow (American poet), The Galaxy

_____

_____

What are your thoughts about the possibility of discovering another Earth-like planet?

Learn About

• the Milky Way Galaxy and the Sun's place in it
• the four components of the galaxy
• other planetary systems
• black holes
• the center of the Milky Way Galaxy

Astronomy has come a long way since the days when people believed the Earth was the center of the universe. Since the mid-1990s scientists have discovered hundreds of planets revolving around stars other than the Sun. The Solar System is not unique. In fact, exoplanets—*planets that orbit stars outside the Solar System*—are fairly common.

But the ones that have been discovered so far are giants, with a mass similar to that of Jupiter and Saturn. They are unlikely to support life. Some of these giant exoplanets, though, may have smaller neighbors more like Mars and Earth.

NASA is embarking on a series of missions to find these new worlds. To this end, the agency will send out some of the most sensitive instruments ever built, able to reach beyond the Solar System's limits.

These will include the Terrestrial Planet Finder. It will provide scientists with images, for the first time, of nearby planetary systems. Scientists will analyze the atmospheres of distant worlds to look for three things: carbon dioxide, water, and ozone. The substantial presence of all three would suggest that life is present. It would be convincing evidence that planet Earth is not the only home to life in the universe.

The Milky Way Galaxy and the Sun's Place in It

Human beings have come fairly recently to understand that they live in a galaxy made up of hundreds of billions of stars. Like so many basic discoveries in astronomy, this one started with Galileo. Four hundred years ago he turned his telescope to the Milky Way and found it made up not of haze, but of stars too numerous to count. The Milky Way appears from Earth as heavenly mist because its stars blur together. The naked eye cannot make them out as individual stars.

The Makeup of the Milky Way

The Milky Way appears as a faint band of light that stretches around the sky. It encircles Earth at an angle of about 63 degrees with respect to the celestial equator. The ancient Greeks called it *galaxies kuklos*, the milky circle. The Romans called it *via lactae*, the milky way.

The Sun is one of about 200 billion stars that make up the Milky Way Galaxy. From the perspective of Earth, the center of the galaxy is in the direction of Sagittarius. Estimates of the galaxy's diameter range from about 100,000 to about 160,000 light-years across—a light-year being *the distance that light travels in a vacuum in one year (about 5.9 trillion miles or 9.5 trillion km).*

The Shape of the Galaxy

Most of the galaxy's stars are arranged in a wheel-shaped disk that circles around a bulging center. The accompanying figures give an idea of what this looks like.

Vocabulary

- exoplanet
- light-year
- parsec
- galaxy disk
- nuclear bulge
- halo
- globular cluster
- galactic corona
- neutrino
- brown dwarf
- pulsar
- binary star system
- astrometry
- microlensing
- accretion disk
- black hole
- Schwarzschild radius
- nova

Star POINTS

The ancients had an easier time seeing the Milky Way than you will today. But you can still get a good look at the galaxy at night in places away from the bright lights of cities. It's definitely something to keep an eye out for the next time you visit such a place. And it might be worth making a special trip!

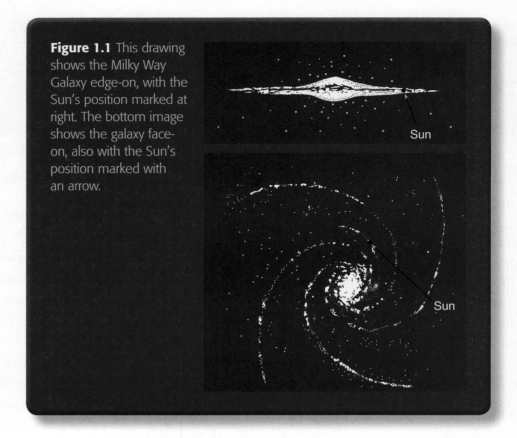

Figure 1.1 This drawing shows the Milky Way Galaxy edge-on, with the Sun's position marked at right. The bottom image shows the galaxy face-on, also with the Sun's position marked with an arrow.

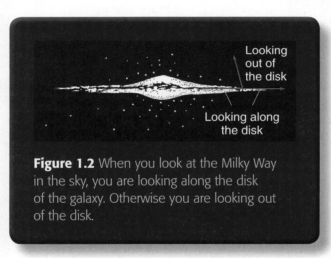

Figure 1.2 When you look at the Milky Way in the sky, you are looking along the disk of the galaxy. Otherwise you are looking out of the disk.

The Sun's Location in the Milky Way

The Sun is about one-third to one-half of the way out from the galaxy's center, depending on the figure you use for the galaxy's size (Figure 1.1). This is approximately 26,000 light-years, or 8,000 parsecs—*an astronomical unit equal to 3.26 light-years*. Astronomers use terms such as parsecs and light-years to deal with distances too great to measure in AU. The galaxy's diameter in parsecs is 31,000 to 50,000 (as opposed to 100,000 to 160,000 light-years).When you look at the Milky Way, you are looking edge-on into the wheel-shaped galaxy. Otherwise, you are looking out of the wheel (Figure 1.2).

Not too long ago—even into the early twentieth century—scientists thought that the Sun was at least fairly near the galaxy's center. Astronomers tried to scientifically locate the Sun within the galaxy, but their reasoning was incorrect.

Back in the 1780s William Herschel, discoverer of Uranus, and his sister Caroline made systematic studies of 700 regions of the sky. They assumed that if they found more stars in one direction than another, that direction would be closer to the center. They observed that the stars' density seemed about the same throughout the sky. That led them to conclude that the Sun was probably at the center of the galaxy (Figure 1.3).

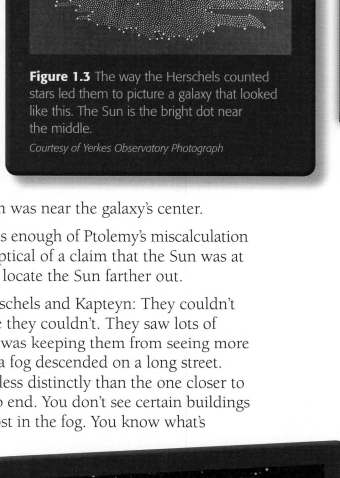

Figure 1.3 The way the Herschels counted stars led them to picture a galaxy that looked like this. The Sun is the bright dot near the middle.

Courtesy of Yerkes Observatory Photograph

In the early twentieth century, Dutch astronomer Jacobus Kapteyn applied a different study technique to the same problem and came up with the same conclusion: The Sun was near the galaxy's center.

But in this same century, scientists were conscious enough of Ptolemy's miscalculation of the Solar System that they were naturally skeptical of a claim that the Sun was at the center. Further studies have led scientists to locate the Sun farther out.

Here's the problem that tripped up both the Herschels and Kapteyn: They couldn't see the edge of the galaxy, and they didn't realize they couldn't. They saw lots of stars, but they didn't know that interstellar dust was keeping them from seeing more distant ones. Here's a way to picture it: Imagine a fog descended on a long street. You may be able to see several blocks, each one less distinctly than the one closer to you. But at some point the street simply seems to end. You don't see certain buildings at all. If you know the street, you know what's lost in the fog. You know what's there, even if you can't see it. But earlier astronomers didn't have that knowledge.

Andromeda, a Galaxy Similar to the Milky Way

Almost everything you can see in the sky without a telescope is in the Milky Way Galaxy. The Magellanic Clouds visible from the Southern Hemisphere are one exception. And in the Northern Hemisphere, the Andromeda Galaxy is another. It's visible in the autumn sky as a fuzzy patch in the constellation Andromeda. It's a spiral galaxy (that is, a galaxy with a swirled and coiled shape) about 2.9 million light-years from the Milky Way. The two galaxies are probably similar in appearance.

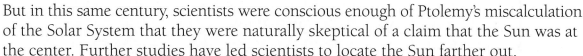

In the autumn sky of the Northern Hemisphere, the Andromeda Galaxy is visible to the naked eye as a fuzzy spot in the constellation Andromeda.

Courtesy of NOAO/AURA/NSF

The Four Components of the Galaxy

Scientists describe four main parts of the Milky Way Galaxy. These consist of the disk, the nuclear bulge, the halo, and the galactic corona. The disk is where the Sun orbits. The nuclear bulge is the crowded neighborhood at the galaxy's center. The halo is home to the clusters of stars that helped scientists figure out where to place the Sun within the galaxy. And the corona may explain why there seems to be more mass in the galaxy than scientists can see.

The Galaxy Disk

A galaxy disk is *the large, flat part of a spiral galaxy, rotating around its center*. The disk contains individual stars, clusters of stars, and almost all the gas and dust found in the galaxy. Most of the stars in the disk are young, that is, in their early stages. The disk as a whole has a bluish cast. This seems to have something to do with the formation of new stars. The disk is densest right at its plane, where it is about 1,000 parsecs thick. Above and below it, stars are fewer and farther between. You can visualize the disk's shape by imagining two compact discs stacked one atop the other, or two Frisbees stacked bottom to bottom.

The Central Region of the Galaxy

The nuclear bulge is *a spiral galaxy's central region*. The Milky Way Galaxy's nuclear bulge is about 2,000 parsecs, or 6,500 light-years, in diameter. To get a sense of its thickness, imagine a peanut stuck in the middle of a pair of CDs. The nuclear bulge is packed more densely with stars, dust, and gas than any other part of the galaxy. The bulge has a reddish cast because it contains so many red giant and supergiant stars.

The Halo

A galaxy's halo is *the outermost part of a spiral galaxy, nearly spherical and lying beyond the spiral*. An important feature of the halo is globular clusters. A globular cluster is *a spherical group of up to hundreds of thousands of stars, found primarily in a galaxy's halo*. Globular clusters orbit a galaxy's nucleus, but they don't remain within the plane of the disk. They pass through the disk twice during each orbit.

Star POINTS

Harlow Shapley (1885–1972) quit school after the fifth grade, returned at age 16, and earned a PhD in astronomy at Princeton at age 27.

Globular clusters provided astronomer Harlow Shapley with the data he needed to place the Sun correctly some distance from the center of the Milky Way Galaxy (Figure 1.4). Whereas earlier researchers had counted individual stars and found them consistently throughout the galaxy, Shapley counted clusters. He found that they concentrated and centered on a point in the constellation Sagittarius, some 50,000 light-years from the Sun. From this he concluded that the Sun was at a greater distance from the galaxy's center than astronomers had originally thought.

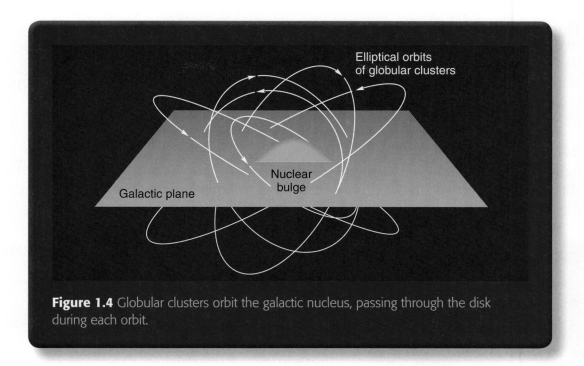

Figure 1.4 Globular clusters orbit the galactic nucleus, passing through the disk during each orbit.

The Galactic Corona

Beyond the galaxy's halo lies the galactic corona, or *outer halo*. This subject presents one of the unresolved problems in galactic astronomy. As astronomers watch the movements of galaxies, they see more evidence of gravitational force at work than seems proportional to the amount of mass they see—by a factor of six.

To explain this, scientists posit the existence of an outer halo perhaps two or three times the radius of the disk and halo. They don't know what the corona consists of, but think it may include small black holes (defined and explained later in the lesson), cool dwarf stars, large numbers of neutrinos, and other exotic particles. (A neutrino is *an uncharged, or electrically neutral, particle believed to have very little or no mass.*) Scientists think the corona may extend to perhaps two or three times the radius of the disk and halo.

Other Planetary Systems

The presence of so many visible stars in the sky, the understanding that the Sun is a star just like them, and the knowledge that many other galaxies are out there naturally causes people to wonder: Just how unusual is the Solar System? Leaving aside the question of life on other planets, is it common for other stars even to have planets? These have been difficult questions to answer until recently. In this section you will read about scientists' new techniques for spotting exoplanets, also known as extrasolar planets.

The Obstacles to Discovering Exoplanets

It's hard to see planets around other stars by direct observation. Their companion stars overwhelm them not only with their size but also with the brightness of their light. Planets in the Solar System are as visible as they are only because they are, in astronomical terms, such near neighbors.

Infrared technology, measuring electromagnetic waves below the range of visible light, has been a boon to astronomers. It modifies the imbalance between a bright star and a dim planet. In the infrared, the brightness of a planet peaks, while that of a star declines. This makes planets easier to spot.

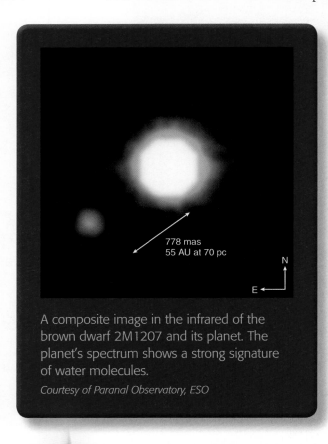

778 mas
55 AU at 70 pc

N
E

A composite image in the infrared of the brown dwarf 2M1207 and its planet. The planet's spectrum shows a strong signature of water molecules.

Courtesy of Paranal Observatory, ESO

The Major Methods for Discovering Exoplanets

Scientists today have developed a number of different methods for finding exoplanets:

Direct observation/infrared companion: The European Southern Observatory's Very Large Telescope obtained the first direct image of an exoplanet in April 2004. The planet is five times as massive as Jupiter and orbits a brown dwarf— *a star-like object that gives off light but lacks sufficient mass for nuclear reactions in its core.* This particular system, about 230 light-years from the Sun in the direction of the constellation Hydra, is only about 8 million years old. Infrared technology made it possible to find.

Dust disks: Astronomers have found disks of dust and gas the size of the Solar System around several stars. Infrared studies of β Pictoris, the No. 2 star in the constellation Pictor, in the Southern Hemisphere, suggest that this star contains fine dust with an icy consistency. Such dust generally either gets blown out of a star system or fuses to the star. That this dust is there all the time suggests that collisions among planetesimals orbiting the star probably keep replacing it. Gaps in the dust imply the presence of a planet orbiting the star.

Star POINTS

β is the Greek letter *beta*. Stars are assigned names by several methods. The brightest stars in each constellation are given a Greek letter according to their brightness. Thus β Pictoris is the second-brightest star in the constellation Pictor. The constellation is visible only from the Southern Hemisphere.

Pulsar companions: A pulsar is *a rotating neutron star that emits beams of radio waves that, like a lighthouse beacon, are observed as pulses of radio waves with a regular period.* In 1992 astronomers reported finding variations in the rate of signals from the pulsar PSR 1257 + 12. Such variations can signal the presence of one or more companion objects, such as a planet.

Binary systems and visual wobble: Two bodies held together by gravity revolve around their common center of mass. As you have read, this is the case with Pluto and its moon, Charon. A binary star system is *a pair of stars that revolve around each other.* Sometimes only one of a pair is visible to astronomers. But they can deduce the existence of the other by looking for a bit of wiggle or wobble in the body they can see. By the same method, astronomers can detect the presence of a very large planet in orbit around a star. This method works best for nearby stars whose motion is easier to observe.

It's not easy to detect the wobble of a star by carefully measuring its position in the sky relative to other stars. Astrometry is *the branch of astronomy dealing with measurement of the positions and motions of celestial objects.* Scientists must be extremely accurate with their measurements, and they must make them over long periods of time to capture the orbital period of a star's motion. The advantage of this technique is that it also lets astronomers find a planet's mass.

Binary systems and Doppler wobble: Another way to detect an exoplanet is to measure the Doppler shift of a star's spectrum (Figure 1.5). That is, scientists measure changes in the light a star gives off as the star alternately wobbles toward and away from Earth. The star wobbles because of an orbiting planet's gravitational pull on it. Astronomers have found that this is a very successful method for identifying planets. This system also tends to provide a lot of information about the planets it helps detect. As of early 2009 scientists had discovered about 335 planet-like objects orbiting about 285 stars.

The Doppler Effect

You may not know the term *Doppler effect* (also called the *Doppler shift*), but you've experienced it many times. When a fire truck passes you, the pitch of its siren rises as the truck approaches and then falls after it passes. The Doppler effect is the change in the apparent frequency of a wave as observer and source move toward or away from each other. The waves get compressed, as the source and observer come closer together—leading to a higher pitch. The waves are pulled apart as the source and observer move away from each other—leading to a lower pitch.

The Doppler effect occurs in other kinds of waves, too. Light from a star moving away from an observer shifts to red. A star moving towards the observer shifts to blue. How much the waves shift can tell astronomers how fast the star is moving.

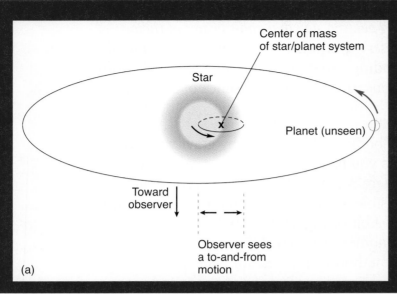

(a)

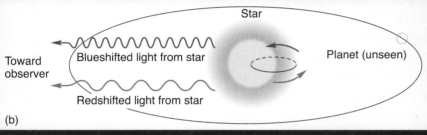

(b)

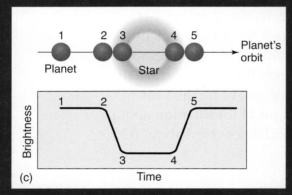

(c)

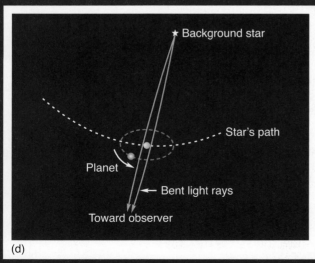
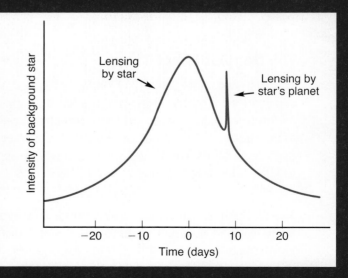

(d)

Figure 1.5 These drawings illustrate four different methods for finding exoplanets. Drawing (a) illustrates how scientists infer a planet's existence from the way it makes its companion star wobble. Drawing (b) illustrates the Doppler wobble. An observer can tell if a star is moving toward or away from Earth depending on whether the wave is shifting to blue or red. Drawing (c) illustrates how scientists detect the presence of a planet when it blocks the light from its star. Drawing (d) illustrates microlensing.

Stellar occultation: When a star seems to dim, that can be a sign of occultation—of a planet passing in front of the star and blocking its light.

Gravitational microlensing: According to the general theory of relativity, when a massive object passes between an observer and a star, the object's gravity will bend the light. The massive object serves as a magnifying lens that multiplies the intensity of the light from the star for a period of time. Microlensing is the term for this *temporary brightening of the light from a distant star orbited by an object such as a planet.* The planet's gravity acts like a lens.

How Planetary Systems Form

As if detecting exoplanets weren't challenging enough, scientists are also at work figuring out how planets form in the first place. Scientists today have two models for the formation of planets: the core-accretion model and the disk-instability model. The core-accretion model assumes that planets start out as small chunks of rock, dust, and debris. After a series of collisions among planetesimals and the accretion of dust and gas, they eventually turn into a planet.

This model explains Earth-like planets better than Jupiter-like ones. A planet like Jupiter needs to collect a lot of rocks for its core, and that process can take tens of millions of years. But an accretion disk—*a rotating disk of gas orbiting a star, formed by material falling toward the star*—tends not to last long enough for that. It typically evaporates in the stellar wind or as a result of influences from nearby stars in just a few million years.

The competing disk-instability model suggests that dense regions forming in the disk accrete (or collect) more and more material and suddenly collapse to form one or more planets. Under this model, it can take only a few hundred years to form a gas giant planet. But the disks in such cases must be quite massive—10 percent of the star's mass. Scientists have not commonly observed this.

Many Jupiter-sized exoplanets are observed close to their parent stars. But it seems unlikely that they formed there, amid the high temperatures, and with few raw materials. A likelier scenario is that they formed some distance away from their star and have moved inward.

Black Holes

A scientific puzzle easily as demanding as planet formation is the black hole. A black hole is *an object whose escape velocity exceeds the speed of light.* (Escape velocity is the speed at which an object must travel to "escape" a star's gravity.)

Shortly after Einstein introduced his theory of general relativity, the German physicist Karl Schwarzschild recognized that a star could collapse so completely, with such a powerful gravitational pull, that nothing could escape from it, not even light.

He calculated that when a star collapses to a dimension equal to or less than a certain radius, light is unable to escape. That radius is now named after him. The Schwarzschild radius is *the radius of a sphere around a black hole from within which no light can escape.*

Star POINTS

The American physicist John Archibald Wheeler usually gets credit for coining the term "black hole." But he later explained that an audience member at one of his talks suggested the phrase. The term in use at the time was "gravitationally completely collapsed object." It was tedious. "How about black hole?" someone in the audience asked. "Suddenly the name seemed exactly right," Wheeler said later.

What a Black Hole Is

The theory behind black holes is complex. But black holes themselves are relatively simple. Science can describe any black hole using only three numbers: one for its mass, one for its electric charge, and one for its angular momentum. Size, texture, color—all those properties of ordinary objects have no meaning for a black hole. Whatever the properties of the material that formed the black hole, that information is forever removed from the universe.

The Three Measurements of a Black Hole

In principle, it would be easy to measure a black hole's mass. You could do this just as you measure the mass of any other object. You would observe an object in orbit around the black hole and apply Kepler's third law to calculate the total mass of the black hole and the orbiting object.

In theory a black hole might have an electric charge, either positive or negative. But whichever it had, it would draw charges of the other type until it became neutral. Thus, scientists do not expect that black holes have electric charges. And they generally don't consider electric charges when discussing black holes.

The third property that a black hole may have is angular momentum. As an object shrinks, its rotation rate speeds up. Therefore a rapidly spinning black hole would drag nearby space-time around with it. Light passing near a rotating black hole on one side would behave differently from light passing by on the other side. On one side the light would be moving along with space-time, and on the other side, light would be moving against the motion of space-time.

Since black holes have just these three properties, astronomers say, "A black hole has no hair." Any properties of objects that fall into it are forever gone from the universe. All that's left is mass, electric charge, and angular momentum.

How Scientists Detect Black Holes

Scientists first predicted the existence of black holes in the 1930s. But a prediction of something as dramatic as a black hole cries out for verification by observation. On the other hand, there's no hope of directly seeing something from which no light escapes. That has turned astronomers' thoughts to indirect observation.

If matter were falling into a black hole, some of that matter would orbit the black hole the same way matter falling onto a white dwarf orbits it before falling in, creating a nova in the process. A nova is *a star that grows brighter than usual for a time and then returns to its original state.* Because the gravitational field near a black hole would be so strong, any matter nearby would orbit very fast. Particles would collide randomly, as regular orbital motion gave way to random thermal motion. Temperatures would reach hundreds of millions of degrees. This hot material would radiate energy. It would appear as an X-ray source (Figure 1.6).

Cygnus X-1, the first X-ray source in the constellation Cygnus, was the first candidate for a black hole. Scientists discovered it in the 1960s. In 1989 the star V404 Cygni, a star in the Milky Way Galaxy, drew attention to itself by erupting with a powerful X-ray flare. Astronomers have observed black-hole candidates with masses ranging from as small as three times the mass of the Sun to billions of solar masses at the cores of galaxies. Since the early 1990s orbiting X-ray observatories have been providing important data to help astronomers tell the difference between black holes and other space phenomena.

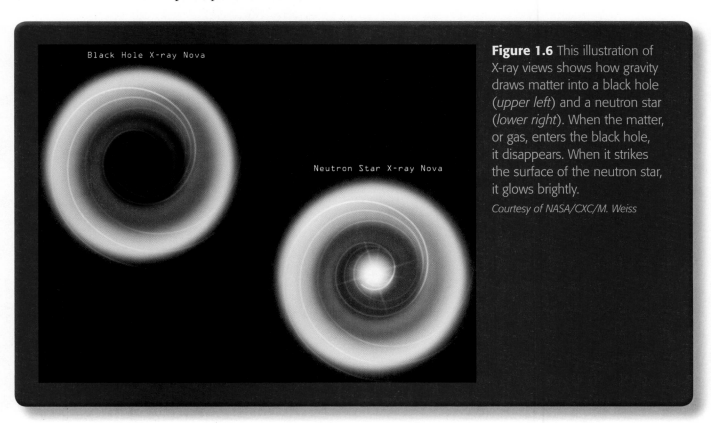

Black Hole X-ray Nova

Neutron Star X-ray Nova

Figure 1.6 This illustration of X-ray views shows how gravity draws matter into a black hole (*upper left*) and a neutron star (*lower right*). When the matter, or gas, enters the black hole, it disappears. When it strikes the surface of the neutron star, it glows brightly.
Courtesy of NASA/CXC/M. Weiss

The Center of the Milky Way Galaxy

It took astronomers until the early twentieth century to get a good understanding of the Milky Way Galaxy's nucleus. Dust and gas in the galactic plane dim the visible light from the nucleus. But the development of nonoptical telescopes opened new windows on the galaxy's center.

How Scientists Observe the Galaxy's Nucleus

To observe the galactic nucleus, scientists use wavelengths in the infrared/radio part of the spectrum or in X-rays and gamma rays. These observations have helped them construct a picture of the Milky Way Galaxy's nucleus. In the process, they discovered that the nucleus has a massive black hole. Findings from these nonoptical scopes have also helped scientists piece together some of the galaxy's violent history.

Some other spiral galaxies also have tremendously powerful energy sources at their center. The massive-black-hole hypothesis seems to be the best explanation for the energy source in the Milky Way Galaxy and other similar galaxies.

But scientists are a long way from understanding the center of the galaxy. They need many more observations and much more data to get a clearer picture. In the meantime, they have learned much in recent years about other galaxies and objects in the universe. The next lesson will introduce you to some of them.

CHAPTER 4 Deep Space

CHECK POINTS

Lesson 1 Review

Using complete sentences, answer the following questions on a sheet of paper.

1. The Sun is one of how many stars that make up the Milky Way Galaxy?

2. How are the stars of the galaxy arranged?

3. Why were twentieth-century scientists skeptical of claims that the Sun was near the center of the galaxy?

4. What do the Magellanic Clouds and the Andromeda Galaxy have in common?

5. What are most of the stars in the disk of the galaxy like? Why does the disk have a bluish cast?

6. Why does the nuclear bulge of the galaxy have a reddish cast?

7. What did Harlow Shapley find when he counted globular clusters?

8. What do scientists think the galactic corona includes?

9. Why has infrared technology, measuring electromagnetic waves below the range of visible light, been such a boon to astronomers?

10. What is gravitational microlensing?

11. What is the core-accretion model of planetary formation?

12. Scientists can describe any black hole using which three numbers?

13. Why would a black hole be expected to spin rapidly? What effect would it have on nearby space-time?

14. Why would astronomers use indirect observation for their work on black holes rather than direct observation?

15. How do scientists observe the galactic nucleus?

16. What is currently the best explanation of the energy source of the Milky Way and other spiral galaxies?

APPLYING YOUR LEARNING

17. Astronomers discovered interstellar dust in 1930. How did failure to understand this phenomenon earlier lead to incorrect estimates of the Sun's position on the disk of the Milky Way Galaxy?

Quick Write

What other example can you give, either in history or from your own life, of an important discovery made during the search for something else?

Learn About

- other galaxies and their classifications
- five types of space objects
- the electromagnetic spectrum
- the big bang theory

I n 1967 Jocelyn Bell (later Burnell) was a graduate student at Cambridge University, looking for quasars—extremely distant starlike objects giving off enormous waves of energy. But she ended up finding something else instead: pulsars, a kind of pulsating radio source.

The first sign of what would turn out to be a new discovery appeared as merely "noise" in the data. As she wrote afterward, "Six or eights weeks after starting the survey, I became aware that on occasions there was a bit of 'scruff' on the records, which did not look exactly like a scintillating source, and yet did not look exactly like man-made interference either."

It's been said that the most critical moments in science are not the times of textbook-perfect lab data, or of great flashes of inspiration, but the moment when someone thinks, "Hmm, that's funny...."

Other Galaxies and Their Classifications

In the early twentieth century, scientists were trying to make sense of something astronomers had discovered in the late 1700s, but whose nature had remained a mystery: the so-called spiral nebula—nebulae, in the plural. At Mount Wilson Observatory in California, Adriaan van Maanen was convinced that these nebulae were vast clouds of gas. But in 1924 Edwin Hubble proved conclusively that they were separate galaxies like the Milky Way Galaxy.

That fact established, Hubble began a serious study of galaxies. As is common in science, his first steps were observation and classification. When scientists come to a group of objects about which they know little or nothing, they typically respond by observing them and grouping them into classes on the basis of what they note about them.

The Three Types of Galaxies

Hubble's own scheme for classifying galaxies became the basis for the one still used today. He divided galaxies into three major groups: *spiral*, *elliptical*, and *irregular*. Each of these groups has subdivisions. More recently, astronomers have discovered objects too peculiar even to qualify as "irregular."

The Characteristics of Spiral Galaxies

A spiral galaxy, which you first read about in Chapter 4, Lesson 1, is a disk-shaped galaxy with arms in a spiral pattern. Hubble divided galaxies of this group into two smaller groups: ordinary spiral galaxies and barred spiral galaxies. A barred spiral galaxy is *a spiral galaxy in which the spiral arms come from the ends of a bar through the nucleus, rather than from the nucleus itself.* Astronomers designate ordinary spirals with a capital S. They mark barred spirals with an SB. The Milky Way is a barred spiral galaxy.

Science further subdivides both groups into groups a, b, and c, depending on how tightly their arms wrap around their nucleus. The more tightly wound galaxies have the most prominent nuclear bulges. Galaxies designated "c" tend to have more gas and dust than those designated "a."

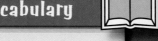

Vocabulary

- barred spiral galaxy
- variable star
- lenticular galaxy
- elliptical galaxy
- irregular galaxy
- apparent magnitude
- luminosity
- proper motion
- optical double
- fluorescence
- emission nebula
- reflection nebula
- dark nebula
- missing mass
- dark matter
- dark energy
- electromagnetic spectrum
- big bang

The Man Who Expanded the Universe

Edwin Hubble (1889–1953) entered the University of Chicago at 16. He studied law after he won a Rhodes scholarship to the University of Oxford in England. By his mid-20s, though, he had settled on astronomy. He began graduate study at the Yerkes Observatory in Wisconsin. Early in his time there, he heard a presentation by Vesto M. Slipher on a hot topic of the day: whether spiral nebulae were part of the Milky Way or separate galaxies in their own right. Slipher presented data that pointed in the latter direction. The question remained open, however.

Soon after, Hubble, perhaps inspired by Slipher's talk, started photographing nebulae with the 24-inch reflecting telescope at the observatory. His doctoral thesis grew out of this research. It was leading him to the conclusion that the nebulae were outside the galaxy.

World War I interrupted his work, though. He enlisted three days after receiving his PhD, although the Armistice was declared before his division reached Europe.

After the war he went back to his research. By 1924 he observed the Andromeda "nebula" and identified six distinct variable stars within it. A variable star is *a star that appears to brighten or dim either because of changes going on within the star itself or because something has moved between it and an observer on Earth*. Hubble's calculations weren't exact—they were off by a factor of more than two. But he showed that the nebulae he had studied were too far away to be within the Milky Way Galaxy's known limits.

Soon after this, Hubble started investigating the idea that the universe is expanding. In 1929 he published a paper stating what's known as the Hubble law of redshifts. It indicates that the universe is expanding because the velocities of galaxies increase at increasing distances from any chosen point. Hubble's work has influenced other astronomers and, in fact, scientists named the famed Hubble Space Telescope after him.

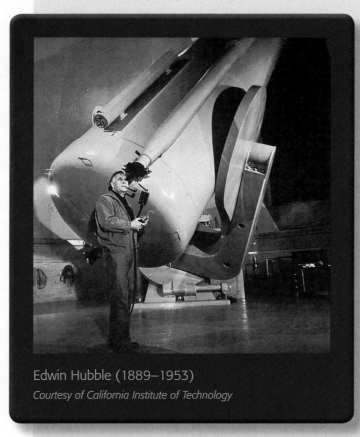

Edwin Hubble (1889–1953)
Courtesy of California Institute of Technology

CHAPTER 4   Deep Space

This photo shows an example of an ordinary spiral galaxy, M77, in the constellation Cetus.

Courtesy of NOAO/AURA/NSF

This photo shows an example of a barred spiral galaxy known as NGC1365, about 60 million light-years from the Milky Way. Note the horizontal bar through the nucleus. The two prominent arms stem from either end of the bar, rather than directly from the nucleus.

Courtesy of Todd Boroson/NOAO/AURA/NSF

About two-thirds of spiral galaxies have bars. A few seem to have the nuclear bulge and disk of a spiral galaxy, but no arms. Hubble called these lenticular galaxies. A lenticular galaxy is *a galaxy with a flat disk like a spiral galaxy, but with little spiral structure, and a large bulge in the nucleus.* Lenticular means shaped like a lens. Hubble designated such galaxies as S0.

The Characteristics of Elliptical Galaxies

An elliptical galaxy is *a galaxy with a smooth spheroidal shape.* That is, round but not perfectly round.

Astronomers classify elliptical galaxies by their degree of eccentricity. But it's hard to tell from Earth just what the shape of a given galaxy is. Is it extremely eccentric—or is it just the angle from which people on Earth are viewing it? Because scientists can't really answer that question with the current level of technology, science classifies elliptical galaxies according to how they appear from Earth, from round (E0) to very elongated (E7).

This photo shows M59, a type E5 elliptical galaxy in the constellation Virgo.

Courtesy of NOAO/AURA/NSF

The Large Magellanic Cloud, the second-nearest galaxy to the Milky Way at only 160,000 light-years away, is generally classified as an irregular galaxy. It is visible primarily in the Southern Hemisphere.

Courtesy of NOAO/AURA/NSF

The Characteristics of Irregular Galaxies

An irregular galaxy is *a galaxy of irregular shape that cannot be classified as spiral or elliptical.* Fewer than one-fifth of all galaxies fall into this group. They tend to be small, with normally fewer than 25 percent as many stars as are in the Milky Way.

The Magellanic Clouds are usually classified as irregular galaxies. Some astronomers, however, think the Large Magellanic Cloud is a barred spiral that has been disrupted by being so close to the Milky Way— and perhaps by a past collision with the Small Magellanic Cloud.

Scientists calculate that collisions between galaxies are not unusual because relatively small distances separate them—only 20 times their diameter. Stars *within* a galaxy, on the other hand, are relatively farther apart when their much smaller diameters are taken into account. This leads to few collisions between individual stars. When galaxies collide, they actually pass through each other or merge— at least, this is what computer simulations suggest. The distances involved are so great and the apparent motions so small that scientists have to rely on simulations rather than direct observation.

Five Types of Space Objects

Galaxies are made up of many objects, which have different properties. These objects, along with the galaxies themselves, are in constant motion. Human beings have known about some of these objects, such as stars, since they first looked at the night sky. Others have only recently been discovered. This section will discuss five types:

1. Stars—specifically their brightness, luminosity, and motion

2. Optical doubles and binary star systems

3. Interstellar clouds and nebulae

4. Pulsars

5. Dark matter.

The Brightness, Luminosity, and Motions of the Stars

Earlier in this lesson you read about the classes of galaxies based on a system Edwin Hubble devised in the twentieth century. The system for classifying stars goes back to the ancient Greeks. The astronomer Hipparchus, who lived in the second century BC, compiled a list of some 850 stars and put them into six groups, according to their brightness. First-magnitude stars were the brightest. Sixth-magnitude stars were the dimmest. Today's astronomers can use photographic and electronic techniques to measure brightness. But they still use a version of Hipparchus's system.

In discussing the brightness of a star, it's important to distinguish between its apparent brightness and the energy it actually emits. Scientists call a star's apparent brightness apparent magnitude. This is *the amount of light received from a celestial object*. Apparent magnitude depends partly on how far away the star is from Earth and on Earth's position in the galaxy. This measure contrasts with luminosity—*the rate at which electromagnetic energy is emitted from a celestial object*. That is, luminosity measures how much energy a star releases, rather than how bright it appears from Earth (apparent magnitude).

Motions of the Stars

Everything in the universe is moving, and that includes the stars. You may occasionally hear people speak of "fixed stars"—as if they were permanently attached somewhere. (That's what "fixed" means in this sense.) The stars do move relative to the Solar System, but they do so very slowly, which makes them appear fixed in place. Thus, the constellations known to the ancients have retained their shape to the present day. The star we identify as the North Star is the same one people identified 2,100 years ago. In 1718, however, Edmund Halley discovered that stars also move with respect to one another. That means that constellations do change shape over time. The North Star (Polaris) has, in fact, moved closer to true north than it was in the year 1 AD.

Barnard's star is one of the closest stars to the Sun. It shows the greatest motion as observed from Earth. Proper motion is the term for *the angular velocity of a star as measured from the Sun*. (*Proper* in this sense means the motion that actually "belongs to" a star, as property is something that "belongs to" someone.) Proper motion is opposed to *observed* motion that is due to Earth's movement.

> ### Star POINTS
>
> Remember that the constellations are stars that appear to be grouped together when seen from Earth. If you could look at the same stars from a different vantage point, the pattern would be completely different.

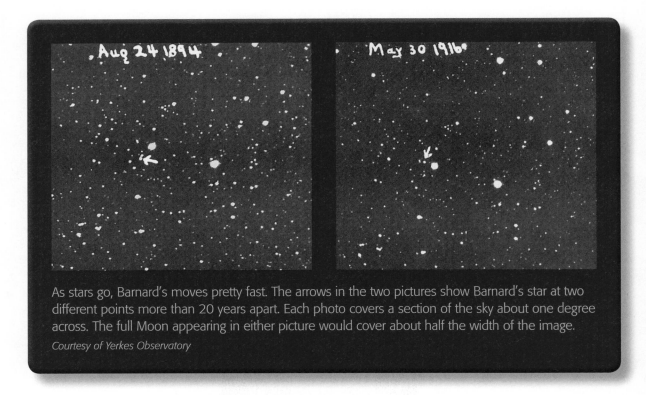

As stars go, Barnard's moves pretty fast. The arrows in the two pictures show Barnard's star at two different points more than 20 years apart. Each photo covers a section of the sky about one degree across. The full Moon appearing in either picture would cover about half the width of the image.

Courtesy of Yerkes Observatory

Optical Doubles and Binary Star Systems

In Chapter 4, Lesson 1, you read about binary star systems: systems of two stars gravitationally bound so that they orbit one another around a common center. Over time astronomers have come to see these as much more common than they used to think. Now they believe that more than half of what appear to be single stars are actually multiples of two or more stars.

There's an important distinction between true binaries, or other multiples, and so-called optical doubles. An optical double is *two stars that, from Earth, appear to be very close but are not actually gravitationally bound.* As the (not-to-scale) drawing in Figure 2.1 suggests, the apparent closeness may be a matter of sightlines.

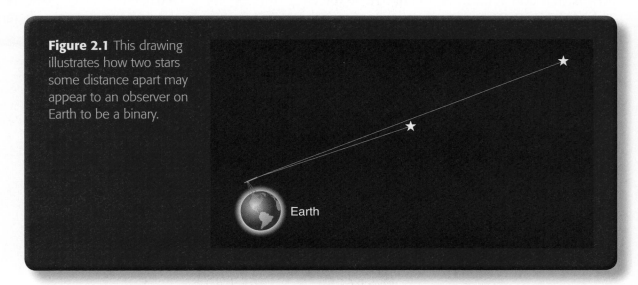

Figure 2.1 This drawing illustrates how two stars some distance apart may appear to an observer on Earth to be a binary.

Earth

To put it in human terms: An optical double is like two trees in a field, one close to you, the other at the field's far end. If both trees are in your line of sight, they may *appear* to be close together from your perspective when in fact they are far apart and unrelated.

Interstellar Clouds and Nebulae

The dust and gas of the galaxies are not spread evenly around. They tend to clump together and form into interstellar clouds. Astronomers describe these clouds with a term borrowed from weather science: cirrus clouds. Cirrus clouds on Earth are the thin wispy kind, made up mostly of ice crystals. Interstellar cirrus clouds are even wispier than their counterparts on Earth. They spread over vast distances. Scientists reckon the total mass of one of them might be equivalent to that of a small-to-average star.

Scientists have classified three types of interstellar clouds, or nebulae. If a cloud is near a hot star, the ultraviolet radiation from the star causes the cloud to fluoresce. Fluorescence is *the process of absorbing radiation of one frequency and re-emitting it at a lower frequency.* An emission nebula is *a cloud of interstellar gas receiving ultraviolet radiation and fluorescing as visible light.*

A reflection nebula is *a cloud of interstellar dust that becomes visible because it refracts and reflects light from a nearby star.* A dark nebula is *an interstellar molecular cloud whose dust blocks light from stars on the other side of it.*

Star POINTS

Interstellar material is in a constant state of flux. New stars use it up by including the material as the star is born, whereas dying stars replenish it.

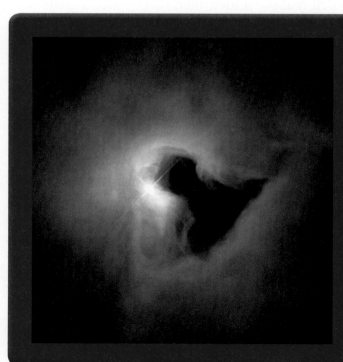

This image illustrates two phenomena at once.

Just left of center it shows a bright young star, which lights up the reflection nebula NGC 1999 in the constellation Orion. To the right in the photo, and in front of NGC 1999, is a dark nebula blocking the star's light. A dark nebula is a condensation of cold molecular gas and dust so thick that it blocks light.

Courtesy of Hubble Heritage Team (STScI)/ NASA

The Discovery of Pulsars

As you read earlier in this lesson, Jocelyn Bell was a graduate student who stumbled upon pulsars in 1967 while searching for quasars—small, intense celestial sources of radiation with a very large redshift. The radio telescope she was using had no giant dish. Rather, it looked more like a field of clotheslines. Its purpose was to pick up faint radio sources and see quick changes in their energy.

Soon she was picking up an unexpected, and unexplained, new source of radio waves. The signal pulsed rapidly every 1.3 seconds. This was much faster than any stellar source had previously been known to pulsate. So she and her team thought the signal perhaps had a terrestrial source, and wasn't really from outer space after all. But a check of local radio transmitters failed to turn up any that could be the source of the signal.

What's more, they picked up the signal four minutes earlier each night. They knew that any given star sets four minutes earlier each night because of the way the Earth moves around the Sun. So the researchers concluded that the signal was coming from outer space, with no human source.

Was it a signal from some extraterrestrial beings? That was the next tantalizing possibility to consider. In fact, the researchers referred to the source as "LGM"— for "Little Green Men," a reference to the way the extraterrestrials of science fiction are sometimes known.

But the pulsations continued in a very regular fashion. A message from space aliens would have some variety in its tones—something like the dots and dashes of Morse code, the researchers reasoned. That made them think the signal wasn't a message.

More convincing, however, was that Bell had soon identified three more signals. One message from an alien race was going to be a stretch. Four extraterrestrial civilizations trying to contact Earth at the same time was beyond belief. The "LGM" soon had a new name: pulsar. As you read in Chapter 4, Lesson 1, a pulsar is a pulsating radio source with a regular period, between a millisecond and a few seconds, believed to be associated with a rapidly rotating neutron star. A neutron star is one that, in essence, has collapsed.

Theories About Dark Matter in Space

Astronomers have several ways to measure the mass of galaxies. All are limited in precision, primarily because the distances involved are so great. Scientists can't wait for even a piece of a galaxy to complete a revolution. Instead, they use Doppler shift data to reach some conclusions about the speed of part of a galaxy, and combine what they find with Kepler's law to come up with an estimate of the mass of the part of the galaxy they are studying. Despite the lack of pinpoint accuracy, though, astronomers do have some ideas about the mass they're dealing with.

CHAPTER 4 Deep Space

Sisterhood of the Telescopes

Up to this point, you've read about the contributions of male scientists to our understanding of the universe. But going back at least to the eighteenth century, women have made important contributions, too. As you read in Chapter 4, Lesson 1, Caroline Herschel worked with her brother William in efforts to understand the shape of the Milky Way Galaxy and to locate the Sun in it. She discovered eight comets and was one of the first two women elected to honorary membership in Britain's Royal Astronomical Society.

Annie J. Cannon was a member of the Harvard College Observatory for almost 50 years.
Courtesy of Harvard College Observatory

Maria (pronounced ma-RYE-a) Mitchell (1818–1889) learned astronomy from her schoolteacher father and her readings while she worked as a librarian. In 1847 she discovered a comet while observing the sky from her rooftop. This led to her becoming the first woman elected to the American Academy of Arts and Sciences.

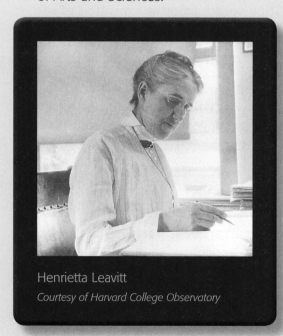

Henrietta Leavitt
Courtesy of Harvard College Observatory

Mitchell later taught astronomy at the newly established Vassar College. One of her students, Antonia Maury (1866–1952), went on to become one of a trio of notable women astronomers at Harvard College Observatory, along with Annie Jump Cannon (1863–1941) and Henrietta Leavitt (1868–1921).

In fact, the observatory's director, Edward Pickering, hired a group of some 40 women to work there, starting in the 1880s. Many people at that time still saw science as an inappropriate field for women. Pickering, for his part, was apparently motivated less by progressive principles than by the idea that women would probably work for less money than men.

In fact, the big issue in all these galactic calculations is what scientists call "missing mass." Remember what you read in Chapter 4, Lesson 1: Scientists see more pulling and tugging going on in the heavens than they can readily explain in terms of the celestial objects they can see.

Missing mass is *the difference between the mass of clusters of galaxies as calculated from Keplerian motions and the amount of visible mass.* Scientists know there must be something "out there" to account for all that mass. They just don't know what it is. By one reckoning, this amounts to 80 percent of the "ordinary matter" of the universe. Dark matter is the scientists' term for *matter that can be detected only by its gravitational interactions.* It makes up 20 percent of the universe, while normal matter makes up only 4 percent.

Astronomers are trying to understand dark matter better. They've offered several ideas to explain what it might consist of:

Ordinary nonluminous matter: Common examples include white dwarf stars, which no longer generate their own energy; brown dwarfs, which have never generated enough light to be easily observable; or planets, meteors, comets, and interstellar dust clouds.

Hot dark matter: This refers primarily to neutrinos. Even if these particles have only a small amount of mass, as experiments suggest, so many of them travel at high speeds around the universe that they could account for much of the missing mass.

Black holes: These come in all sizes, and astronomers speculate that they could contribute to the dark matter.

> ### *Star* POINTS
> Astronomer Vera Rubin had a few words about missing mass. She said, "Nature has played a trick on astronomers. We thought we were studying the universe; now we know we are studying only the small fraction that is luminous."

Cold dark matter: This is the term for a group of particles whose existence theories suggest must be there, but which scientists have not yet detected. These particles go by the name of *photinos*, *axions*, or *neutralinos*. The idea is that these particles are moving through space relatively slowly, in contrast with the way that "hot" neutrinos zip around the universe.

Along with dark matter in space is a force scientists call dark energy. This is *an exotic form of energy whose negative pressure speeds up the expansion of the universe.* Dark energy makes up about 70 percent of the universe.

Scientists are trying to understand the mystery of dark energy, because the fate of the universe depends on it. If the universe expands forever, the Solar System will slowly lose contact with the rest of the universe beyond the local supercluster of stars. If the strength of dark energy decreases with time or becomes attractive, gravity could lead to a big collapse into an incredibly dense ball. If dark energy gets stronger with time, it will eventually overcome all other forces and lead to a big rip, tearing apart everything in the universe.

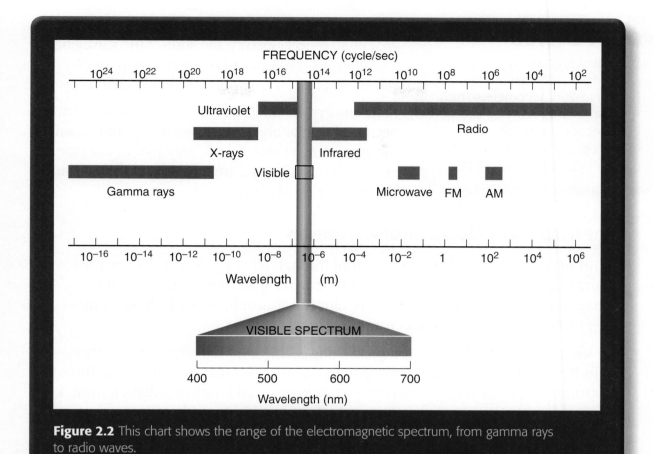

Figure 2.2 This chart shows the range of the electromagnetic spectrum, from gamma rays to radio waves.

The Electromagnetic Spectrum

Astronomers can't see dark matter with their eyes, although they can study its effects using instruments. They also can't see a large part of the electromagnetic spectrum, which affects people here on Earth every day. The electromagnetic spectrum is *the entire array of electromagnetic waves.* All the light you can see— visible light—consists of waves. Visible light takes up only a relatively narrow slice in the middle of the spectrum, however. Many other types of waves and rays exist besides visible light (Figure 2.2).

The Types of Electromagnetic Waves

The electromagnetic spectrum ranges from gamma rays, X-rays, and ultraviolet waves on through visible light and then infrared radiation and various forms of radio waves, including FM and AM. But all these waves are essentially the same phenomenon. They differ in wavelength, so they differ in properties. But they are all waves, and they do many useful things for people. This includes sending messages of all kinds, such as cell phone calls and television programs.

Wavelengths are directly related to energy levels. The greater an object's energy, the shorter the wavelength. Radio waves result from far less energy than do gamma rays at the other end of the spectrum.

Astronomers are interested in electromagnetic waves because celestial objects emit waves up and down the spectrum. Visible light is only part of the story. Space scientists learn a great deal from the invisible radiation celestial objects emit.

The Atmospheric Absorption of Wavelengths

Most of this radiation does not pass well through air. Therefore it does not reach Earth's surface. Visible light and part of the radio spectrum are the exception. They can cut through the atmosphere. But the atmosphere blocks most of the rest of the electromagnetic spectrum to some degree.

In some cases, this blockage is a good thing. Ultraviolet radiation, for instance, kills living cells (including your skin cells), so it's good that not much gets through. Astronomers, however, are interested in picking up all kinds of radiation from outer space. They can get above enough of the Earth's atmosphere on mountaintop observatories and in aircraft to study some of the infrared spectrum. And they send up balloons and satellites to get instruments well up into the atmosphere, or even beyond it.

The Big Bang Theory

Waves and rays allow mankind to transmit messages to one another, just as information gathered from the study of space imparts messages to astronomers about Earth, the Milky Way Galaxy, and what lies beyond. These messages include answers about the universe's beginnings.

The motion of the galaxies has led scientists to think there was a time when all the matter in the universe must have been packed closely together. As they ran the imaginary "movie" of the universe's expansion backward in their minds, they decided an initial explosion must have jump-started that expansion. The big bang is *the theoretical initial explosion that began the universe's expansion.*

Understanding the Big Bang

In trying to understand the big bang, you must be careful not to think of the material of the big bang as being at a certain place in the universe. It *was* the universe—the whole thing.

The big bang represents the highest concentration of energy ever reached in the history of the universe. After the initial explosion, the steady cooling of the universe allowed nuclear particles to form into atoms of low mass (hydrogen and helium). These atoms then clustered into stars, clusters of stars, galaxies, and clusters of galaxies. The clusters of galaxies are still moving apart.

The Prediction and Discovery of Cosmic Background Radiation

The evidence for the expansion of the universe was what led to the idea of the big bang. But scientists have to consider: What evidence is there to support the theory? The first evidence was found quite by accident when a prediction made earlier— by two independent groups of people—was found to be correct.

As scientists learned more about nuclear processes in the 1940s, they began to form some idea of what the universe was like shortly after the big bang. The material of the big bang was originally extremely hot. But it cooled as it expanded outward. And once it cooled to about 3,000 K, it became transparent to radiation—in other words, radiation could pass through unblocked. The universe has remained transparent to radiation since that time, and so scientists proposed that this radiation should still exist. If this radiation were detectable, it would be coming from far back in time and therefore from a great distance.

Star POINTS

K stands for kelvin, which is a way to measure temperatures down to what's called absolute zero, where energy is as low as it will go. The science community often relies on the Kelvin scale, named for Scottish scientist William Thomson, Lord Kelvin.

In 1948 scientists theorized that this cosmic microwave background radiation (CMB) would be striking Earth from all directions but that it should be very faint. At the time, scientists couldn't test the idea because they hadn't yet invented a way to detect such weak waves. So the idea was set aside temporarily.

Then in the mid-1960s some Princeton University physicists were studying the big bang. Once again, they posited the idea of CMB radiation—unaware others had already proposed the idea 15 years before. This group was confident they could build a radio receiver to detect the radiation, and they set about putting it together.

But they were too late. Coincidentally, just a few miles up the road at Bell Telephone Laboratories, Arno Penzias and Robert Wilson were doing some applied research on microwave transmission. They wanted to improve message transmission. They were frustrated, though, by some low-intensity radio waves reaching their receiver from all directions.

It was the CMB radiation predicted years before, as well as by the Princeton physicists, on the basis of the big bang theory. Penzias and Wilson won the Nobel Prize in 1978 for their discovery of something they were not looking for and of whose larger implications they were unaware. Their discovery established the big bang theory as the accepted explanation for the beginning of the universe.

Other Evidence for the Big Bang Theory

The big bang theory is the best current model for the universe's beginnings. In addition to the CMB radiation, other evidence supports it. The darkness of the night sky is another argument for a universe that had a starting point. The sky is dark at night because the universe is finite—limited—in size and age. A finite universe doesn't have enough stars to light up all of space. Hubble's findings about the expanding universe support the big bang theory as well.

The observed proportions of light chemical elements (deuterium, helium, and lithium) are consistent with the big bang theory. These proportions are consistent with a process of nuclear fusion in the first few minutes of a hot young universe.

As scientists continue to explore the universe's origins, they may find more supporting evidence for the big bang theory. Who knows what other discoveries they may make as well. As you've read, some of the most surprising scientific revelations over the years have been the most unexpected. The future should only grow more fascinating.

CHECK POINTS

Lesson 2 Review

Using complete sentences, answer the following questions on a sheet of paper.

1. What became of Hubble's system of classifying galaxies?

2. What is a lenticular galaxy?

3. How does science classify elliptical galaxies?

4. Why do scientists think that collisions between galaxies are not unusual?

5. Who was Hipparchus, and how did he contribute to scientists' understanding of the brightness of stars?

6. What is an optical double?

7. What is the difference between a reflection nebula and a dark nebula?

8. When Jocelyn Bell and her Cambridge University colleagues were trying to figure out the source of an unexplained pulsating radio signal, why was it significant that they detected the signal four minutes earlier each night?

9. What is "missing mass"?

10. Why are astronomers interested in electromagnetic waves?

11. Which types of electromagnetic waves can cut through Earth's atmosphere?

12. After the big bang's initial explosion, what did the steady cooling of the universe allow to happen?

13. What prediction did scientists make in 1948 that would lead to confirmation of the big bang theory?

14. How do the observed proportions of light chemical elements support the big bang theory?

APPLYING YOUR LEARNING

15. Explain why you think the big bang theory is difficult for people to understand.

UNIT 2

Apollo 14 lunar module pilot Edgar Mitchell stands by a US flag in the Moon's Fra Mauro region. Mission commander Alan Shepard, casting a deep black shadow, took this photo during the crew's first spacewalk.
Courtesy of NASA

Exploring Space

Unit Chapter

CHAPTER 5 Exploring, Living, and Working in Space

Mission specialist Kathryn Thornton replaces the Hubble Space Telescope's solar arrays during *Endeavour* mission STS-61 in 1993. NASA launched the mission for the sole purpose of repairing the telescope.
Courtesy of NASA

Exploring, Living, and Working in Space

Chapter Outline

> Astronomy compels the soul to look upwards and leads us from this world to another.
>
> *Plato,* The Republic

Quick Write

_____

_____

What similarities do you see between the Lewis and Clark Expedition and space exploration? What differences?

Learn About

• the historical benefits of exploration
• the US strategic plan to explore space
• the current costs of exploring space
• the practical benefits of space exploration

In 1804 President Thomas Jefferson sent Capt. Meriwether Lewis and Capt. William Clark on a voyage of discovery. Their mission was to explore the territory the United States had bought from France in the Louisiana Purchase. With them was a group of volunteers from the US Army and some civilians, including Sacagawea, one civilian's Native American wife.

Jefferson was not only a politician, he was a scientist. He wanted to explore the new American territory and prevent the British and Spanish from claiming part or all of the land. But he also wanted to make contact with the Indian tribes living in the region and study the geography, animals, and plants.

Among the 32 soldiers and civilians who made up the party, there were interpreters, hunters, cooks, river men, saltmakers, carpenters, blacksmiths, tailors, and trackers. Jefferson arranged for Lewis to receive additional schooling in botany, natural history, medicine, celestial navigation, mapmaking, and geology from some of the best minds in the American scientific community.

The "Corps of Discovery" left St. Louis in the spring of 1804. It traveled up the Missouri River to the Dakotas, Montana, Idaho, and down the Columbia River, the border between Washington State and Oregon. The men reached the Pacific Ocean in November 1805. They wintered there and began their return trip in March 1806, finally reaching St. Louis on 23 September the same year. Only one expedition member had died.

Over the course of two years, four months, and 10 days, the Corps of Discovery traveled approximately 7,689 miles (12,374 km). It was out of communication with the US government and beyond help during most of the trip. It brought back invaluable geographic and scientific data, including 178 new plants and 122 previously unknown species and subspecies of animals. When the party finally returned, its journey had already captured the American people's admiration and imagination.

The Historical Benefits of Exploration

As societies today think about whether and how to explore outer space, they can look for reasons and inspiration in the great period of European exploration. Journeys to distant lands were as enticing, exotic, and challenging to people back then as space travel is to mankind in the twenty-first century.

The modern period of European exploration began in the fifteenth century with the voyages of Christopher Columbus and others. It went on to include explorers from Spain, Portugal, France, England, and the Netherlands. Their voyages led to, among other things, the settlement of the "new worlds" of North and South America and the founding of the United States.

How Exploration Strengthens Nations

Successful civilizations are often the most likely to undertake the dangers and costs of exploring strange lands. This has been a rule of history. In part this is because as their populations grow and prosper, they need more room.

The Soyuz TMA-17 rocket launches from the Baikonur Cosmodrome in Kazakhstan on 20 December 2009 carrying a crew to the International Space Station.
Increased knowledge and the development of new technologies benefit exploring nations.
Courtesy of NASA/Bill Ingalls

Historically, such civilizations sought access to distant riches—or reliable sources of spices, such as cloves, that could lead to riches. Indeed, spices were an important goal when Ferdinand Magellan sailed around the globe from 1519–1522. The more that exploring nations learned about other places, the more they wanted to visit them.

And exploring nations wanted to share their values and culture with other peoples. The French had a term for this. They described their colonial activities as their *mission civilisatrice*—their "civilizing mission" to the rest of the world. The British similarly spoke of seeking to spread "the British way of life." Other peoples have formed similar goals. Of course, the people at the receiving end of these activities often had a different view of them. Today such colonization is not without controversy. But the ability to carry it out can be seen as a sign of a successful civilization.

Of course exploration exposes people to dangers, whether it's the explorers or the people they "discover." But it has led to increased knowledge and the development of new technologies that have strengthened the exploring nations. These benefits offset the risks (Figure 1.1).

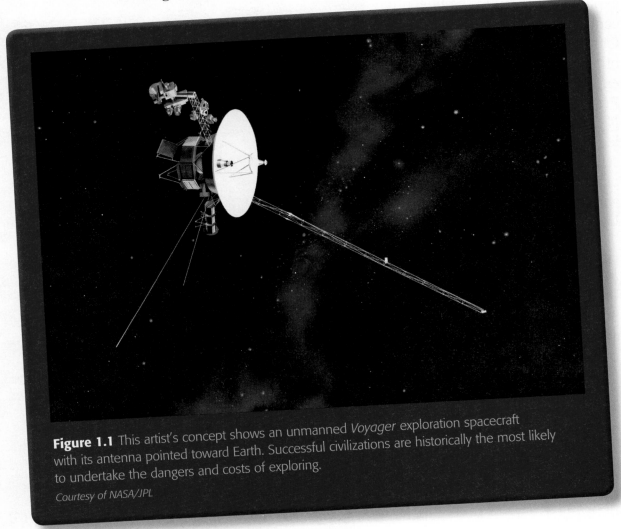

Figure 1.1 This artist's concept shows an unmanned *Voyager* exploration spacecraft with its antenna pointed toward Earth. Successful civilizations are historically the most likely to undertake the dangers and costs of exploring.

Courtesy of NASA/JPL

CHAPTER 5 Exploring, Living, and Working in Space

China as an Exploring Nation

Some scholars point to China as a negative example of exploration. In seven voyages from 1405 to 1433, a Chinese mariner named Zheng He sailed to Southeast Asia, the Indian Ocean, Arabia, and the east coast of Africa. But he accomplished much less than the European explorers. He established no colonies, opened no trade routes, and formed no alliances. His voyages did nothing to help educated classes back in China get a better grasp of their world.

That's because a new Ming emperor decided to call a halt to the whole business. It cost too much and delivered too little in return, the imperial court decided. Bureaucrats destroyed the records of Zheng's voyages. And China withdrew from the world for the next four centuries. During that time, the country made few, if any, technological advances. Other civilizations surpassed it in both technology and knowledge. China opened up again only when European traders arrived and started banging on its doors.

"Fully equipped with the technology, the intelligence, and the national resources to become discoverers," writes historian Daniel J. Boorstin, "the Chinese doomed themselves to be discovered."

How Knowledge Gained Through Exploration Has Enhanced Society

The European voyages of discovery led to colonies and ultimately dozens of whole new nation-states. You may think of "globalization" as a new thing of your own time. But the voyages of the explorers led to the establishment of regular trade routes that tied much of the world together in a web of commerce. Raw materials and finished goods crossed the oceans regularly in Yankee clippers and other vessels.

Exploration also had its effects at an individual level. The establishment of colonies opened new worlds of possibility to countless men and women in Europe. They could cross the ocean in search of a better life. But even those who stayed behind benefited from a larger understanding of their world.

Historical Examples of Technological Advancements Through Exploration

The voyages of oceangoing explorers led to improvements in sailing techniques, shipbuilding, and navigation. The rediscovery of Ptolemy's mapwork, the invention of the printing press, and the explorers' reports led to more-accurate maps of the world. The Portuguese invented the quadrant to aid in navigation, improved the compass, and built better ships—not only to find new places, but to get back with the information. To develop and implement such new inventions and technologies, Prince Henry the Navigator brought together the basic elements of a modern research institute. Progress in navigation, in particular, was tied closely to progress in astronomy. Sailors navigated by the stars, and their observations, in turn, helped inform the work of astronomers.

One of the biggest hurdles sailors had to overcome was the problem of longitude—east-west location on the globe. Efforts to solve this problem eventually led to the marine clock and the portable watch. The need for even more accurate clocks later led to improvements in metallurgy and machining, such as the invention of the lathe.

Exploration also resulted in all manner of New World food and plant products arriving in European ports. These new foods took their place so firmly in Europe that people forget that they aren't native fare there. But "French fries" are made of potatoes, a New World product. "Dutch chocolate" is based on a product Spanish explorers found in Mexico. The sauce you enjoy on your "Italian" pasta includes tomatoes from South America.

These new foods, and others like corn and peanuts, led to new industries and new diets. But other plant products, such as rubber and tobacco, had a similarly revolutionary effect in the Old World. All these New World contributions made it possible for the Old World to support larger populations.

The US Strategic Plan to Explore Space

Just because adventurers have by now visited almost every corner of Earth, and scientists using satellite photos can map the planet down to its slightest detail, doesn't mean the age of exploration is over. On the contrary, people look to the heavens as a still largely unexplored frontier.

On 14 January 2004 President George W. Bush announced a new vision for American efforts in space. Less than a year after the tragic loss of the shuttle *Columbia* and its crew, the president committed the nation to a long-term program to explore the Solar System, first with robots and then with astronauts. Its first phase would consist of a return to the Moon. Ultimately, it would lead to the exploration of Mars and other destinations.

The US Vision for Space Exploration

Key to this vision would be a new spaceship to carry astronauts to other worlds. NASA, along with private industry, is developing the new spacecraft, a so-called crew exploration vehicle named *Orion*. It is to have its first manned mission no later than 2014. NASA will also use *Orion* to transport astronauts and scientists to the International Space Station once the agency retires the space shuttle.

Star **POINTS**

The *Orion* spacecraft is the first of its kind the US has built since the *Apollo* command module.

Visitors at the John C. Stennis Space Center in Mississippi view a full-scale mockup of the *Orion* crew exploration vehicle. NASA plans to use *Orion* to transport astronauts and scientists to the International Space Station once the agency retires the space shuttle.

Courtesy of NASA/SSC

According to President Bush's original plan, the United States would return to the Moon as early as 2015 and no later than 2020. From the Moon, astronauts would embark on even more ambitious missions. A series of robotic—*machine- or robot-based*—missions to the Moon, like those of the *Spirit* Rover on Mars, would explore the lunar surface and prepare for future human exploration. Humans would carry out extended lunar missions. Their goal would be to have humans living and working on the Moon for longer and longer periods. This will help them prepare for missions to other worlds.

These long-stay missions would also help astronauts develop new technologies, those who support the plan say. In addition, astronauts would have the opportunity to explore the Moon's natural resources.

Moreover, the Moon would make a better place than Earth from which to launch a mission to Mars, advocates say. Lunar gravity would be much easier for a spacecraft to escape than Earth's. So it would take much less energy (and fuel) to launch a mission to Mars from the Moon rather than from Earth. This is far into the future, of course. Until there are bases or settlements on the Moon with manufacturing ability, it would still be cheaper to launch from Earth.

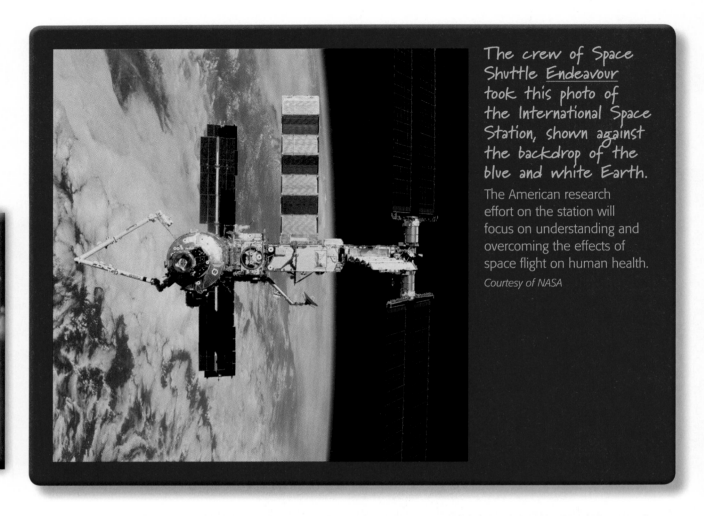

The crew of Space Shuttle _Endeavour_ took this photo of the International Space Station, shown against the backdrop of the blue and white Earth.

The American research effort on the station will focus on understanding and overcoming the effects of space flight on human health.

Courtesy of NASA

Experience and knowledge gained on the Moon would provide the foundation for missions to more distant destinations. NASA also expects to make more use of robotic exploration as a means of learning new facts about the Solar System in preparation for these long journeys. Probes, landers, and other unmanned vehicles can provide enormous amounts of data to help in planning future human missions.

How the International Space Station Will Advance Space Exploration

The United States is set to complete its construction work on the International Space Station by 2010. The research effort on the space station will focus on understanding and overcoming the effects of space flight on human health. What scientists learn in this research will help make future space missions safer for astronauts.

The Long-Term Goal of Exploring Mars

After the Moon, the current plan calls for a manned mission to Mars. The red planet has fascinated humanity at least since nineteenth-century astronomers thought they saw "canals" on its surface. At this writing, President Barack Obama has proposed major changes to his predecessor's plan. But it's not clear what Congress will decide.

But most of the necessary technology is already available to get humans to Mars, according to experts such as Norman Augustine. He's a retired aerospace executive. President Obama put him in charge of a commission to study NASA and make recommendations about future American space missions. As his commission concluded its work, Augustine said that the timing of a mission to Mars will be determined mostly by NASA's budget—*a sum of money set aside to spend for a specific purpose*.

The Current Costs of Exploring Space

Americans have long wondered whether the exploration of space was worth the money. Since the Apollo program's glory days, they have been asking: Are there other things we should spend our money on? And how much are we spending, anyway? It's probably fair to say that most people have no idea of the amounts involved.

NASA's Budget as a Percentage of the US Budget

NASA's entire budget accounts for around six-tenths of one percent of the federal budget. That's six dollars out of every thousand, or sixty cents out of every hundred dollars. NASA is a relatively high-profile agency, though, and people tend to imagine it costs more than it does. They imagine that NASA spending might somehow be on par with that of the Defense Department, or the Social Security Administration or other large federal agencies. The amount of NASA's budget has gone up and down over the past few decades. But since 1975 it has not exceeded 1 percent of the total federal budget.

NASA's defenders argue that if the agency's budget were simply to disappear, the resulting savings would be too small to matter. Shutting down NASA would do almost nothing, they say, to trim the national debt. These defenders charge as well that it might also damage the country's overall national defense.

In 2009 the Augustine commission, appointed by President Obama, issued recommendations for the US space program. It advised that Congress increase NASA's budget by $3 billion annually to about $18 billion.

Spending on Space Exploration Versus Other Spending

Americans spend more on pet food than they do to explore outer space. They spend more on cosmetics—lipsticks and mascara and the like—than they do on space. And the Gillette razor people reportedly spent more to develop and market one of their new shavers than it cost to fly the space shuttle for a year. The comparisons can go on and on.

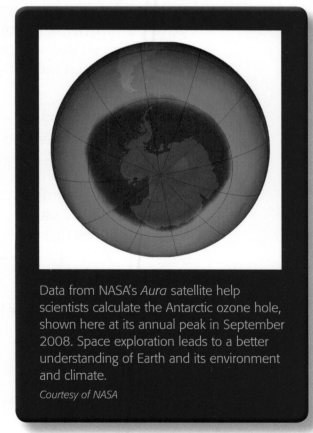

Data from NASA's *Aura* satellite help scientists calculate the Antarctic ozone hole, shown here at its annual peak in September 2008. Space exploration leads to a better understanding of Earth and its environment and climate.

Courtesy of NASA

The United States is a large economy, and so the amounts Americans spend on even trivial items like pet food or cosmetics or razors add up to numbers that sound impressive. But you have to figure that Americans don't spend that much, in proportional terms, on these items. And if you understand that NASA's budget is an even smaller sum, it helps you get a sense of where space exploration fits in the national economy.

Pet food, cosmetics, and razors are examples of consumer goods. They are things that ordinary people buy for their ordinary needs. When they consume them (or, in the case of the pet food, when the pets eat it), there's nothing left. Space exploration, however, is arguably more like an investment. Money spent on space tends to have lasting impacts.

The Practical Benefits of Space Exploration

A further look into the benefits of space exploration may surprise you. Some of the inventions and discoveries pinpointed here touch your everyday life. The way of science is not often a straight line from Point A to Point B, however. A discoverer doesn't know what he's going to discover. An inventor doesn't know what she will invent. And so the practical benefits of space exploration are often byproducts or chance discoveries.

But the efforts of America's space scientists, as astrophysicist Neil deGrasse Tyson has written, "have the capacity to improve and enhance all that we have come to value as a modern society."

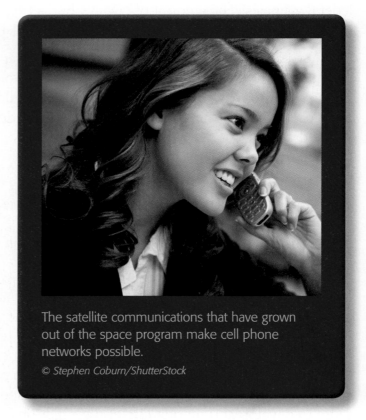

The satellite communications that have grown out of the space program make cell phone networks possible.
© Stephen Coburn/ShutterStock

Here are some examples: Handheld infrared cameras of the type used on the space shuttle have helped firefighters on the ground identify hot spots in brush fires. The satellite communications that have grown out of the space program are now a routine part of daily life. They allow news organizations to broadcast live from anywhere. They are the foundation of the cell phone networks that keep families, friends, and businesses in touch. They support automatic teller networks (ATM machines) that give people access to their money around the clock, and not just during "bankers' hours."

Star POINTS

Observers estimate that more than 1,300 NASA and other US space technologies have contributed to US industry.

How Space Exploration Encourages the Study of Science, Technology, Engineering, and Mathematics (STEM)

"If you build it, they will come." That paraphrased line from a movie has become proverbial. It's used to suggest that if you create something or start a process, you will draw others along.

Space exploration is like that. NASA and its programs, developed in response to President Kennedy's call for the United States to put a man on the Moon, have "built" an establishment of space scientists. Over the years NASA has continued to draw new talent. Space exploration requires a broad range of scientific disciplines. These include astrophysics, biology, physiology, chemistry, engineering, and planetary geology.

President Kennedy first called for putting a man on the Moon in a historic speech to a joint session of Congress in May 1961.

Courtesy of NASA

Unit 2: Exploring Space

Astronaut Buzz Aldrin poses next to the American flag during the Apollo 11 mission. Space missions have encouraged many young people to study the subjects they would need to become space scientists.

Courtesy of NASA

This broad range—what the astrophysicist Tyson calls "the cross-pollination of disciplines"—is part of what makes space exploration such a fertile field for innovation and discovery.

The first astronauts had their roots in conventional aviation. But now that humans have been exploring space for some decades, NASA is full of people who dreamed of going to space when they were children. They knew they wanted to be space scientists when they grew up. Without the space program, thousands of highly educated specialists would not have jobs, and the United States would have far less scientific expertise.

The Mercury, Gemini, and Apollo programs encouraged young people to study the subjects they would need to pursue careers as space scientists—science, technology, engineering, and mathematics. Astronauts were heroes, and so they made science and math "cool." Many young people have followed in their footsteps.

How Science and Technology Fuel Economic Growth

Besides encouraging more students in the sciences and producing byproducts such as ATMs, science and technology have always fueled economic growth. That was true in the days when railroads and the telegraph—that object of Henry David Thoreau's scorn—were the leading edge of technology. And it's certainly true today.

Technologies such as the personal computer or the cell phone can make people more productive. They allow people to share information quickly and easily across vast distances and to be less tied to their desks. This productivity growth leads to economic growth.

Other new technologies make whole new kinds of human activity possible. Aviation technology let humans fly, and led to the aircraft industry as well as airlines. Both have important roles in the national economy and have employed enormous numbers of people.

Photography is another example of a new kind of activity made possible by new technologies. Think how taking pictures has changed from the days of glass plates in the nineteenth century to the digital images you snap on your cell phone today.

But some people worry that there aren't enough Americans pursuing studies and careers in today's leading-edge technologies. Specifically, they worry that not enough Americans are studying math and science.

The critics fret that Americans are willing to benefit from new technologies but aren't playing enough of a role in making them happen. They worry that without ongoing investment—of money and human resources in science and technology— Americans won't be able to maintain their standard of living.

The image processing used in MRI technology came from technology first developed to enhance pictures of the Moon for the Apollo program.

Courtesy of Chief Warrant Officer 4 Seth Rossman/US Navy

How Space Exploration Has Led to Medical Breakthroughs

Space science not only lifts Americans economically, it's a boon as well to American health. Advocates of space exploration can point to several important medical breakthroughs that have been byproducts of NASA and other programs:

- The image processing used in so-called MRI technology in hospitals around the world came from technology first developed to enhance pictures of the Moon for the Apollo program

- NASA developed a chemical process that led to the invention of kidney dialysis machines

- Insulin pumps were developed from technology used on the Mars *Viking* spacecraft

- NASA's satellite electrical systems led to programmable heart pacemakers in the 1970s

- Fetal heart monitors were developed from technology originally used to measure airflow over aircraft wings

- Special lighting technology developed for plant growth experiments on space shuttle missions led to the development of surgical probes used to treat brain tumors in children.

How the Hubble Space Telescope Helped in the Fight Against Breast Cancer

The greatest explorer of outer space today isn't even human. It's the Hubble Space Telescope. For years now, the Hubble, orbiting Earth outside the atmosphere, has been humanity's picture window on the cosmos.

The Hubble didn't get off to a very good start, however. A blunder in the design of its optics meant that when it was launched in 1990, it gave blurred images from space. You might say the telescope needed "glasses."

Once they realized the problem, NASA scientists set to work to correct it. In December 1993 astronauts made a very expensive service call to the Hubble. After their repair, Hubble's images were as good as its designers intended.

But during the early 1990s, as the service mission was being readied, other scientists were trying to find a way to make use of the defective images. Astrophysicists at Baltimore's Space Telescope Science Institute wrote advanced image-processing software to help identify stars in Hubble's fuzzy images.

And meanwhile, medical researchers at the Lombardi Cancer Research Center at the Georgetown University Medical Center in Washington, DC, recognized that the astrophysicists' problem was like what doctors face when they look for tumors in mammograms. Doctors, too, were looking for important points in a fuzzy image.

Soon doctors started using the techniques developed for the Hubble's defective imagery to detect breast cancer in its early stages. Medical experts believe the new approach has saved the lives of countless women.

The Hubble Space Telescope hovers at the boundary between Earth and space.

Courtesy of the STS-82 Crew/HST/NASA

There's no telling what other benefits may arise from future space programs, since such discoveries often happen by chance. But these are exciting times in science and math—times Galileo and Newton could only dream of. Still, it will take a lot of hard work and risks, and a lot of study by future space scientists, to achieve tomorrow's great scientific and practical breakthroughs.

There's one other reason to explore space—humanity's survival could one day depend on it. Humans' knowledge of what's out there could someday be crucial in defending the planet from an asteroid strike or other event of the kind that may have wiped out the dinosaurs. Humans may never be masters of their fate, but space exploration could give them a better chance at survival.

A nation may decide to explore space. But getting there is another matter. It takes a huge support team to assemble, launch, and support even an unmanned mission, let alone a human in space. The next lesson will discuss what's involved in putting such a mission together.

CHAPTER 5   Exploring, Living, and Working in Space

Lesson 1 Review

Using complete sentences, answer the following questions on a sheet of paper.

1. What role did cloves play in the European voyages of discovery?

2. How did European explorers lead to an early form of globalization?

3. How did sailors and early astronomers help each other out?

4. Why would the Moon make a better place than Earth from which to launch a mission to Mars?

5. What will be the focus of the American research effort on the International Space Station?

6. Is it possible to send astronauts to Mars? What do people say is the main constraint?

7. NASA's budget accounts for what share of overall US federal spending? What has the peak been since 1975?

8. Identify two types of consumer goods on which the American people spend more annually than they do on NASA's budget.

9. How does space exploration encourage the study of science and technology?

10. How do technologies such as the personal computer or the cell phone lead to economic growth?

11. How is MRI technology a byproduct of the space program?

APPLYING YOUR LEARNING

12. Explain what you think would happen if the United States stopped exploring space while other nations continued to do so.

Quick Write

_____

_____

What example can you give from your own experience of having a "click" moment, like Wiseman's, when you realized there was something in particular you wanted to do? What was that like? If you can't think of an example from your own life, give one from the life of someone you know or know about.

"I am not sure there was a definitive moment," Gregory R. Wiseman responded when someone asked him when he first knew he wanted to become an astronaut. NASA chose the Navy flier for its Astronaut Class in the summer of 2009. "I grew up watching space shuttle launches on television and my parents took me to see the Blue Angels perform at Annapolis each spring," he recalled. "These two amazing displays of aviation certainly provided me with the motivation to pursue this path. When I saw my first space shuttle launch from the side of a road in Cocoa Beach in 2001, my ambition was sealed. There is nothing more exhilarating than watching the most complex machine on Earth accelerating downrange."

Learn About

- how NASA plans and implements space missions
- the essential components of a space mission
- the selection and training of astronauts

Lieutenant Commander Gregory R. Wiseman enrolled in Astronaut Class in 2009. "When I saw my first space shuttle launch from the side of a road in Cocoa Beach in 2001, my ambition was sealed," he said.

Courtesy of NASA

How NASA Plans and Implements Space Missions

Vocabulary

- aeronautical
- mission directorate
- launch vehicle
- payload
- flight simulator
- cosmonaut

To put together a space mission, you have to have someone to plan, assemble, and run it. You need money to pay for all the people and equipment that will be involved. And you need public support.

NASA—the National Aeronautics and Space Administration—oversees the US space program. The agency has a broad range of duties. It reaches out into the heavens, but has a strong anchor here on Earth. NASA's Earth-related activities include such things as basic Earth science and efforts to improve the air transport system. For instance, NASA is trying to come up with technology to help airports handle up to three times the number of flights they manage today.

In addition, the types of projects that NASA tackles tend to take a long time to plan and complete. Therefore, when requesting money from the federal government for its annual budget, NASA must look years down the road to predict how much it will require. And as you have probably realized by now, anything involving outer space gets into distances and time frames that are, well, astronomical!

NASA's goals and missions reflect a mix of science, technology, and politics. The agency's plans reflect its relationships with the executive and legislative branches of government, as well as its outreach to the public.

NASA's Mission

Congress established NASA in 1958 with the support of President Dwight Eisenhower. The space agency grew out of the National Advisory Committee for Aeronautics, established in 1915 to conduct and promote aeronautical research. Aeronautical means *anything related to the science, design, or operation of aircraft.*

NASA today conducts cutting-edge research to help transform the nation's air transportation system. It's goals are to:

- improve airspace capacity and mobility

- improve aviation safety

- improve aircraft performance while reducing noise, emissions, and fuel burn.

President Dwight D. Eisenhower (*center*) commissions T. Keith Glennan (*right*) as NASA's first administrator and Hugh L. Dryden (*left*) as deputy administrator in 1958. Congress established NASA the same year with Eisenhower's support.

Courtesy of NASA

NASA plans and carries out many air and space missions. But its overriding mission is to pioneer the future in space exploration, scientific discovery, and aeronautics research. For more than half a century, thousands of people at NASA have been working around the world, and in outer space, to answer some basic questions: What's out there? How do we get there? What will we find? And what can we learn there—or learn just by trying to get there—that will make life better here on Earth?

NASA's Four Principal Directorates

NASA is divided into four mission directorates—*the four main organizations through which NASA carries on its work.*

- *Aeronautics* is the directorate that pioneers and proves new flight technologies. These improve human abilities to explore space. But they also have practical applications on Earth.

- *Exploration Systems* is the directorate focused on new capabilities and spacecraft for affordable, sustainable exploration. This includes robotic and human missions.

- *Science* is the directorate that explores the Earth, the Moon, Mars, and beyond. It charts the routes of space missions. And through its Applied Sciences Program, it helps society at large enjoy the benefits of discoveries in space science.

- *Space Operations* is the directorate that provides technical support for the rest of NASA through the space shuttle, the International Space Station, and flight support.

In these early years of the twenty-first century, NASA's reach extends across and beyond the Solar System. The two Mars Exploration Rovers, named *Spirit* and *Opportunity*, arrived on the red planet in 2004 and at this writing are still operating. The *Cassini* orbiter is circling Saturn. The Hubble Space Telescope is transmitting images of the deepest reaches of the cosmos back to Earth. Thirty years after their launch in 1977, Voyagers 1 and 2 were three times farther from the Sun than Pluto.

Closer to home, the International Space Station is extending the human presence in space. NASA satellites are sending back floods of data on Earth's oceans, climate, and other features. NASA's aeronautics team is working with other parts of government and with universities and industry to improve air transport. It's also striving to help maintain American leadership in global aviation.

Among NASA's contributions to aeronautics are improvements in aircraft wings, safer casings for jet turbine fans, better icing detection and deicing fluids, and runway grooves to channel away water. The agency also developed "winglets"— vertical extensions to wingtips that improve airflow and fuel efficiency. And it partnered with the US Army to research and improve rotor designs for helicopters.

NASA's Human Robotic Systems Project—part of the Exploration Systems Mission Directorate—tests spacesuits and lunar robots in the rough, sandy terrain at Moses Lake, Washington.

NASA's four mission directorates provide technical and other support to space missions.

Courtesy of NASA/Sean Smith

How NASA Funds Its Space Missions and Programs

NASA gets the money for all these activities from Congress. As an arm of the federal government, NASA makes a budget request every year. That request goes to the White House, along with similar requests from other parts of the federal government. The president and his staff review and modify these requests. Then they incorporate them into the official White House budget they submit to the Congress. Under the American system, the Congress has the power of the purse—control over government spending. The federal government doesn't spend money, in other words, unless and until Congress appropriates it.

This control means that elected representatives ultimately decide the nation's agenda for space. And members of Congress answer to the public. If public support for space missions appears to be lacking, congressional leaders will hesitate to give NASA the money for them.

But the White House, too, has historically staked out a strong role in setting the national agenda for space. President Kennedy's Cold War call to put a man on the Moon by the end of the 1960s is the best example of this. More recently former President George W. Bush made a bold call to reenergize America's efforts in space. And even more recently, President Barack Obama endorsed an increase in NASA's budget and programming.

President Barack Obama—joined by former astronaut Senator Bill Nelson (in front) of Florida and a group of schoolchildren—places a call to astronauts aboard the International Space Station in March 2009. The president endorsed an increase in NASA's budget and programming that same year.

Courtesy of White House/Pete Souza

NASA's Siblings Around the World

NASA may be the premier space agency in the world. But it's not the only space agency.

The European Space Agency is NASA's counterpart in Europe. It describes itself as "Europe's gateway to space." Its mission is "to shape the development of Europe's space capability and ensure that investment in space continues to deliver benefits to the citizens of Europe and the world." Europe consists of a number of affluent but small countries. By pooling its members' resources, the ESA can undertake projects far beyond the scope of any single one of them.

Russia established its Russian Federal Space Agency, known as Roscosmos, in 1992. The Soviet Union's prestigious space program was part of its Cold War effort against the United States. But after the Soviet Union's collapse in the early 1990s, Russia went through hard times and had little money for space exploration. This prompted a push to excel in commercial satellite launches and space tourism. Space science has not been Russia's long suit lately. But Roscosmos has done well with its Mir space station and other space activities. In late 2009 Roscosmos announced a plan to cooperate in space with the Japan Aerospace Exploration Agency.

China also has a space program. Communist leader Mao Zedong became convinced China needed one after the Soviet Union's successful launch of a satellite in 1957. In 1970 China launched its first satellite. In 1984 it sent its first communications satellite aloft. And then in 2003 it sent a man into space—only the third nation on Earth to do so.

The Indian Space Research Organization (ISRO) first launched a satellite into orbit in 1980. It launched the lunar orbiter Chandrayaan-1 in 2008. The orbiter returned data until August 2009. ISRO also builds boosters.

In Japan, the Japan Aerospace Exploration Agency (JAXA) combined three organizations to create a new agency in 2003. It sent several Japanese astronauts on the space shuttle and to the International Space Station. JAXA has launched several research satellites and cargo resupply craft for the space station. It, too, has its own launch vehicles.

The national conversation among Congress, the president, the public (especially professors and scientists), and NASA determines America's role in space. Together these players work out a space program that reflects scientific, technical, and political goals.

One of the challenges in this process is that as political priorities change, NASA projects may also be changed or canceled. This can make it difficult to complete projects that take several years to develop. Sometimes NASA has to cancel projects that show promise simply because it doesn't have the money to finish them.

The Essential Components of a Space Mission

With funding in hand, NASA can turn its attention to specific missions. A space mission—a journey into space—starts with an idea, a goal, or set of goals. Once the mission's goals are in place, NASA must think about what combination of spacecraft plus launch vehicle—or *rocket*—can best accomplish these goals. If a manned mission is called for, the agency must hire and train astronauts. It must prepare them for space travel in general and their own mission in particular. Communications and ground support round out the list of essential components of a mission into space.

The Process of Research and Development for a Mission

NASA's Science Mission Directorate manages a complex list of research goals by means of what's known as the Science Plan. The current plan covers the years 2007 through 2016, although it is likely to change. The plan identifies and prioritizes space missions—whether manned space flight, robotic exploration, or observatories in Earth orbit or deep space. The plan lays out the research program in detail. And it spells out just what's needed to achieve NASA's goals for both space and Earth science. The plan further identifies the research that a given mission will require. It also details exactly what advanced technology, data management, and other related aspects it will involve.

The Task of Building a Launch Vehicle

To get from a Science Mission Directorate concept to actual execution demands lots of work and lots of high-tech parts. One important step when pulling off a space mission is to figure out how to propel a spacecraft into orbit around the Earth, or even beyond Earth's gravity. A spacecraft like the *Apollo 11* capsule that carried Neil Armstrong, Buzz Aldrin, and Michael Collins to the Moon in 1969 can't get into outer space all by itself. It needs a rocket.

Star POINTS

A rocket can produce more power for its size than any other kind of engine.

A rocket is a type of engine that pushes itself forward or upward by producing thrust. Unlike a jet engine, which draws in outside air, a rocket engine uses only the substances carried within it. Because it carries its fuel with it—either liquid or solid—a rocket can operate in outer space, where there is little or no air.

NASA currently uses several different types of rockets. Weight and the mission's objective determine the choice of rocket. The heavier the payload—*the cargo the rocket is to carry aloft*—the bigger the launch vehicle it requires.

From a high perch in the Vehicle Assembly Building at Kennedy Space Center, people watch a crane lift the *Ares I-X* Super Stack 4, which will become part of a solid rocket booster's upper stage. The *Ares I-X* is a test vehicle for the next US mission to the Moon. NASA contracts with different large companies to build its rockets, but a fair bit of actual assembly work is done at Kennedy.

Courtesy of NASA/ Dimitri Gerondidakis

NASA contracts with different large companies to build its rockets under the agency's supervision. These are typically built in pieces at various locations around the country. From there they are shipped by barge or rail to the Kennedy Space Center and assembled on site. A fair bit of actual assembly work is done at Kennedy.

Launch Sites

Despite frequent thunderstorms, it made good sense to establish the site for most US rocket launches on Florida's east coast. It's relatively close to the equator, which helps propel rockets into space because the Earth spins faster the closer it gets to the equator. There's an ocean to the east, so NASA could launch its rockets eastward to take advantage of Earth's spin in that direction and still be over water. And when NASA chose the site, Florida had relatively few people. It did have good roads, however, because of the strong military presence there.

NASA launched its Interstellar Boundary Explorer (IBEX) with an Orbital Sciences Pegasus XL rocket on 19 October 2008 from the Reagan Test Site on the Kwajalein Atoll in the Marshall Islands. IBEX will image and map the Solar System's outer limits.

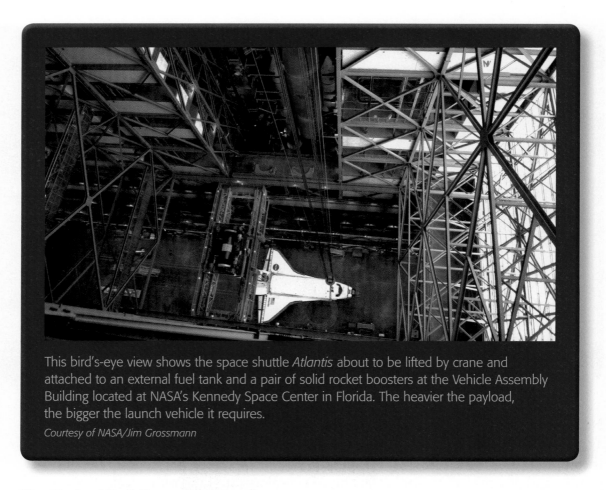

This bird's-eye view shows the space shuttle *Atlantis* about to be lifted by crane and attached to an external fuel tank and a pair of solid rocket boosters at the Vehicle Assembly Building located at NASA's Kennedy Space Center in Florida. The heavier the payload, the bigger the launch vehicle it requires.

Courtesy of NASA/Jim Grossmann

The Task of Building a Spacecraft

Just as vital as the launch vehicle to any mission is the payload, or spacecraft. The building of a spacecraft and the design of a mission are intimately connected. NASA has to build a craft to do what the mission demands. But sometimes the available technologies can hamper the mission.

The Apollo program illustrated how all this worked. Early on, the plan was for the entire assembly that left Earth's orbit to land on the Moon. Then the assembly's upper stage that landed on the Moon would have to be relaunched from the lunar surface, enter lunar orbit, and return to Earth.

This scenario would require the propellants and associated hardware for the return trip to Earth to be carried to lunar orbit, taken out of orbit, and then soft-landed on the Moon. Propellants and hardware would then have to be brought back up into lunar orbit and burned to send *Apollo* back to Earth.

Under this scenario, the mission would need two *Saturn V* booster rockets to get all this into low Earth orbit. This scenario would also have required an Earth-orbit rendezvous at the beginning of the mission.

The way it all actually worked, though, was much different. NASA mission planners suggested using a lunar orbit rendezvous (LOR) instead. That meant that they needed only one *Saturn V* booster rocket. Under the LOR scenario, NASA parked the return capsule, with its engine and return propellants, along with one of its astronauts, in lunar orbit.

The spacecraft that actually made it to the Moon's surface was the combined Lunar Descent Module and Lunar Ascent Module. It needed fewer propellants to deorbit and land than the original scenario envisioned. And once the ascent module had done its work of lifting astronauts back off the Moon, the astronauts left it to crash back onto the lunar surface.

Astronaut Training

Another important piece of many space missions is the astronaut. Those chosen to be NASA astronauts are men and women of great accomplishment. (You'll read more later on in this lesson about the requirements to apply for these jobs.) But astronaut candidates typically go through 20 months of additional training. This is on top of all the training and experience that got them into the space program in the first place.

Their training includes sessions in the classroom and in flight simulators—*machines that duplicate what it's like to operate an airplane or a spacecraft*—as well as learning to fly the T-38 high-performance training aircraft. (Mission specialists don't train as pilots.) They learn to use the ejection seat and their parachutes. They practice getting out of an aircraft that is sinking underwater.

They get training in scuba diving. This is useful in preparing them for walking in space.

Astronaut Leroy Chiao trains in a pool. Scuba diving training prepares astronauts for spacewalks.
Courtesy of NASA

As with anything connected with aviation, astronaut training involves lots of checklists. There's a fair bit of memorization involved, so astronauts resort to the use of something you may remember from grade school: flash cards!

Astronaut candidates get training on the computers and software they will use on their missions. They learn about rocket engines. They get to know their spacecraft intimately. They master the radios they will be working with.

The astronauts' training includes wilderness survival skills as well: learning to trap wild animals for food, to find water, to build shelters, and to deal with medical emergencies. It even includes learning to find their way in the woods. But one of the most important outcomes of the training, the astronauts say, is that it teaches each class of candidates to bond as a team.

Preparing Communications

When the European explorers set off on their great voyages, they knew they wouldn't be able to keep in touch with their home ports. Today's astronauts, however, expect more in the way of communication.

NASA relies on the Tracking and Data Relay Satellite System (TDRSS). TDRSS is a communication-signal relay system that provides tracking and data communications between low-Earth-orbiting spacecraft and control facilities on the ground. The system can transmit to and receive data from spacecraft during at least 85 percent of the spacecraft's orbit.

The system relies on six Tracking and Data Relay Satellites (TDRS). Three satellites are available for operational support at any given time. The other TDRSs provide ready backup in the event of a failure.

The TDRSS ground segment is located near Las Cruces, New Mexico, at the White Sands Complex. Forward data is sent up from the ground segment to the TDRS and from the TDRS to the spacecraft. Return data is sent down from the spacecraft via the TDRS to the ground segment and then on to the designated location for collecting data.

Apollo 13 Flight Directors (left to right) Gerald D. Griffin, Eugene F. Kranz, and Glynn S. Lunney cheer Command Module *Odyssey*'s successful splashdown in the South Pacific Ocean on 17 April 1970. Kranz rallied his ground support staff to figure out how to bring the stranded *Apollo 13* astronauts safely home after things went seriously wrong aboard their vulnerable spacecraft thousands of miles from Earth.

Courtesy of NASA

The Importance of Ground Support Operations

Astronauts are at the top of a vast support structure. But they'd be without critical help if not for ground support personnel. Have you ever seen the movie *Apollo 13*? It told the true story of the rescue of three astronauts whose spacecraft suffered an explosion on their way to the Moon.

One of the story's big heroes was Eugene F. Kranz, the mission flight director. He never left Earth. But after things went seriously wrong aboard a vulnerable spacecraft thousands of miles from Earth, he rallied his ground support staff to figure out how to bring the stranded astronauts safely home. Ground support for a NASA mission includes everything from the tracking stations around the globe to the grief counselors on standby at places such as Edwards Air Force Base in case of accident.

The Selection and Training of Astronauts

Before someone gets to sit atop a rocket or even go through training, men and women who dream of soaring into space must endure an extraordinarily tough screening process. Still, it all begins with that desire to navigate the stars. In fact, *astronaut* comes from Greek words meaning "one who sails among the stars." NASA uses the term to refer to all those who have been launched as crew aboard NASA spacecraft bound for orbit and beyond. NASA also identifies as astronauts those selected for the NASA corps of astronauts. These people have made "star sailing" their profession, and they get the designation of "astronaut" even before they launch.

Star POINTS

After astronauts complete their training, they each receive a silver pin. Once they actually go into space, they exchange their silver pin for a gold pin.

The former Soviet Union identified its star sailors as cosmonauts—sailors amid the universe, or the starry firmament. The Soviet Union is no more, but the term cosmonaut is still in use to refer to *astronauts in the space program of the Soviet Union and its successor state, Russia.*

Qualifications Required to Become an Astronaut

Every manned spacecraft that NASA launches carries a crew made up of astronauts from each of four different categories: commanders, pilots, mission specialists, and payload specialists.

Each of these roles has its own basic qualifications. These qualifications may sound relatively simple, but make no mistake. These are extremely competitive positions, and those who win them are remarkably accomplished individuals.

To become an astronaut pilot or commander, someone must have:

- a bachelor's degree in engineering, biological science, physical science, or math. An advanced degree is desirable.

- at least 1,000 hours of pilot-in-command time in jet aircraft. Flight test experience is highly desirable.

- the ability to pass a NASA space physical. This is similar to a military or civilian flight physical. It requires:

  - good distant vision: 20/100 or better uncorrected, correctable to 20/20 in each eye.

  - blood pressure: 140/90 measured in a sitting position.

  - height between 64 and 76 inches.

Astronaut candidates from NASA's class in 2004 experience reduced gravity in a KC-135 flying in parabolas over the Gulf of Mexico. Pilot candidate Randolph J. "Randy" Breznik and mission specialist candidate Shannon Walker tumble around in the foreground. Astronaut jobs are extremely competitive positions, and those who win them are remarkably accomplished individuals.

Courtesy of NASA

All so-called "mission applicants" for the astronaut candidate program (everyone but the payload specialists, that is) must be US citizens.

Requirements for Mission Specialists

The academic requirements for becoming a mission specialist are like those for becoming an astronaut pilot: a degree in engineering, biological science, physical science, or mathematics. Mission specialists don't face the same requirement for pilot-in-command time in jet aircraft. But they must have either professional experience or an advanced degree. Their vision requirements aren't as strict as the pilots': They can be only 20/200 uncorrected. And they don't have to be as tall as pilots, either. They can qualify at only 58.5 inches.

What Exactly Is a Payload, Anyway?

Payload originally referred to the load of goods on a truck or similar vehicle. It was the load someone paid a trucker to haul. Later *payload* referred to the load a rocket carried into space—a communications satellite, for instance, or even a manned space capsule. This was an extension of the earlier usage. *Payload* also refers to the charge in a guided missile—explosives or chemicals or even biological agents. In NASA lingo, a payload is any "cargo" on a spaceflight. Typically a payload is the equipment related to an experiment or other similar task to be carried out in space. The astronauts in charge of such payloads are known as payload specialists.

Requirements for Payload Specialists

Payload specialists aren't part of the astronaut candidate program. But they must have the appropriate education and training related to the payload or experiment they are responsible for. All applicants must meet certain physical requirements. They must also pass NASA space physical exams, although standards will vary depending on how a given applicant is classified.

How NASA Selects Astronauts

Have you ever wondered whether you have the "right stuff" to become an astronaut? The first astronauts came from the armed services. They were a group of seven military pilots, all men, chosen in 1959. Since these "Original Seven," NASA has chosen 18 more groups of astronauts. They are no longer all male, and their backgrounds are somewhat more diverse. But they are still a pretty exclusive group. Thousands of applicants seek to join the Astronaut Candidate training program. But only 321 have ever been chosen, including the "Original Seven."

NASA developed the process for selecting astronauts to select highly qualified individuals for human space programs. It selects astronaut candidates as they are needed. Both civilian and military personnel may apply for the program. Applicants must meet the minimum requirements mentioned earlier. Once an astronaut candidate's application has survived the screening process, the candidate undergoes a weeklong process of personal interviews, medical screening, and orientation. After NASA makes its final selections from among the finalists, the agency notifies all applicants of the outcome.

Duties of Commanders and Pilots

Pilot astronauts serve as both commanders and pilots of the space shuttle and the International Space Station. During flight, the commander is responsible for the vehicle, the crew, the mission's success, and safety. In other words, he or she is in charge of getting everyone back to Earth unharmed. The pilot helps the commander control and operate the spacecraft. The pilot may also help deploy and retrieve satellites. To this end, the pilot uses the remote manipulator system. The pilot also may have a role in spacewalks or in other payload operations.

Duties of Mission Specialists

Mission specialist astronauts, working with the commander and pilot, manage shuttle crew activity planning. They keep track of "consumables"—food, water, oxygen. And they have roles carrying out experiments aboard space missions.

Mission specialists must have detailed knowledge of their spacecraft and its systems. They also have to understand the operational characteristics, mission requirements and objectives, and supporting systems and equipment for each payload element on their assigned missions.

Mission specialists sometimes work outside their spacecraft. NASA calls this "performing extravehicular activities." Most people call this going for a spacewalk. Mission specialists are the ones who use remote manipulators—special tools to handle payloads outside the spacecraft.

Astronaut Doug Wheelock, STS-120 mission specialist, picks out a meal in the space shuttle *Discovery*'s galley. Among many other duties, including spacewalks and conducting experiments, mission specialists keep track of consumables—food, water, oxygen.

Courtesy of NASA

Duties of Payload Specialists

Payload specialists are those other than NASA astronauts who have specialized onboard duties. When foreign nationals take part in NASA missions into space, they do so as payload specialists.

Payload specialists may be added to shuttle crews when the mission's activities require more than the minimum crew of five. Qualified NASA mission specialists are first in line when a place becomes available for an additional crew member on a mission. When payload specialists are required, they are nominated by NASA, the foreign sponsor (in the case of a foreign national), or the designated payload sponsor.

As you can see, putting a space mission together is a complicated and time-consuming task. It's made even more complicated by an unavoidable fact: Space is a dangerous place for spacecraft and human beings alike. The next two lessons will explore those dangers and what is necessary to allow both hardware and people to survive, carry out their missions, and return home.

CHECK POINTS

Lesson 2 Review

Using complete sentences, answer the following questions on a sheet of paper.

1. What is NASA's mission? What are NASA's four mission directorates?

2. List four NASA contributions to aeronautics.

3. What path does NASA's budget request take to get to Congress?

4. What is the Science Plan and what does it do?

5. Why is a rocket able to function in outer space?

6. In the original scenario for the Apollo missions to the Moon, how many booster rockets did it call for and what was the rocket name?

7. Why do astronauts get training in scuba diving?

8. What is TDRSS and where is its ground element located?

9. Who is Eugene F. Kranz and what is he known for?

10. What academic degree must an astronaut pilot have?

11. Who were the "Original Seven"?

12. What is the commander of a space flight responsible for?

13. What are "consumables," and what responsibility does a mission specialist have with regard to them?

14. When foreign nationals take part in a NASA mission, what role do they play?

APPLYING YOUR LEARNING

15. How do the Congress, the president, and the public work with NASA to set the nation's space agenda? Would you change this if you could? Why or why not?

LESSON 3 The Hazards for Spacecraft

Quick Write

_____

_____

How do you think NASA scientists should protect against the hazards of space impact? Are collisions in space an acceptable part of the price to pay for space exploration?

A funny thing happened in 1967 on *Mariner 4*'s trip home from a flyby of Mars: It got caught in a cloud of dust. This wasn't just any dust. It was a cloud of space dust, and it was a pretty intense experience, even for NASA controllers who were monitoring it all from millions of miles away.

As Bill Cooke of the Marshall Space Flight Center Space Environments Team recalled later, "For about 45 minutes the spacecraft experienced a shower of meteoroids more intense than any Leonid meteor storm we've ever seen on Earth."

The impacts ripped away bits of insulation and turned *Mariner 4* around. But this was minor damage, sustained on the return journey after a mission accomplished. It could have been worse. Interplanetary space holds many uncharted dust clouds like the one that clobbered *Mariner 4*. Some of them are probably quite dense.

Learn About

- the threat caused by high levels of radiation
- the hazard of impact damage to spacecraft
- the threats associated with surface landings
- fire hazards in space

The Threat Caused by High Levels of Radiation

Vocabulary

- solar radiation
- solar storm
- plasma
- microgravity
- oxidant

When you're planning a space mission, you have to understand the space hazards your spacecraft will face and develop defenses against them.

Of the many hazards to spacecraft, radiation exposure is one of the biggest. Outside the protection of Earth's atmosphere and magnetic field, spacecraft and their crews are exposed to solar radiation—*solar flares*—to cosmic radiation from elsewhere in the universe (called galactic cosmic rays), and to belt radiation trapped in Earth's magnetic field. Accordingly, one of the biggest challenges of long-distance space travel is to re-create, in some form, Earth's "safety blankets" to protect spacecraft (and astronauts) from long exposure to radiation.

This shot, taken over northwestern Africa, shows Earth's atmosphere—composed of 78 percent nitrogen, 21 percent oxygen, and 1 percent other elements. One of the most important things the Earth's atmosphere does is to protect the planet from radiation—whether from the Sun or from other sources in outer space.
Courtesy of NASA/JPL/UCSD/JSC

The Threat of Solar Storms to Spacecraft in High Orbit

Electrical storms can cause plenty of trouble on Earth by overloading power lines and leading to widespread blackouts. A similar threat to hardware in space is a solar storm, or geomagnetic storm—*an episode of violent space weather resulting from particles streaming outward from the Sun.*

The solar particles themselves don't cause severe space weather, but they become energized when the solar magnetic field becomes oppositely directed to Earth's and reconnects in a different way.

Star POINTS

This new discovery about holes in the Earth's magnetic field came from NASA's THEMIS mission. THEMIS stands for Time History of Events and Macroscale Interactions during Substorms. Earlier missions had sampled the enormous layer of solar particles within the Earth's magnetic field. But the THEMIS fleet consists of five spacecraft—enough to span the whole layer and make definitive measurements.

Researchers have recently discovered that Earth's magnetic field often develops two large holes, one at high latitude over the Northern Hemisphere, and one at high latitude over the Southern Hemisphere. The holes form over Earth's daylit surface, on the side of the magnetic shield facing the Sun (Figure 3.1).

Scientists think this new discovery will likely allow them to better predict the severity of solar storms. They compare it to predicting storms on Earth. For example, a weather forecaster knows to expect more-intense hurricanes at a time when the oceans are warmer.

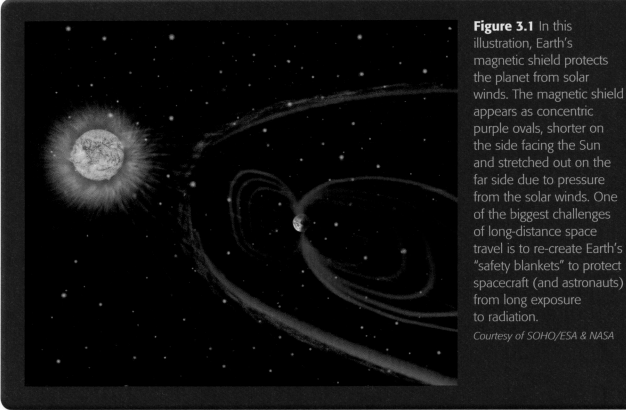

Figure 3.1 In this illustration, Earth's magnetic shield protects the planet from solar winds. The magnetic shield appears as concentric purple ovals, shorter on the side facing the Sun and stretched out on the far side due to pressure from the solar winds. One of the biggest challenges of long-distance space travel is to re-create Earth's "safety blankets" to protect spacecraft (and astronauts) from long exposure to radiation.
Courtesy of SOHO/ESA & NASA

NASA's TRACE spacecraft records solar activity, such as this solar flare from 21 April 2002. Outside the protection of Earth's atmosphere and magnetic field, space crews will be exposed to solar and cosmic radiation.

Courtesy of NASA/Goddard Space Flight Center, Scientific Visualization Studio

How Radiation Can Damage Machines and Cause Computer Failure in Space

This radiation from the Sun and elsewhere in space poses several threats to space hardware. It can cause metals and materials to deteriorate, lose strength, and become less flexible—similar to what happens to plastics on Earth that sit in the Sun for a long time. How a spacecraft and its materials withstand radiation determines the craft's lifetime. So NASA studies different materials to determine which ones hold out best against the radiation in space.

In addition, very high doses of radiation can cause machines to fail by triggering computer malfunctions—and in the space program just about every machine relies on a computer. For example a large solar event bombarded the Martian Radiation Environment Experiment (MARIE) instrument aboard the Mars Odyssey spacecraft with radiation in 2003. NASA engineers and scientists believe a solar particle crashed into MARIE's computer board, damaging a chip. MARIE was unable to collect data after that event.

Electronics can break down when cosmic rays pass through critical parts and short the circuits. But that's not the only kind of trouble that radiation can make. Ions or electromagnetic radiation can affect the output or operation of an electronic device. The term for this is a single event upset, or SEU. An SEU is considered a "soft error" that doesn't permanently damage the device. But it can produce faulty data transmission, or give a false reading on a navigational device, causing a spacecraft to miss its destination, for instance.

To protect against these dangers, critical computer systems need the ability to detect and correct errors. The space shuttle, for instance, has four computers that "vote" on each action before they make a decision. This protects against the possibility of a computer "memory lapse" because of exposure to charged particles.

The Hazard of Impact Damage to Spacecraft

Another serious threat to spacecraft is colliding with another object in space. But one problem many people think of is actually *not* much of a risk: collision with asteroids. Popular movies may make it seem otherwise. But lots of space lies between asteroids. Thousands of them may roam the Solar System, and astronomers may spot several hundred new ones every year. Yet these asteroids are spread over a ring more than half a billion miles around, more than 50 million miles wide, and millions of miles thick.

Nonetheless, plenty of other impact hazards, ranging from space junk to space dust, fill outer space.

The Damage That Meteoroids Can Cause to Spacecraft

Asteroids may not be much of a problem, but meteoroids definitely are, even in Earth's own backyard. The tiniest meteoroids can pose a risk for Earth-orbiting satellites. They can blast holes in solar panels. They can pit surfaces. And they can short out electronics. Both the space shuttle and the *Mir* space station have suffered broken windows because of strikes by tiny objects in Earth's orbit.

This is apparently what happened to the European Space Agency's *Olympus* communications satellite in 1993. When a meteoroid hits a spacecraft, the meteoroid can disintegrate, creating a cloud of plasma—*an electrically charged gaseous form of matter distinct from solids, liquids, and normal gases.* Under the right (or arguably, the *wrong*) conditions, this plasma cloud can set off a chain reaction, causing a massive short circuit and zapping a satellite's delicate electronics. This is what scientists think happened to *Olympus*.

Star POINTS

Windows are a major issue for spacecraft designers. It's obviously desirable for astronauts to be able to see out. But the fragility of windows makes them a real vulnerability.

When scientists forecast meteoroid showers, NASA and other authorities take precautions, such as pointing vulnerable systems away from the incoming meteoroids. They also may power down sensitive instruments to minimize damage in the event of a collision.

CHAPTER 5 Exploring, Living, and Working in Space

Mission Specialist Marsha S. Ivins, along with astronauts Kenneth D. Cockrell and Mark L. Polansky, looks through a window in the space shuttle *Atlantis* at one of her spacewalking colleagues who snapped this photo in 2001. Windows on both the space shuttle and the *Mir* space station have suffered damage from strikes by tiny objects in Earth's orbit.

Courtesy of NASA

The crew of any mission to Mars would have to be alert to risks from meteoroid strikes. The odds of such a thing happening are not large, but the damage would be devastating if it did. Even a small particle of debris can do a lot of damage if it travels on a collision course with a spacecraft. At this point, an object has to be about the size of a baseball to show up on NASA's radar.

How Junk in Space Can Damage Spacecraft

Besides natural threats to spacecraft, there's a class of manmade threats as well: space junk.

What goes up doesn't always come down. Satellites and humans have been going into space for half a century. Some spacecraft have returned to Earth, touching down at a landing strip or dropping more or less harmlessly into the ocean. Others have mostly burned up in the Earth's atmosphere. Some objects sent into space, like the two *Voyager* craft, are still at work years after their launch.

But many items no longer serving a useful purpose remain in orbit. They keep going round and round the Earth—posing a potential collision hazard to everything else in space that still does have a useful purpose. Eventually items in Earth orbit reenter the atmosphere, where most of them burn up.

Space junk can be as small as a dot of paint that has flaked off a spacecraft. Or it can be as big as a satellite that has stopped working. The United States tracks as much space junk as it can. Radar finds a lot of it. And scientists have ways of estimating what they can't see with radar.

Space junk moves pretty fast. That makes it all the more dangerous. As with meteoroids, even a small object can do a lot of damage. The International Space Station is relatively strong and can stand up to strikes from space junk. The space shuttle's advantage is maneuverability—it can do more to get out of the way of a threat. But it still returns to Earth with tiny craters and even cracks in its windows.

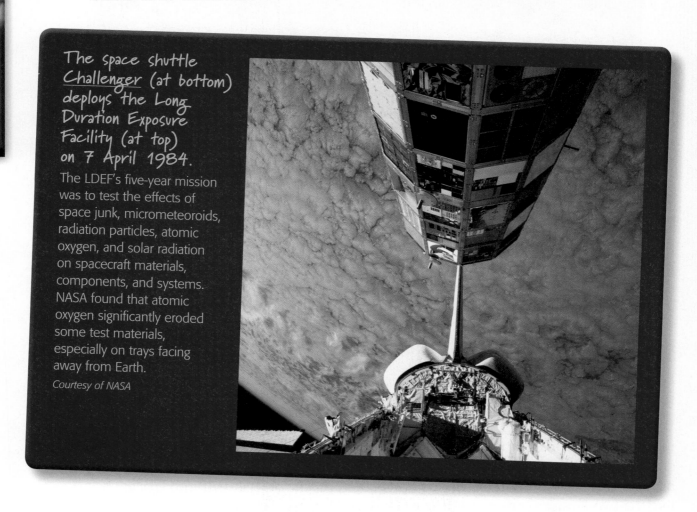

The space shuttle Challenger (at bottom) deploys the Long Duration Exposure Facility (at top) on 7 April 1984.

The LDEF's five-year mission was to test the effects of space junk, micrometeoroids, radiation particles, atomic oxygen, and solar radiation on spacecraft materials, components, and systems. NASA found that atomic oxygen significantly eroded some test materials, especially on trays facing away from Earth.

Courtesy of NASA

The Threat Caused by Clouds of Space Dust

A similar danger is space dust. *Mariner 4*'s encounter with a dust cloud, which you read about at the beginning of this lesson, was an example of this impact hazard in space. As it happened, that spacecraft was the only one of NASA's Mars spacecraft to be sent off with a micrometeoroid detector. As expected, it registered occasional impacts from interplanetary grains of dust. These are harmless in small numbers. But when *Mariner 4* encountered the cloud, the impact rate increased 10,000 times.

Think of yourself in the desert. If the wind throws a single grain of sand in your face, that's no big deal. But if you're caught in a sandstorm, suddenly thousands of grains of sand are hitting you. That's a very different story—for one thing, it hurts!

As NASA's unmanned missions to Mars continue, this kind of encounter with a dust cloud is likely to be repeated. That's why it's a priority for NASA to map these dust clouds and calculate their orbits.

Much of this work involves computer modeling of streams of cometary debris. These are long rivers of dust shed by comets as they orbit the Sun. NASA scientists study how clumps form within the streams and how the gravity of planets, especially Jupiter, deflect them. Scientists also watch the sky for meteor outbursts on Earth. It's a good way for them to test their models and discover new streams of debris to track.

This much-enlarged photo shows a grain of space dust 10 microns across. A U-2 aircraft in the stratosphere captured it. Dust clouds are another type of impact hazard in space.

Courtesy of NASA

The Threats Associated With Surface Landings

After all the planning and designing and training that goes into a mission, and all the stresses involved in launching a rocket, the next step is engineering the landing on the Moon or another planet. This raises an entirely different set of new issues.

How Autonomous Landing and Hazard Avoidance Technology Helps Spacecraft Avoid Hazards

"The Apollo landing missions were an unqualified success," Chirold Epp of NASA's Johnson Space Center in Houston said in 2008. Every single time a lunar module headed toward the surface, it landed successfully and its crew got out and did a great job. "But that doesn't mean there were not close calls," Epp went on. "On four of the six Apollo landings, conditions were such that it gave us pause."

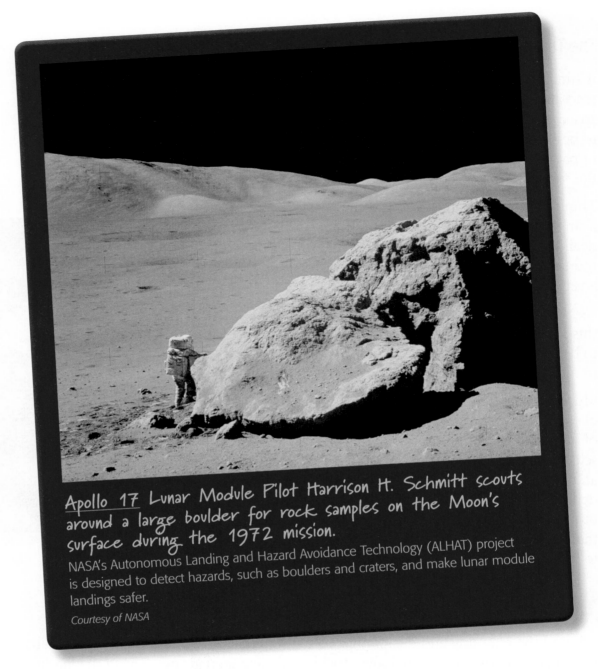

Apollo 17 Lunar Module Pilot Harrison H. Schmitt scouts around a large boulder for rock samples on the Moon's surface during the 1972 mission.

NASA's Autonomous Landing and Hazard Avoidance Technology (ALHAT) project is designed to detect hazards, such as boulders and craters, and make lunar module landings safer.

Courtesy of NASA

Epp is the manager of NASA's Autonomous Landing and Hazard Avoidance Technology (ALHAT) project. The project's aim is to provide NASA's next Moon crews with the data that could make the difference between a good day and a very bad day. The ALHAT system is designed to automatically detect hazards—such as craters and boulders—then direct landers to the safest spot available for touchdown.

It's a job that will have to be done on the fly. Like the Apollo crews, Moon crews on future missions will have limited fuel supplies. They will have to act quickly. Things move fast during a Moon landing. But ALHAT will be able to guide spacecraft precisely and repeatedly to certain designated landing sites—anywhere on the Moon's surface.

How Surface Boulders, Craters, and Sloping Hillsides Threaten Spacecraft

The advantages that ALHAT provides will be particularly important for future Moon expeditions because the locations future crews will explore will be more hazardous than those the Apollo astronauts visited. The terrain and lighting will be notably more challenging.

One such site is near the Moon's south pole, near the rim of a cavity called the Shackleton Crater. It's nearly 12 miles (20 kilometers) wide and about 7.5 miles (12 kilometers) deep. Sunlight can throw shadows that minimize or hide surface features that an astronaut pilot will want to see when trying to bring a lunar lander down safely.

Any single boulder, crater, or sloping hillside could be enough to threaten a spacecraft's safe landing. Failing to see such an obstacle, a pilot might land the craft at an awkward angle, or even cause it to topple over.

Another lunar safety hazard that's just as important is the blinding dust that the lunar lander's rocket exhaust kicks up. That dust can block an astronaut's view of boulders, craters, and other features.

But NASA is confident that ALHAT is the answer to all these challenges. The system provides light and helps astronauts figure out the nuances of the surface terrain from far enough away that dust is not an issue.

Fire Hazards in Space

Another space hazard is one familiar to everybody—fire. In outer space there's no neighborhood firehouse or 911 to call. There's no "normal" gravity either. This causes fire to behave differently in space.

But astronauts rely on fire—in the form of rocket engines—to get around. They have no choice but to learn to manage fire hazards in space. As a NASA official put it, "The safety of NASA's space crews and vehicles can depend on our knowledge of combustion in space."

The Threats Associated With Propulsion Maneuvers in Space

Spacecraft rely on enormous rockets to get up into space in the first place, and then on smaller rockets to maneuver once they leave Earth. Spacecraft even use rockets as braking systems. The reverse thrusters that softened the landings of the Apollo program's lunar modules were a type of rocket.

Any time a spacecraft pilot fires up the rockets to maneuver into or out of orbit, he or she is playing with fire. At any point, something could go wrong, as happened in the shuttle *Challenger* disaster. And so at every moment astronauts must pay attention to safety and to minimizing threats.

The biggest threat from a propulsion system is the possibility that it won't work properly—leaving the spacecraft unable to maneuver in space.

The Special Concerns for Fire Prevention in a Microgravity Environment

The familiar teardrop shape of a candle flame is the result of gravity. A process called gravity-driven buoyant convection carries soot to the flame's tip, making it yellow. But this rule applies only on Earth. In microgravity—*a condition of gravity so low that weightlessness results*—no convective flows exist, and flames are spherical, soot-free, and blue.

The presence of gravity and the effects of air or gas movement, plus the type of fuel and oxidant (*a chemical substance that mixes with oxygen*), determine many things about a flame:

- its shape
- its temperature
- its burn rate
- its burn pattern
- its soot production and deposition
- the speed with which you can or can't extinguish it.

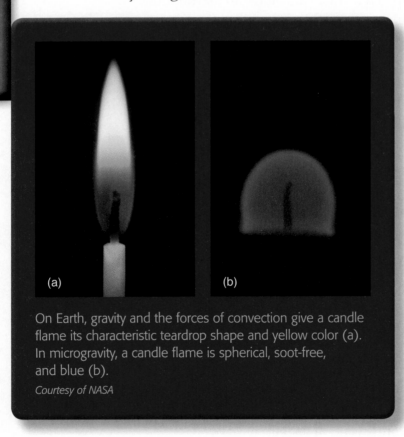

On Earth, gravity and the forces of convection give a candle flame its characteristic teardrop shape and yellow color (a). In microgravity, a candle flame is spherical, soot-free, and blue (b).
Courtesy of NASA

No wonder NASA is studying fire in microgravity! The scientists have plenty of variables to consider. All this is interesting just as a matter of pure research. But as a practical matter, scientists also need to understand how a fire that broke out aboard a spacecraft would burn. What if fabric caught fire? What if chemicals or electronics did? It is vital to know what makes flames start and stop in low gravity, and how flames in such conditions behave while burning.

Research Into Spiral Flames

Some important research on how flames behave in microgravity has focused on creating spiral flames. By studying these peculiar flames in the comfort of the lab, researchers hope to mitigate fire hazards on spacecraft and gain new insights about complex systems in nature.

Two researchers at the University of California at San Diego experimented by taking plastic disks about the size of an ordinary compact disc and setting them on fire with a blowtorch. With the disk stationary, starting a fire at the center produces a flame front that moves outward in a circle. It fades as the fire consumes the disk (the fuel). If the disk is spinning, however, the circular disk flames become spiral flames under certain circumstances.

The researchers thought that spiral flames could be common in microgravity. It was a big surprise, though, to find they could create spiral flames under the conditions of normal Earth gravity.

All this is interesting because scientists want to understand better the phenomenon of *fire whorl*, or "devil wind," and they think they can learn from studying spiral flames. A fire whorl is like a tornado of fire, or a spike of flame that towers over the rest of a fire. It is a terrifying phenomenon when it occurs on Earth, typically as part of a wildfire.

Star POINTS

The spiral is an important shape throughout the natural world. Galaxies often have a spiral shape. So do tornadoes and hurricanes and the shells of mollusks. Even within the human body, the heart's bioelectric impulses move in a spiral pattern. Brain waves seem to flow along the neurons and down the spinal cord in a spiral pattern.

Spiral flames burn counterclockwise on a disk that is spinning clockwise.

Scientists are studying these types of flames in hopes of preventing similarly dangerous fires on spacecraft.
Courtesy of NASA

Scientists hope the study of spiral flames will help them understand fire whorls, tornadoes of fire that often appear in wildfires.

© Nancy G. Fire Photography, Nancy Greifenhagen/Alamy Images

But scientists worry what it could do to a spacecraft. And so combustion studies are a key aspect of spacecraft safety research.

As you've read, spacecraft face hazards from radiation, collision with space objects and dust, surface landings, and fire. You can imagine, then, that if space is risky for spacecraft and hardware, it's even riskier for humans. In the next lesson, you'll read about those risks and the ways in which scientists are studying how to overcome them.

CHECK POINTS

Lesson 3 Review

Using complete sentences, answer the following questions on a sheet of paper.

1. What is a geomagnetic storm?

2. What is a single event upset, or SEU?

3. What is thought to have happened to the communications satellite *Olympus*?

4. What advantage does the space shuttle have over the International Space Station with regard to avoiding space junk?

5. What piece of equipment did *Mariner 4* have that none of NASA's other Mars probes had, and what did it do?

6. What is the ALHAT system designed to do?

7. How will the lunar destinations of future Moon missions be different from those of the Apollo program?

8. What is the risk inherent in all propulsion maneuvers in space?

9. If you lit a candle in a microgravity environment, what would its flame look like?

10. Why are NASA researchers studying spiral flames?

APPLYING YOUR LEARNING

11. If Earth's atomosphere were to let more radiation from the Sun penetrate to ground level, what would be some of the effects, based on what you have read in this lesson?

Quick Write

_____

_____

What would be some ways to protect astronauts against radiation in space?

Learn About

- how the microgravity of space travel affects the human body
- the threat of radiation to astronauts traveling in space
- the study of space biomedicine

n 6 August 1972, a solar flare erupted. Billions of tons of protons raced toward Earth. The particles, moving nearly as fast as light, swarmed around and past the planet. They enveloped Earth in one of the five biggest space radiation storms ever recorded. On the planet below the event passed mostly unnoticed. Earth's atmosphere and magnetic field warded off the radiation, and no one was harmed.

But what if an astronaut had been "out there?"

Only four months later the astronauts of Apollo 17 were. In December 1972 they spent 12 days cruising to the Moon and back, protected only by the aluminum shell of their spacecraft. If they had left on 6 August, Gene Cernan, Ron Evans and Harrison Schmitt, would have absorbed, in a single hour, about 100 times more radiation than a person would living on Earth at the top of a high mountain—in a lifetime.

This is a story often told to illustrate the radiation dangers of space travel. A solar flare can erupt at any moment, peppering astronauts' bodies with light-speed particles they rarely encounter on Earth. It sounds bad.

"In fact, those Apollo astronauts would have been OK," says Frank Cocinotta, the chief scientist for NASA's Space Radiation Health project at the Johnson Space Center. "The command module was shielded well enough to ward off a strong solar flare."

Radiation doses to humans are usually described in *rem*—short for R. E. M. A rem measures the amount of damage to human tissue from a dose of ionizing radiation.

"The August '72 radiation storm would have delivered a sudden dose of 30 to 40 rem inside an Apollo command module," says Cocinotta. For comparison, the threshold for radiation sickness—nausea, loss of appetite and fatigue—is about 75 rem. "Fatalities don't begin until 300 rem," says Cocinotta. "[At that dose,] without medical care, the death rate is 50 percent."

So the Apollo astronauts wouldn't have gotten sick. And they wouldn't have died.

Vocabulary

- atrophy
- deconditioning
- vestibular system
- physiological
- ionize
- cataract
- telemedicine

How the Microgravity of Space Travel Affects the Human Body

Some 50 years of US and Russian space missions have taught researchers important lessons about the dangers to human bodies in space—even when they're protected by their space capsule or a space station.

If you've ever seen television news reports of astronauts in space, you've surely seen videos of them "floating" around their spacecraft. It looks relaxing. But medical experts see the microgravity of space as a real challenge to human health.

The human body and all its parts need exercise to remain in good condition. Bones must bear weight. Otherwise they lose their density and strength. Muscles need to push or pull against something to stay in shape. But floating around in space is so "easy" that it's not good for astronauts.

The rule "use it or lose it" applies to the human body on Earth. And it applies even more in outer space. Exercise programs to keep astronauts fit are a key part of safe and productive space travel. For example, the International Space Station has a variety of exercise equipment. The exercise complement includes a resistance device and a cycle ergometer, which measures the amount of work different

Mission Specialist Joseph Acaba exercises on space shuttle Discovery's stationary bike (called an ergometer) in 2009.
Exercise programs to keep astronauts fit are a key part of safe and productive space travel.
Courtesy of NASA

muscles are doing. The station also features two treadmills. The treadmills are different in many ways, but both share the need for an exercise harness. Astronauts must use a harness to attach themselves to the treadmill while running in space due to the lack of gravity. The harness prevents them from floating off the machine and provides friction against the treadmill belt as they run. It also exerts an external load, or force, on their body to simulate the resistance of gravity that a terrestrial workout would naturally provide.

There's another thing to keep in mind as you read about the challenges of microgravity. Astronauts going to Mars via the Moon will not have to adjust only from Earth gravity to the microgravity of space flight. They will also experience lunar and Martian gravity.

> ### *Star* POINTS
>
> An astronaut who weighs 175 pounds on Earth would weigh 65.9 pounds on Mars but only 29 pounds on the Moon.

The Effect of Space Travel on the Human Heart

The heart is a muscle—the most critical one in the human body. Medical experts know that astronauts lose heart mass and exercise capacity after long periods in microgravity. They suspect this could lead to impaired heart function once astronauts return to Earth—low blood pressure and even fainting. But they believe they still have a lot to learn. They aren't sure, for instance, whether a heart in space simply loses muscle mass the way a body builder does when he takes a break from the gym, or actually becomes scarred.

NASA scientists expect to answer these questions. They also hope to identify risk factors that will help flight surgeons pick the best candidates for long space missions.

A two-year study began in July 2009 on the International Space Station to determine how, how much, and how fast the heart deteriorates during long-duration space travel. The study will examine 12 different astronauts before, during, and after their time in space. It will use equipment such as echocardiograms to watch how the heart relaxes and fills. Magnetic resonance imaging (MRI) will allow researchers to determine how much the heart weakens and whether it scars or is infiltrated by fat.

The team is also studying the effects of heart weakening on crew members' ability to exercise and on the likelihood of their developing unusual heart rhythms—both on the space station and after returning to Earth. In addition, the researchers will look closely at other cardiovascular issues, such as how blood pressure responds to the reintroduction of gravity at the levels experienced on Earth, the Moon, and Mars.

It may also turn out that the heart does well in space. The strategies now used to keep astronauts in shape overall may turn out to be enough to keep their hearts functioning well, too.

Astronaut Story Musgrave aboard the shuttle *Challenger* in 1985 attends to the blood samples he drew from his crew members to study the effects of prolonged weightlessness on bones. Researchers have found that in the microgravity of space, men and women of all ages lose up to 1 percent of their bone mass per month.

Courtesy of NASA

The Effect of Space Travel on Bones and Muscles

If heart function in space turns out to be not much of a problem, flight surgeons will still have plenty of other issues to think about. These include the effect of space travel on bones.

Astronauts in space lose calcium from their bones. The absence of the effects of Earth's gravity disrupts the process of bone maintenance. Bones support body weight. With no gravity there is no weight and hence no bone maintenance.

Researchers have found that in the microgravity of space, men and women of all ages lose a percentage of their bone mass each month. For example, they lose about 2 percent of bone mass in the heel and 1 percent in the pelvis, lower hip, and spine. The loss of calcium in certain bones for one year in space is equivalent to 10 years of loss in a person starting at 50 and going to 60 years. The term for this is disuse atrophy. Atrophy is *a wasting away or shrinking of a body part, typically from lack of use*. This disuse atrophy is similar to the osteoporosis that troubles older women in particular. It's not yet clear whether astronauts' losses in bone mass will continue as long as they are in space. It may level off in time. Scientists are looking for the signals that permit bone tissue to adapt to either a microgravity environment or Earth's gravity.

Deconditioning is space researchers' term for *the weakening of heart and lung function in microgravity*. This weakening again illustrates the "use it or lose it" principle. It also shows how fighting against Earth's gravity toughens humans up. As in the case of bone loss in space, scientists are looking for clues as to whether heart-lung deconditioning continues as long as an astronaut's mission, or whether it levels off in time. Research on the astronauts aboard NASA's *Skylab* space station, launched in 1973, showed that after four to six weeks, deconditioning stabilized.

The body gets rid of the extra calcium released from bone mass loss through the kidneys. So people who have a history of kidney stones are not permitted to fly into space—their body's disposal of calcium would increase in microgravity.

Space researchers are studying the effects of reduced gravity on other muscles as well. NASA studies have suggested that after a long stay in space, an astronaut could lose up to 40 percent of his or her overall muscle function. This is obviously a concern for astronauts once they return to Earth. But loss of muscle function to this degree could put astronauts at greater risk of injury. It could also leave them less able to fulfill their missions in space.

How Space Travel Affects the Sense of Body Weight and Movement

Human beings have evolved to cope with the gravity they experience on Earth. But the microgravity of space travel is a whole new thing. Human systems no longer work as they do on Earth. In microgravity it's hard at first for astronauts even to know the orientation of their own bodies. This is because they have no sense of weight to guide them.

About two-thirds of astronauts feel some space motion sickness during their first experience of microgravity. This is nausea similar to that of carsickness or seasickness.

After almost a year of training, these astronauts are getting their first taste of weightlessness aboard the NASA KC-135.

Nearly all astronauts feel some space motion sickness during their first experience of weightlessness.

Courtesy of NASA

The human brain has learned how to determine which way is up because of the force of Earth's gravity. In space, the signals are different.

Within the inner ear is the vestibular system—*the system that helps people maintain balance and a sense of which way is up.* Information from this organ, along with information from muscles, joints, and the senses of touch and vision, lets the brain identify and determine the body's position relative to the pull of gravity. But in space, with virtually no gravity, the signals from the vestibular organ and other body sensors give conflicting information.

Microgravity also fools the nerves that tell astronauts where their arms and legs are. "The first night in space when I was drifting off to sleep," recalled one Apollo astronaut, "I suddenly realized that I had lost track of … my arms and legs. For all my mind could tell, my limbs were not there. However, with a conscious command for an arm or leg to move, it instantly reappeared—only to disappear again when I relaxed."

Another astronaut from the Gemini program reported waking in the dark during a mission and seeing a disembodied glow-in-the-dark watch floating in front of him. Where had it come from? He realized moments later that the watch was around his own wrist.

These sorts of mismatches between what the eyes can see and the body can feel trigger nausea. Fortunately, after a few hours or days astronauts typically overcome their space sickness.

Space Travel and Body-Fluid Shift

While on Earth, gravity causes most of the body's fluids to be distributed below the heart. In contrast, living in space with less gravity allows fluids in the body to spread equally throughout the body.

When astronauts first travel into space, they feel as if they have a cold and their faces look puffy. Many astronauts talk about not feeling thirsty because of this fluid shift. The body records this shift as an increase in blood volume. The body takes care of this fluid shift by eliminating what it thinks are extra fluids as it would normally—through the kidneys—resulting in visits to the restroom. Once this "extra fluid" is flushed from the body, astronauts adjust to space and usually feel fine.

Puffy faces and chicken legs are short-term changes that astronauts feel. Within three days of returning to Earth, the fluid level of the astronauts return to normal, and the body is "back to normal."

The Effects of Space Travel on the Immune System

Various tests of organisms in space and of astronauts' blood provide evidence that microgravity may also affect astronauts' immune systems. In addition some bacteria's ability to cause disease seems to be increased. That's not a good combination!

Scientists need to understand more about how and why this happens. They're conducting experiments to gain that deeper understanding. In the meantime, exercise, which helps keep people healthy on Earth, may be a good preventative measure.

The Psychological Challenges Associated With Space Travel

You read earlier in the lesson about some of the physiological challenges of space travel—*having to do with the internal organs and how they work.* These challenges include heart health, cardiovascular weakening, and the loss of bone and muscle mass. NASA researchers are also working on the mental, or psychological, challenges of long space flights. Sleep loss and anxiety can affect the health, safety, and productivity of space crews. So can communication woes and the problems of team dynamics. Researchers are looking for ways to keep astronauts motivated and productive, with good morale and team cohesion.

Star POINTS

When choosing crews for long-duration spaceflights, NASA looks at the skills needed for the mission, the ability to work with others under stress, and the compatibility of team members. NASA also looks for teams whose members complement one another—balance one another's strengths, in other words. The agency chooses leaders for their ability to coordinate and motivate crews, and to function under normal conditions and in a crisis.

Astronaut Richard M. Linnehan tries to get some shuteye aboard a 1998 space shuttle *Columbia* mission while wearing a sleep cap that monitors and measures electrical impulses from the brain, muscles, eyes, and heart. Sleep loss can affect the health, safety, and productivity of space crews.

Courtesy of NASA

NASA tries to help the astronauts by sending up "care packages," including books, comfort food, CDs, and so forth. Calls to family members and the ability to send e-mail are also morale boosters. As space missions get longer, the emphasis on maintaining the right mental outlook, no matter what, will only increase.

Outer space is no place for a bad mood. NASA chooses astronauts, in part, for their good interpersonal and team skills. Crew members receive a variety of tools to help them maintain what NASA calls good "behavioral health" and mental function. These include extensive training to help them identify early warning signs of trouble and make self-assessments.

The Threat of Radiation to Astronauts Traveling in Space

One of NASA's mission focuses—space radiation—deserves a closer look. In fact, NASA, along with other federal researchers and universities, devotes a good amount of attention to radiation's effects on astronauts.

In the previous lesson, you read about the damage radiation in space can cause to electronic equipment. But radiation—high-energy electrons and protons, gamma rays, and X-rays—poses an immense threat to human health, too, especially on long space voyages. Short-term exposure to radiation can cause nausea and decreased blood counts. In the long term, exposure can cause cancer, cataracts, and death. And children whom astronauts conceive after flying in space may have a larger risk of birth defects.

The Cancer Risk From Exposure to High Levels of Radiation in Space

When highly charged particles come into contact with living tissue, they ionize molecules of water or oxygen (that is, they *gain either a positive or negative electric charge as a result of gaining or losing electrons*). This creates what are known as free radicals. These can damage cells. When free radicals affect cellular DNA, this can lead to uncontrolled cell division and ultimately cancer.

Scientists calculate that a typical healthy, nonsmoking 40-year-old man already has a 20 percent chance of eventually dying of cancer. This is the "baseline" risk of death from cancer, in other words. They feel sure that exposure to radiation on long-haul space voyages will necessarily mean more cancer risk for astronauts.

The question is, how much more risk? Scientists don't really know. In 2001 they conducted a study of people exposed to large doses of radiation— for example, survivors of the nuclear attack on Hiroshima and cancer patients who have undergone radiation treatments. The researchers concluded that the additional risk of a 1,000-day Mars mission, aboard an aluminum spacecraft similar to other NASA vehicles, would be somewhere between 1 percent and 19 percent. The most likely percentage of increased risk is 3.4 percent, NASA researchers say. But the margin of error in this study is wide. The scientists can't say how likely the "most likely" number is to be the real number.

Workers at Kennedy Space Center's Spacecraft and Encapsulation Facility check out their *Mars Odyssey Orbiter*. In 2001 NASA launched the orbiter, which carries the Mars Radiation Environment Experiment to gather data on near-space radiation and the risks it poses to human health. Scientists feel sure that exposure to radiation on long-haul space voyages will necessarily mean more cancer risk for astronauts.
Courtesy of NASA

If the real number were not 3.4 percent but 19 percent, then the average healthy 40-year-old astronaut would have a nearly 2 in 5, or 40 percent, chance of eventually dying of cancer. That would be unacceptable.

It will take a lot more research to narrow the margins of error in estimating radiation hazards. Scientists know they need to simulate the conditions of actual space travel to get more accurate data than they could draw from studies of bomb survivors and cancer patients.

Flying Through Space in a Ship Made of Plastic?

All other things being equal, NASA would like to go to Mars in a spacecraft made of aluminum. Yes, that would be rather like flying through the heavens in an enormous and very expensive soda-pop can. But aluminum is a strong, lightweight material. The whole aerospace industry has accrued a lot of experience building aircraft and spacecraft with it. This metal offers a certain level of protection against radiation.

It may not offer enough, though. If ongoing research determines that a spacecraft bound for Mars will have to do a better job shielding astronauts from all those particles zipping and zapping around in space, NASA may have to turn to other materials.

Plastic, for instance. A plastic spaceship may sound even crazier than a flying soda-pop can. But plastics are rich in hydrogen, which is good at absorbing cosmic rays. Polyethylene—the stuff from which garbage bags are made—absorbs 20 percent more cosmic rays than aluminum. The Marshall Space Flight Center has developed a reinforced polyethylene that's 10 times the strength of aluminum, and lighter, too. It can also be cut and shaped by machine.

If it can be made cheaply enough, the space industry could make whole ships out of it. If not, NASA and its partners could at least use plastic to shield crew quarters, as on the International Space Station.

If the plastic doesn't prove strong enough against radiation, another idea would be to use pure liquid hydrogen. Some advanced designs for spacecraft call for big tanks of liquid hydrogen fuel. Engineers could wrap the fuel or water tanks around the crew's quarters to protect them.

NASA opened its Space Radiation Laboratory (NSRL), at Brookhaven National Laboratory in New York, in 2003. The NSRL has particle accelerators that can simulate cosmic rays. Researchers can expose cells and tissues to the particle beams and then see what damage occurs. The goal is to narrow the margin of error in risk estimates to only a few percent by 2015.

The Risk of Cataracts Associated With Space Travel

The Apollo astronauts saw things that no humans had ever seen before. They saw Earth as a bright blue disk against the black of space. They saw the far side of the Moon. And they saw flashes of light *inside* their eyeballs.

Astronauts aboard later space missions reported the same phenomenon. Scientists explain it as the result of subatomic particles zapping their retinas. The brain interprets these as flashes of light, and so that's how astronauts see them.

As you might guess, these flashes are not good for anyone's vision. Years after they returned to Earth, many of these astronauts developed cataracts. A cataract is *a clouding of the lens of the eye*. The lens focuses light onto the retina.

Astronauts aboard <u>Apollo 7</u> took this photo of the Sun's reflection on the Gulf of Mexico and the Atlantic Ocean during their 134th orbit around Earth on 20 October 1968.

In addition to breathtaking views such as this, astronauts also see flashes of light *inside* their eyeballs during missions. Scientists explain it as the result of subatomic particles zapping their retinas.

Courtesy of NASA

A NASA study done in 2001 found that at least 39 former astronauts have suffered some form of cataracts after flying in space. Of those, 36 had flown on high-radiation missions—outside Earth orbit—such as the Apollo Moon landings.

Scientists have long known of a link between radiation and cataracts. But they've never fully understood it. They don't know just how radiation clouds the eye's lens. They also don't know whether some astronauts' genes make them more likely to suffer the problem.

The Potential Risk of Birth Defects Caused by Exposure to Radiation

The greatest threat astronauts would face on a trip to Mars is galactic cosmic rays—GCRs. These rays consist of particles accelerated by supernovas (exploding stars) far outside the Solar System. These particles move at nearly the speed of light. The energetic nuclei of iron atoms are a particular concern. They can barrel like tiny cannonballs through the skins of spacecraft and of space travelers alike. They break the strands of DNA molecules, damaging genes and killing cells. Scientists' research on Earth tells them that cells have not evolved to be able to repair such damage when they sustain it. An astronaut—man or woman—whose genetic material has been damaged this way is seen to be at greater risk of producing children with birth defects.

How Diet Can Help Defend Against Radiation

Studies have shown that an astronaut's diet may help defend against radiation's effects in space. Foods rich in antioxidants (vitamins E, C, and beta-carotene), fish oils rich in Omega 3, and fruit and vegetable fiber may prevent or even repair radiation damage.

Getting enough calories, vitamins, and minerals is as important for astronauts as it is for people living on Earth. Astronauts need the same number of calories for energy during spaceflight as they need on the ground.

The Study of Space Biomedicine

Besides potential long-term effects from radiation, astronauts must prepare for medical emergencies in space that require immediate attention. When you're in the deep reaches of space and far from any hospital, you have to carry your own doctor with you. NASA expects that on any mission to Mars, the crew will include at least one physician and one medically trained assistant. Already, all astronauts are trained in first aid. That will continue. The doctor aboard a mission to Mars will have medicines and equipment to monitor heart rate, blood pressure, and other vital signs. The space doctor will also have to be able to test blood, administer medication, and even perform minor surgery.

Using data from the US and Russian space programs, from military aviation, and from experience aboard submarines and in Antarctica, NASA figures it should peg its expected rate of significant illness or injury at about 0.06 per person per year. (By "significant" NASA means an illness or injury requiring a visit to an emergency room or admittance to a hospital.)

For a crew of six heading off on a $2^1/_2$-year trip to Mars, this means an incidence of about 0.90, or about one person per mission. That is, planners expect about one significant medical episode per mission. The expected incidence of serious illness or injury is about 0.02 per person per year, or one incident every third mission.

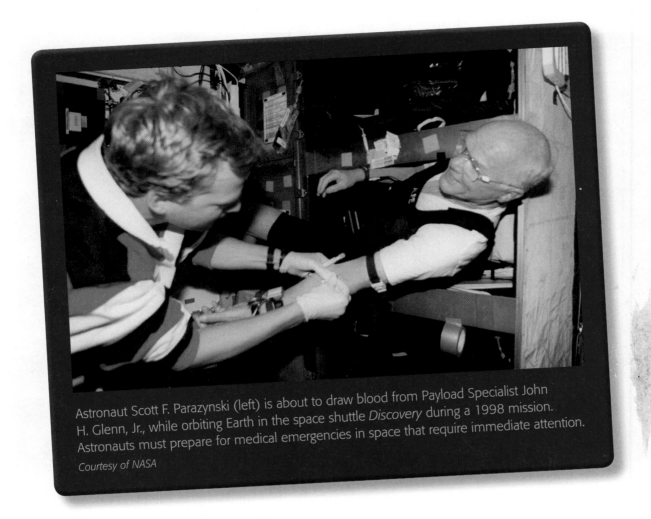

Astronaut Scott F. Parazynski (left) is about to draw blood from Payload Specialist John H. Glenn, Jr., while orbiting Earth in the space shuttle *Discovery* during a 1998 mission. Astronauts must prepare for medical emergencies in space that require immediate attention.

Courtesy of NASA

Of course, hospitals don't exist in outer space—not even neighborhood emergency clinics. NASA medical planners expect that a serious illness will occupy the medical doctor's time at some point in the journey. It will also require additional support from the ground.

Communication may be close. But it won't be instant. A call from a base on Mars would take between seven and 40 minutes to get an answer, depending on the distance between planets at any given time. Telemedicine is the term for *the delivery of medical support to remote locations*. This is typically done with computer, camera imaging, and audio systems. NASA expects to use telemedicine techniques during the flight to Mars as well as on the planet's surface.

The Benefits of Space Biomedical Research for Health on Earth

All of this biomedical research for the space program will more than likely find applications on Earth. Many useful technologies, including tools for hospitals, have already come from space exploration. Medicine will continue to profit from this kind of study.

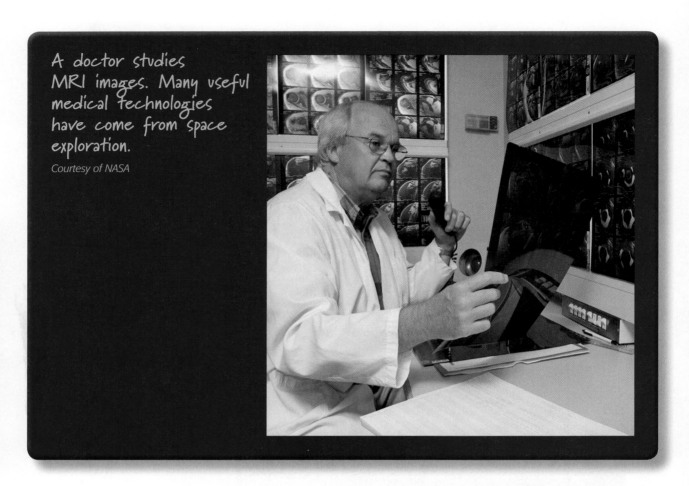

A doctor studies MRI images. Many useful medical technologies have come from space exploration.
Courtesy of NASA

Outer space may be an extreme environment. But many of the health challenges astronauts face in space have their counterparts on Earth. Many older women are under a medical verdict of osteoporosis, a condition in which their bones become porous and fragile.

What's learned about the loss of muscle mass in astronauts in space may benefit sufferers from muscle-wasting disorders back on Earth. Sleep disorders, balance problems, and cardiovascular system troubles are other areas where space research may benefit people on Earth. Space science could even benefit people who have been bedridden for extended periods.

Research being done to understand the link between radiation and cataracts could also benefit large numbers of people who will never see the inside of a space capsule. Medical experts say cataracts seem to be a natural sign of aging. But if science can prevent them, it would be a great boon for many older people.

As with most adventures, space combines excitement with danger. Outer space is inhospitable to human beings, so NASA's job is to make space people-friendly. The agency, along with teams from other federal departments and universities, has come up with and continues to invent new technologies to further mankind's venture into space. It's also probing the risks to human bodies from radiation, microgravity, subatomic particles, and more to protect US astronauts during flights. In the process, it hopes to find solutions to health problems back here on Earth.

CHECK POINTS

Lesson 4 Review

Using complete sentences, answer the following questions on a sheet of paper.

1. Medical experts know that astronauts lose heart mass and exercise capacity after long periods in microgravity. What do they suspect this could lead to?

2. Why do astronauts in space lose calcium from their bones?

3. Where is the vestibular system and what does it do?

4. What are some of the mental, or psychological, challenges of long space flights?

5. What are some of the ways NASA tries to boost astronauts' morale?

6. What tool does NASA's Space Radiation Laboratory have that helps it study the connection between space radiation and cancer?

7. How do scientists explain the flashes of light astronauts have seen inside their eyeballs? To what medical problem are these flashes linked?

8. Why are the energetic nuclei of iron atoms a particular concern to scientists concerned about the health of astronauts?

9. What kinds of foods help defend astronauts against the effects of radiation?

10. Why do older women in particular stand to benefit if biomedical research sheds new light on astronauts' loss of bone tissue in space?

 APPLYING YOUR LEARNING

11. Explain how NASA predicts the level of illness and injury it expects on a Mars mission. How do the NASA predictions make you feel about the wisdom of proceeding with such a mission?

Unit 2: Exploring Space

UNIT 3

Manned and Unmanned Spaceflight

Unit Chapters

Astronaut John Glenn enters his *Friendship 7* capsule shortly before launch on 20 February 1962. He became the first American to orbit Earth. The flight placed the country one step closer to putting a man on the Moon.
Courtesy of NASA

Space Programs

Chapter Outline

> "A day will come when beings, now latent in our thoughts and hidden in our loins, shall stand upon Earth as a footstool and laugh, and reach out their hands amidst the stars.
>
> *H. G. Wells, English author*

Quick Write

_____

_____

Was President Kennedy correct in his prediction about the importance of US space efforts? Why or why not?

Learn About

- the history and accomplishments of Project Mercury
- the history and accomplishments of Project Gemini
- the history and accomplishments of Project Apollo

On 12 September 1962, after the US manned spaceflight program had achieved some of its first triumphs, President John F. Kennedy gave a speech at Rice University in Houston, Texas. He said: "We set sail on this new sea because there is new knowledge to be gained, and new rights to be won, and they must be won and used for the progress of all people. For space science, like nuclear science and all technology, has no conscience of its own. Whether it will become a force for good or ill depends on man, and only if the United States occupies a position of preeminence can we help decide whether this new ocean will be a sea of peace or a new, terrifying theater of war."

The History and Accomplishments of Project Mercury

Vocabulary

President Kennedy's stirring call to send a man to the Moon and bring him back safely was a crystallizing moment in the history of human space exploration. But Project Mercury predated the Kennedy administration. It began under the man who served as president just before Kennedy: Dwight D. Eisenhower.

- suborbital
- extravehicular activity
- combustible

Astronaut John Glenn enters his Friendship 7 capsule shortly before launch on 20 February 1962.

His three orbits around Earth placed America one step closer to putting a man on the Moon. President Kennedy's stirring call to send a man to the Moon and bring him back safely was a crystallizing moment in the history of human space exploration.

Courtesy of NASA

The History of Project Mercury

The idea of space travel has fascinated people throughout recorded history. And so it's hard even for NASA to put an exact date on the actual beginning of American efforts to put a man in space. (And in those days, they really meant "a man," not "a man or a woman.")

But by the mid-twentieth century, the necessary hardware was available to realize the dream of spaceflight. By 1958 tests conducted by scientists in government and industry showed that sending astronauts into space would indeed be feasible. On 7 October 1958 NASA initiated the first national manned spaceflight project, later named Project Mercury. This was just a little more than a year after the Soviets' launch of their *Sputnik* satellite. That event had shaken Americans badly. It left them feeling dangerously behind the Soviets scientifically and, by implication, militarily. A nation that could send a satellite into orbit, people believed, could also put a nuclear weapon on a missile and fire it at the United States.

Project Mercury lasted only a relatively short time: not quite five years. It ran from the time of the first official go-ahead (1958) to the final Mercury mission, Gordon Cooper's 34 hours in orbit around Earth in 1963.

NASA named Project Mercury for the ancient Romans' messenger god. Project Mercury was a time of learning about man's abilities in space. Some of this learning actually happened in activities on the ground: simulations and relentless drills on procedures. This helped make it possible to accomplish Mercury's goals in such a short time.

Star POINTS

Project Mercury, over the course of its not quite five years, drew on the skills, initiative, and experience of more than 2 million people from many major government agencies and from much of the aerospace industry.

The Goals and Accomplishments of Project Mercury

Project Mercury's astronauts made six manned flights between 1961 and 1963. Mercury's goals were quite specific:

- To orbit a manned spacecraft around Earth
- To investigate man's ability to function in space
- To recover both man and spacecraft safely.

The Mercury Team: the 'Original Seven'

As you read in Chapter 5, Lesson 2, the first American astronauts chosen were the "Original Seven" of Project Mercury. NASA presented them to the public at a press conference on 9 April 1959. Their introduction, in civilian clothes rather than military uniforms, put a human face on the space program. NASA told the American people that these pioneers would be known as "astronauts," sailors among the stars, because they would go off into an uncharted new "ocean."

The Original Seven gather at Langley Air Force Base in Virginia to train for their Mercury missions. They are *(from left in front)*: Virgil Grissom, Scott Carpenter, Donald Slayton, Gordon Cooper, *(from left in back)* Alan Shepard, Walter Schirra, and John Glenn.

Courtesy of NASA

The public soon adopted these star sailors as heroes. The astronauts got plenty of attention in the news media, especially glossy magazines. They were on space technology's leading edge. Their work required immense courage. And yet the space agency presented them to the public as men who could be their neighbors—or at least neighbors of the magazine-reading public. The astronauts were trim, healthy, average-sized men. They had wives and children and college degrees, typically in mechanical engineering.

The Mercury Seven's missions would pave the way for the two programs that would follow, as well as for all further human spaceflight.

Alan B. Shepard, the First American in Space

The American artist Andy Warhol once said that in the future, everyone will be world-famous for 15 minutes. Today "15 minutes of fame" is a proverbial expression. Well, Alan Shepard's first big moment really did last 15 minutes.

On 5 May 1961, at 9:34 a.m., as schoolchildren watched on televisions in their classrooms, he launched into space aboard *Freedom 7.* He splashed down in the Atlantic Ocean about a quarter of an hour later. He was the first American in space.

His flight was suborbital—*having a trajectory, or path through space, of less than a complete orbit.* The mission's main goal was to see how well an astronaut would do in space and *what* he would be able to do. NASA was also keen to find out

> ### Star POINTS
>
> Extensive psychological testing was part of the selection process for the Mercury astronauts. Among the questions put to the prospective astronauts were, "Who am I?" and "Whom would you assign to the mission if you could not go yourself?"

A Marine helicopter retrieves Alan Shepard from the Atlantic Ocean shortly after the *Freedom 7* splashed down from its history-making flight. Shepard, who flew the capsule into suborbital flight, was the first American in space.
Courtesy of NASA

Star POINTS

Shepard's mission aboard *Freedom 7* came just 23 days after Yuri A. Gagarin of the Soviet Union became the first man in space. But Shepard still made history. Gagarin had been really just a passenger in his spacecraft; Shepard piloted *Freedom 7*. And whereas the Soviet Union veiled Gargarin's mission in secrecy, NASA made sure Shepard's mission from launch to splashdown and recovery was televised live and seen by millions around the world.

how well Shepard would do at launch and on reentry. Aviators know that takeoff and landing are the most potentially dangerous parts of any flight. What happens in between is relatively easier. The same principle holds true for spaceflight.

Some experts feared that even a few minutes in space, with all the strains of launch and reentry, plus the challenge of weightlessness, would leave Shepard disoriented or worse. But he reported no such difficulties. "It was painless," he said, "just a pleasant ride."

John Glenn, the First American to Orbit the Earth

When Marine Col. John H. Glenn Jr. went up in *Friendship 7* on 20 February 1962, he became the first American to orbit the Earth. NASA's objectives for the mission were to put Glenn into orbit, to observe his reactions to being in outer space, and to return him safely to Earth at a point where he could be easily found.

The destroyer USS *Noa* picked him up 800 miles southeast of Bermuda. Lookouts aboard the ship spotted Glenn's parachute nearly a mile up in the sky from about five miles away. The *Noa* had *Friendship 7* aboard 21 minutes after Glenn splashed down.

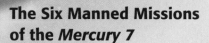

The Six Manned Missions of the *Mercury 7*

Mission No. 1: On 5 May 1961 Alan B. Shepard Jr. made a suborbital flight of a little more than 15 minutes in space in the *Freedom 7*.

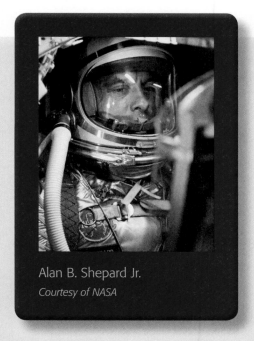

Alan B. Shepard Jr.
Courtesy of NASA

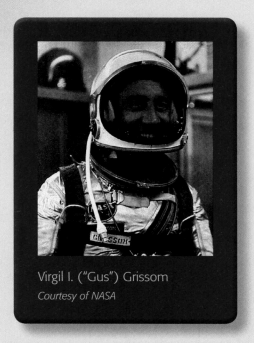

Virgil I. ("Gus") Grissom
Courtesy of NASA

Mission No. 2: On 21 July 1961 Virgil I. ("Gus") Grissom made a second suborbital flight of a little more than a quarter hour in the *Liberty Bell 7*. This follow-up showed that Shepard's very successful flight had not been a fluke. Grissom's mission was marred, however, by the loss of the spacecraft shortly after splashdown.

Mission No. 3: On 20 February 1962 John H. Glenn Jr. orbited Earth three times aboard *Friendship 7*. His mission lasted nearly five hours.

Mission No. 4: The primary goal of M. Scott Carpenter's three-orbit mission in *Aurora 7* on 24 May 1962 was to determine whether an astronaut could work in space. He proved that he could, by carrying out a number of scientific experiments during his flight. This was a major steppingstone toward a lunar landing.

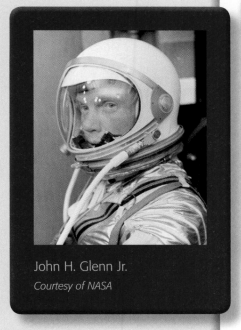

John H. Glenn Jr.
Courtesy of NASA

M. Scott Carpenter
Courtesy of NASA

Mission No. 5: On 3 October 1962 astronaut Walter M. Schirra Jr. orbited Earth six times in his spacecraft, *Sigma 7*. He proved that an astronaut could carefully manage the limited amounts of electricity and maneuvering fuel necessary for longer, more complex flights. Many have termed his precisely engineered flight a "textbook spaceflight."

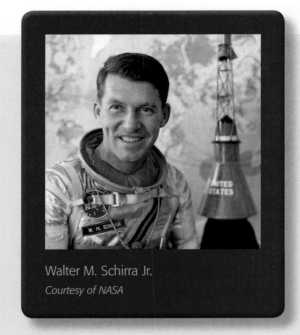

Walter M. Schirra Jr.
Courtesy of NASA

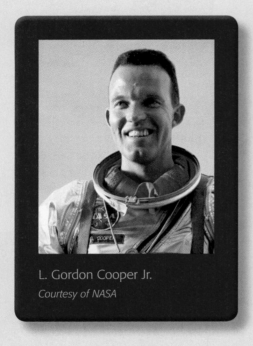

L. Gordon Cooper Jr.
Courtesy of NASA

Mission No. 6: On 15 May 1963 L. Gordon Cooper Jr. took off for the Mercury program's longest mission by far. In his ship, *Faith 7*, he circuited 22.5 times around the Earth. He spent more than 34 hours in flight. He had two primary goals. One was to see how he would hold up on such a long mission. The other was "to verify man as the primary spacecraft system." This is NASA's way of saying that he confirmed the value of sending astronauts, rather than robots or chimpanzees, into space. He also became the first American astronaut to sleep in orbit.

Donald K. Slayton Jr. was the only one of the Original Seven who did not go into space during the Mercury program. On 15 March 1962, just weeks before he would have made Mercury's second orbital flight, NASA announced that he was being grounded by a heart condition. In March 1972 the space agency restored him to full flight status. In 1975 he made his only space flight as part of the Apollo program.

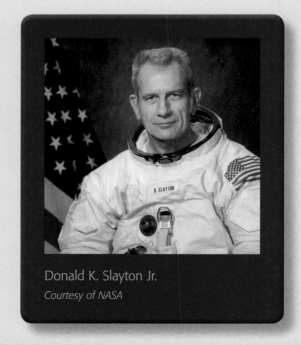

Donald K. Slayton Jr.
Courtesy of NASA

The Gemini VI crew took photos of Gemini VII during a 1965 mission to practice rendezvous maneuvers in space.

NASA decided that Gemini was a needed intermediate step between Project Mercury and the Apollo program, which was meant to put a man on the Moon.

Courtesy of NASA

The History and Accomplishments of Project Gemini

Project Mercury was in the middle of its run when NASA announced a successor program. On 7 December 1961 the agency revealed the development of a two-man spacecraft. A few weeks later, on 3 January 1962, the program received the official designation *Gemini*.

The History of Project Gemini

If you're familiar with the zodiac, you should recognize the name *Gemini* and understand why it was so apt for this program. Gemini, the third constellation in the zodiac, contains the twin stars Castor and Pollux. The name *Gemini* is Latin for *twins*. NASA decided that Gemini was a needed intermediate step between Project Mercury and the Apollo program, which was meant to put a man on the Moon.

The Goals and Accomplishments of Project Gemini

Gemini involved 12 flights, including two unmanned flight tests of the equipment. Like Project Mercury, Gemini had very clear-cut objectives. They were:

- To put men and equipment into space for flights lasting as long as two weeks
- To rendezvous in space and dock with a ship already in orbit—then maneuver, or steer, the two linked ships using the propulsion system of the one that was already on orbit

- To perfect ways of entering the atmosphere and landing at a preselected point on Earth

- To find out more about how weightlessness affects crew members and how their bodies respond to long periods in space.

Gemini was a successful program. The only goal it did not meet was that of bringing a spacecraft back to Earth and having it touch down on dry land, rather than splashing down in the ocean. NASA officials canceled that goal in 1964. They felt, however, that the precision control necessary to make a "land landing," as they called it, had been demonstrated.

Ed White, the First American to 'Walk' in Space

On *Gemini IV*, which was launched on 3 June 1965, Edward H. White II became the first American to "walk" in space. Using a tether, he stepped outside the spacecraft's relative safety. Extravehicular activity (or EVA) is NASA's term for this kind of out-of-capsule experience, otherwise known as *a spacewalk*.

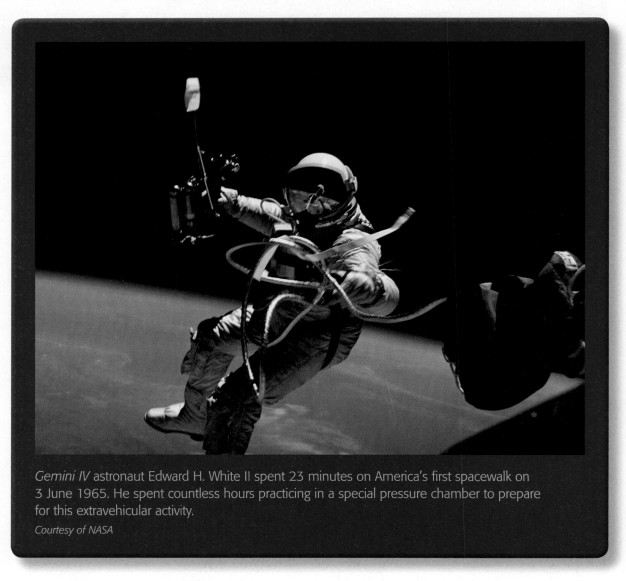

Gemini IV astronaut Edward H. White II spent 23 minutes on America's first spacewalk on 3 June 1965. He spent countless hours practicing in a special pressure chamber to prepare for this extravehicular activity.

Courtesy of NASA

CHAPTER 6 Space Programs

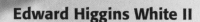

Edward Higgins White II

Edward Higgins White II came from a military family. His father had served in the Air Force and retired as a major general. Young Ed first took the controls of an aircraft at age 12, under his father's supervision. "It felt like the most natural thing in the world to do," he would remember later.

Gemini IV astronauts Ed White (*left*) and James A. McDivitt get a phone call from President Johnson congratulating them on their successful mission. They are on board the USS *Wasp*, which recovered them after their splashdown in the Atlantic Ocean.

Courtesy of NASA

Something else that seemed natural was attending the United States Military Academy at West Point, like his father. But most cadets there are appointed by their representative in Congress. The Whites had moved around so much in the service of their country that no particular member of Congress was "their" representative. So Ed decided to seek an "at large" appointment instead. He went to Washington, DC, and walked the halls of Congress, knocking on doors in search of an appointment. He finally knocked on enough doors to get one.

Another part of the astronauts' work was public relations. Astronauts were sometimes called to speak to employees of spacecraft manufacturers and other segments of the public. Not all of the astronauts were comfortable doing this. White, however, recognized it as part of the job, and joined Toastmasters International to improve his public-speaking skills. At one point he even served as an officer of the organization.

One of the space program's recurring themes was competition with the Soviet Union. You might say the spacewalk was another event in this long-running rivalry. Less than three months earlier, on 18 March 1965, Russian cosmonaut Alexei Leonov became the first human to walk in space after he emerged from *Voskhod II* at the end of a 10-foot tether.

White's spacewalk was considered only a secondary objective of *Gemini IV*. In fact, when NASA introduced the mission's flight plan, the special suit he was to wear for this walk was still on the drawing board. So was the self-propulsion unit he was to use—a hand-held jetpack that would let him maneuver outside the spacecraft. Engineers didn't get the gear certified for use in space until 10 days before the launch. The spacewalk wasn't a confirmed part of the program until a week before launch.

But White spent countless hours practicing in a special pressure chamber to prepare for his spacewalk. Although Leonov made history as the first to *float* in space outside a spacecraft, White gets credit for being the first to use jet propulsion to *maneuver* in space. His EVA was longer than Leonov's, too—23 minutes, compared with about 10 minutes.

The Gemini program's achievements made possible not only Project Apollo and the Moon landing. It also pioneered many of the tasks and procedures necessary for the space shuttle and the various space stations. Gemini showed it was possible to launch two astronauts and to keep them in space for several days. It demonstrated that NASA personnel had the skills necessary to rendezvous in space with another spacecraft and dock with it. Finally, Gemini astronauts, following White, perfected the techniques for leaving a space capsule and maneuvering safely outside it. These were significant successes.

The History and Accomplishments of Project Apollo

The space program's third phase—beyond spacewalks and Earth orbits—was getting astronauts to the Moon. So NASA developed Project Apollo.

"That's one small step for a man, one giant leap for mankind." Those were the first words uttered on the Moon's surface by a human being. The man who spoke them, Neil Armstrong, was an American. But it's notable that he (and the NASA communications people who helped him with his lines) cast his accomplishment in universal terms. It was a national effort that put him on the Moon. But his mission fulfilled a dream as old as humanity.

The History of Project Apollo

The Apollo program overlapped somewhat with the earlier Mercury and Gemini programs. Historians trace Apollo's roots back to the Eisenhower administration. They note that the "space race" between the United States and the Soviet Union helped energize the American quest to go to the Moon.

Abe Silverstein, a NASA official, chose the name "Apollo" for the project. Apollo was the Greek god of music, light, and the Sun, among other things. Silverstein later said that the image of this very versatile god "riding his chariot across the Sun was appropriate to the grand scale of the proposed program."

Apollo had many successes and two important failures. Lessons learned from the first failure, as you will read later in this lesson, helped make the second near-disaster a "successful failure." Historians and others have observed that the Apollo program's success depended on the lessons learned from its early mistakes.

The Goals and Accomplishments of Project Apollo

Project Apollo was about more than just landing Americans on the Moon and bringing them safely back to Earth, however. NASA's other goals for Apollo were these:

- To establish the technology to further other national interests in space
- To achieve preeminence in space for the United States
- To explore the Moon scientifically
- To develop humans' capability to work on the Moon.

On 16 July 1969 a *Saturn V* rocket launches the *Apollo 11* crew to the Moon. The mission fulfilled a dream as old as humanity.
Courtesy of NASA/Stennis Space Center

Astronauts Virgil I. "Gus" Grissom, Edward H. White II, and Roger B. Chaffee were all killed when a fire erupted in their *Apollo 1* capsule during testing.
Courtesy of NASA

Project Apollo comprised 11 manned missions, numbered *Apollo 7* through *Apollo 17*. Six of these were successful Moon landings. They returned a wealth of scientific data and more than 800 pounds (almost 400 kilograms) of lunar samples.

Apollos 7 and *9* were Earth-orbiting missions to test the spacecraft. *Apollos 8* and *10* orbited and photographed the Moon. Apollo also included several unmanned missions, tests of rockets that would lift the Apollo craft into space. Apollo included experiments on soil mechanics, meteoroids, seismic events, heat flow, lunar ranging, magnetic fields, and the solar wind.

The *Apollo 1* Disaster and the Lessons Learned

On 27 January 1967 tragedy struck the Apollo program. It was something everyone in the space program had braced for from the beginning. The surprise was that it happened on the ground.

On that day, *Apollo-Saturn (AS) 204*, also known as *Apollo 1*, sat on the launch pad with three astronauts aboard. They were "Gus" Grissom, Edward White, and Roger B. Chaffee. They were set to be the crew of the *Apollo 1* mission, and they were practicing a mock launch sequence. At 6:31 p.m., after hours of work, a fire broke out in the spacecraft. The pure oxygen atmosphere intended for the flight made the blaze burn intensely. Flames engulfed the capsule. The three astronauts died of asphyxiation. Theirs were the first deaths attributable to the US space program.

The accident left NASA in shock. The next day it appointed a panel to investigate. The panel quickly found that a short circuit in the electrical system had ignited combustible—*flammable*—materials in the spacecraft. The 100-percent oxygen atmosphere, at full sea-level air pressure, had made the fire all the fiercer. Even some metal will burn under such conditions if ignited. The panel also found that the fire could have been prevented. It called for several modifications to the spacecraft, including a move to a less oxygen-rich environment. Changes followed quickly. Within a little more than a year the new spacecraft was ready for flight.

Neil Armstrong, the First Man to Walk on the Moon

Here's how one NASA website summarizes the *Apollo 11* mission: "Perform manned lunar landing and return mission safely. (Achieved)." President Kennedy had set this goal little more than eight years before. On 16 July 1969 *Apollo 11* took off with Neil Armstrong, Edwin Aldrin, and Michael Collins aboard.

On 20 July the radio message went back to the mission's ground controllers: "Houston, Tranquility Base here. The *Eagle* has landed." The lunar module had set down on the Moon, and soon Armstrong and Aldrin would make their first Moon walk—for two-and-a-half hours. As their crewmate, Collins, orbited the Moon,

Neil Armstrong stands by the lunar module. Fellow astronaut "Buzz" Aldrin took the photo during the mission's first extravehicular activity.
While on the Moon, the crew planted an American flag and deployed some sensing instruments.

Courtesy of NASA

Armstrong and Aldrin explored its surface. They planted an American flag and deployed some sensing instruments. They also left a plaque with this inscription: "Here men from planet Earth first set foot upon the Moon. July 1969, A.D. We came in peace for all mankind."

They spent a little less than 22 hours on the Moon's surface and two-and-a-half days in orbit around it.

How NASA Averted Disaster in the *Apollo 13* Mission

Apollo 11 made the phrase "The *Eagle* has landed" part of the national vocabulary. For *Apollo 13*, the key phrase was "OK, Houston, we've had a problem here."

On 13 April 1970 *Apollo 13* was two-and-a-half days into its voyage to the Moon. The crew had just finished a 49-minute TV broadcast to show the American public how comfortably they lived and worked in space. Commander Jim Lovell had just wished his audience a good night.

Then nine minutes later the astronauts heard a "pretty loud bang." The No. 2 oxygen tank on *Apollo 13*'s service module had exploded. Lovell soon saw it venting its precious contents out into space. They and their support team on the ground didn't take long to realize they had "lost the Moon." With so much of the oxygen that their ship's fuel cells needed gone, *Apollo 13*'s mission changed from a lunar landing to bringing the astronauts safely back home.

The crew in space and their handlers on the ground soon had pretty much the same idea: that the astronauts should save themselves by jumping into their lifeboat—their lunar lander. From the early days of planning Moon missions, the so-called lunar module lifeboat scenario had been considered as a life-saving strategy, should the need arise. Their spacecraft was already on a trajectory toward the Moon. They continued toward the Moon, and then "slingshot" around it to return to Earth.

To make it work, they had to manage their "consumables" and to clear out the carbon dioxide gas they were exhaling. The crew actually had plenty of oxygen for breathing—including the backpacks that had been prepared for Moon walks that now they were not going to make. Power and water turned out to be more critical. All three astronauts were seriously dehydrated when they returned to Earth—they lost 31.5 pounds among them.

The days after the accident were marked by constant problem-solving and close teamwork between the astronauts and ground controllers. For example, to vent carbon dioxide, Mission Control devised a system using plastic bags, cardboard, and tape—all materials carried aboard *Apollo 13*. Houston also had to devise a way for the astronauts to manually align the spacecraft by sighting the Sun when they were unable to use the onboard instruments. And ground controllers had to develop a procedure to power-up the Command Module after days of cold "sleep." They accomplished this in three days instead of the usual three months.

Safely back home: Astronauts (*from left*) Fred Haise, Jim Lovell, and Jack Sweigert aboard the USS *Iwo Jima* after their return to Earth. NASA classifies *Apollo 13* as a "successful failure" because, although the mission was aborted, the agency managed to bring all three astronauts home in one piece.

Courtesy of NASA

But they did make it back. NASA classifies *Apollo 13* as a "successful failure" because, although the original mission was aborted, the agency managed to bring all three astronauts home safely. They splashed down gently in the Pacific Ocean near Samoa.

The United States remains the only country to have sent humans to the Moon. Besides *Apollo 11*, astronauts reached Earth's satellite on *Apollos 12, 14, 15, 16*, and *17*. Each one included more time on the lunar surface than the last, with *Apollo 17*'s crew spending a total of 75 hours there. The *Apollo 17* mission was rather long overall as well: It ran from 7 to19 December 1972. The astronauts spent 22 hours in moonwalks and camped out on the Moon for three days total.

Twelve men in all walked on the Moon before Apollo was done. The last three missions featured the *Lunar Rover*, which permitted the astronauts to drive about and explore various terrains too rough for the lunar module to attempt to land on.

Sadly, *Apollo 18, 19,* and *20* were canceled because no funding was available. Apollo spacecraft were used for four later missions—the three long-duration *Skylab* missions and the *Apollo-Soyuz* linkup with the Soviet *Soyuz 19.*

Apollo produced a store of scientific treasure whose value scientists are still discovering. Most significant were the lunar samples the astronauts brought back to Earth. Before 1969 the only samples of extraterrestrial material known on Earth were meteorites. The Apollo samples changed all that.

No other effort to explore the Moon compares with Apollo. In bringing back selected and documented samples from six locations on the Moon, Apollo far surpassed all other efforts. The Apollo samples now stored on Earth allow lunar research to continue, based on new concepts and using better techniques and instruments. Only the limited number of sites from which the astronauts took samples will restrict new concepts of the Moon.

Since the Moon landings, manned space missions have centered on the US space shuttle, US and Russian space stations, and the International Space Station. Now in the twenty-first century, the United States hopes to again send astronauts to the Moon and then beyond. The outcome could be great scientific gains, as well as pure and simple inspiration "for all mankind."

Star POINTS

For the last stages of their return to Earth, the *Apollo 13* astronauts left their lunar module lifeboat to return to the command module. They had to power the command module back up. It was cold and clammy, with droplets of water everywhere. The astronauts worried about short circuits when they powered back up. But there was no problem. Lessons learned from the *Apollo 1* fire meant that the command module was built to avoid such problems.

CHECK POINTS

Lesson 1 Review

Using complete sentences, answer the following questions on a sheet of paper.

1. What happened about a year before NASA initiated the first national manned space-flight project?

2. What were the three goals of Project Mercury?

3. Who were the "Original Seven" and what effect did their introduction to the public have on the space program?

4. What was Alan B. Shepard's achievement in 1961?

5. What did John Glenn achieve in *Friendship 7*?

6. Why did NASA decide Project Gemini was needed?

7. What goal did Project Gemini not meet?

8. Why is Edward White's spacewalk seen as a greater accomplishment than Alexei Leonov's?

9. Who chose the name for Project Apollo? Why did he call it that?

10. What kind of data and samples did the six successful Moon landings of the Apollo program bring back to Earth?

11. What did NASA do on 28 January 1967?

12. Explain the significance of this radio message: "Houston, Tranquility Base here. The *Eagle* has landed."

13. How does NASA classify the *Apollo 13* mission? Why?

APPLYING YOUR LEARNING

14. Given all the risks in connection with spaceflight, was it a good idea for the United States to be as open about its space program as it was? Why or why not?

LESSON 2 The Soviet/Russian Manned Space Program

Quick Write

How do you think you would have felt if you had made a spacewalk like Leonov's? Have you ever had an experience where you felt you were far above, or far away from, ordinary cares and had to be called back in, as he was?

Learn About

- the history and accomplishments of the Russian Vostok project
- the history and accomplishments of the Russian Voskhod project
- the history and accomplishments of the Russian Soyuz Project

When Soviet cosmonaut Alexei Arkhipovich Leonov took his first steps outside his spaceship, *Voskhod 2*, orbiting miles and miles above the Earth, his 4-year-old daughter, Vika, could see it all on live television down below. She didn't know what was going on. She hid her face in her hands and cried. "What is he doing? What is he doing?" she wailed. "Please tell Daddy to get back inside."

Leonov's elderly father, too, was upset. The Soviet space program was "open" enough at that point that its missions were televised, like the American ones. But the civilian public, including the cosmonauts' families, had no idea what to expect. Such was the Soviet habit of secrecy. The senior Leonov, unaware of the plans for the spacewalk, thought his cosmonaut son was simply misbehaving in space. "Why is he acting like a juvenile delinquent?" he shouted in frustration to the journalists gathered at his home. "Everyone else can complete their mission properly, inside the spacecraft."

But then the voice of Soviet President Leonid Brezhnev could be heard delivering a message of congratulations to the cosmonauts: "We are proud of you. We wish you success. Take care. We await your safe arrival on Earth."

Meanwhile, back in outer space, Leonov heard his commander calling him, "It's time to come back in." Leonov later said, "In that moment my mind flickered back for a second to my childhood, to my mother opening the window at home and calling to me as I played outside with my friends, 'Lyosha, it's time to come inside now.' "

The History and Accomplishments of the Russian Vostok Project

Vocabulary

- air lock
- retrofire
- retrorocket

Historians often cast the story of the US and Soviet parallel space programs as the Cold War's space race. It would be wrong to play down the competitive aspect of space exploration. But the Soviet Union made a major effort at space exploration, and this effort made serious contributions to science. In this lesson you will read about ways the two countries' space programs differed, but also how closely they tracked each other.

Cosmonaut Yuri Gagarin, the First Man in Space

Like Project Mercury, Project Vostok was meant to show that it was possible to put a man in orbit, observe his reactions to being in space, and return him safely to Earth at a known point. (*Vostok* is the Russian word for *east*.) NASA wanted to prove that an astronaut was not just a passenger in space but "an invaluable part of the space flight systems as pilot, engineer and experimenter." Soviet designers,

Cosmonaut Yuri Gagarin rides a bus to the launch pad on 12 April 1961 for his famed Vostok 1 spaceflight.

He flew only one mission, but that one made him the first human in space and the first to orbit Earth.

Courtesy of NASA

Yuri A. Gagarin

Yuri A. Gagarin, who would become the Soviet Union's first man in space, was born the son of a carpenter on a collective farm west of Moscow on 9 March 1934. After six years of local schooling he continued his education at vocational and technical schools. In 1955 he joined the Russian Air Force. In 1957 he graduated with honors from the Soviet Air Force Academy. He became a military fighter pilot, and by 1959, his country had selected him for training as one of the first group of Soviet cosmonauts.

Col. Gagarin died on 27 March 1968 when the MiG-15 he was piloting crashed near Moscow. At his death, he was in training for a second space mission.

on the other hand, assigned limited tasks to their cosmonauts. The tradeoff was that the Soviets put a man in space first, and that first flight was longer than either of the first two American manned space missions.

The man whom the Soviets launched into space was Col. Yuri A. Gagarin. He flew only one space mission. But that one made him not only the first human in space but also the first to orbit Earth. On 12 April 1961 his spacecraft, *Vostok 1*, circled the Earth at nearly 17,000 mph (27,400 km per hour). His flight lasted 108 minutes. At his highest point, he was about 200 miles (327 km) above the Earth.

Star POINTS

Although the controls of Gagarin's spacecraft were locked, Soviet space engineers had placed a key in a sealed envelope in case an emergency arose and he needed to take control.

In orbit, Gagarin had no control over his spacecraft. A computer controlled its reentry by sending radio commands to *Vostok 1*. Gagarin's return to Earth was different from that of American astronauts. According to plan, Gagarin ejected from the spacecraft after reentry into Earth's atmosphere and parachuted to the ground.

Valentina Tereshkova, the First Woman in Space

Valentina Tereshkova was an amateur parachutist who happened to be working as an assembly worker in a textile factory when recruiters signed her up for the cosmonaut program. She was one of four women selected for a special woman-in-space program, at the direction of Soviet Premier Nikita Khrushchev. Tereshkova was the only one to complete a space mission. But hers was quite a mission.

She became the first woman in space when she went up aboard *Vostok 6* on 16 June 1963. Her mission lasted nearly three days, in which she orbited Earth 48 times. This meant she had stayed up twice as long, and made more than twice as many circuits, as the Mercury program's marathon man, L. Gordon Cooper Jr., whose mission had taken place a month before hers.

Valentina Tereshkova (1937–)

Valentina Tereshkova, who would become the first woman in space, had a background very typical of the Soviet people of her time. She was born on 6 March 1937 in the Yaroslavl region of Russia, an industrial area not far from Moscow. Her father drove a tractor, and her mother worked in a textile mill. Valentina started school at age 8 and left eight years later to begin work. But she continued her education by correspondence school.

The way Tereshkova became a cosmonaut wasn't very typical, however. She wasn't an air force pilot, for example, as Yuri Gagarin was. While still very young, she developed an interest in parachute jumping. She took up the activity and became good at it. It was this expertise that led the Soviet space agency to select her for its cosmonaut program.

After she completed her space mission, the Soviets honored Tereshkova as a "Hero of the Soviet Union." She also received the United Nations Gold Medal of Peace as a goodwill ambassador for her country.

Cosmonaut Valentina Tereshkova (right) shakes hands with *Expedition 10* commander and NASA International Space Station science officer Leroy Chiao in Kazakhstan in 2004. Tereshkova became the first woman in space when she went up aboard *Vostok 6* in 1963.

Courtesy of NASA/Bill Ingalls

Sergei P. Korolev (1906–1966)

Sergei P. Korolev, the man who designed the rocket that put *Sputnik 1* into space on 4 October 1957, was also a leading advocate of Soviet efforts to put a man on the Moon before the Americans. Korolev was an early experimenter with rockets. In the early 1930s he and his colleagues were testing liquid-fueled rockets of increasing size.

Their work caught the Soviet military's attention. Within a few years Korolev, by now at a military research institute, had designed the RP-318. It was Russia's first rocket-propelled aircraft. But then his career made an abrupt detour: At the peak of the Soviet leader Stalin's purges—campaigns to "cleanse" society of political undesirables—Korolev and a number of other aerospace engineers were thrown into prison. Korolev himself spent a year doing forced labor in the dreaded Kolyma gold mines.

Eventually Stalin realized that rocketry could be useful in the expected war against Nazi Germany. He brought some engineers out of prison. Others he left in place. But he set up a system of "prison design bureaus" to create new weapons for the Red Army.

Korolev was saved when senior aircraft designer Sergei Tupolev, himself a prisoner, asked for his help in one of these bureaus. Some years later Korolev's R-7 rocket launched *Sputnik 1*.

The hardest part of getting a man to the Moon, Korolev and his team found, was building a big enough rocket. The Soviets needed something like the American *Saturn V*. Korolev's design bureau began work on the so-called N-1 rocket in 1962. But even though work continued for years after his death, by the time the project was finally canceled, it had never made a successful flight.

Other 'Firsts' Achieved by the Russian Vostok Project

The cosmonauts of the Vostok program were rightly proud of the "firsts" they racked up in the early 1960s. Among their other achievements:

- In 1961 Gherman Titov became the first person to spend a full day in orbit, aboard *Vostok 2*
- In 1962 *Vostok 3* and *Vostok 4* carried out the first two-spacecraft mission
- On 14 June 1963 the crew of *Vostok 5* carried out the first long-duration mission, five days in orbit.

The History and Accomplishments of the Russian Voskhod Project

For Americans, Project Gemini followed Project Mercury. For the Soviets, the Voskhod Project (which means "dawn" or "sunrise") came after the Vostok Project. In both countries, in fact, engineers were hard at work on the second phase of the quest for the Moon, even as the Mercury and Vostok missions were still being carried out.

The *Voskhod* as the First Three-Man Spacecraft

As the Vostok Project gave way to the Voskhod Project, and Mercury gave way to Gemini, designers in both countries were looking to improve on the spacecraft they had relied on for those first missions. Both Soviet and American designers wanted vehicles that were more flexible and were capable of carrying more than one person.

NASA engineers started out trying to improve the *Mercury* capsule but ended up designing the new *Gemini* craft pretty much from scratch. The Soviets, however, took the *Vostok* spacecraft and modified it to hold two or three astronauts.

The Soviets were apparently eager to beat the goals set for Project Gemini and convinced they didn't have time to design a new spacecraft from the beginning. The *Voskhod* was an attempt to make the most of a tested design.

The *Voskhod*'s First Crew

Voskhod 1, which carried three men into space, was the first "multimanned" flight. Vladimir M. Komarov was the command pilot. Boris B. Yegorov was the physician on board. Konstantin P. Feoktistov, a scientist and spacecraft designer, rounded out the crew. It was Feoktistov who had figured out how to modify *Vostok* to hold more than one astronaut. The Soviet space program rewarded him for his achievement with a trip into outer space himself.

The *Voskhod I* mission was designed:

- to test out the new spacecraft
- to see how well a group of cosmonauts from different professional backgrounds would work together in space
- to conduct physical and technical experiments
- to perform an extensive medical-biological investigation program.

Voskhod 1 cosmonauts (*from left*) Vladimir Komarov, Boris Yegorov, and Konstantin Feoktistov make their way to the launch pad for mankind's first multimanned spaceflight. The crew returned live television pictures from space.
Courtesy of NASA

The crew returned live television pictures from space. And they had enough confidence in the space cabin's life-support system that they could wear overalls rather than cumbersome spacesuits and helmets. (The crowded capsule didn't have room for them.) The Soviets deemed *Voskhod 1* a success.

Voskhod 2's external movie camera captured these photos of Alexei Leonov as he took man's first spacewalk on 18 March 1965. During his brief adventure, which lasted only about 20 minutes, he felt both tension and euphoria.

Courtesy of NASA

The Experiments Conducted During the Voskhod Project

Yegorov was the first trained medical doctor in space. He conducted a number of medical experiments during his flight. He tested the cosmonauts' lung function and their sense of balance. He also took the first blood samples in space. This will confirm that the health challenges of spaceflight you read about in Chapter 5, Lesson 4, have been under study for some time.

Alexei Leonov, the First Man to 'Walk' in Space

On 18 March 1965 the Soviets launched the spacecraft *Voshkod 2* with two men aboard, Pavel I. Belyayev, pilot, and Alexei A. Leonov, co-pilot. On this mission, Leonov made the first spacewalk.

Leonov's spacecraft had an extendable air lock. An air lock is *an airtight chamber, usually located between two regions of unequal pressure, in which air pressure can be regulated.* An air lock lets people flying in space close one door behind them, so to speak, before they open another. On *Voshkod 2*, the air lock let Leonov step outside without having to spill all the air in the main cabin out into space.

Leonov put on a spacesuit to go outside. It was a sign of progress that cosmonauts no longer had to wear spacesuits inside their capsules. He carried a life-support system in a backpack. A TV camera recorded his spacewalk, and he had a handheld movie camera as well. During his brief adventure, which lasted only about 10 to 20 minutes, he felt both tension and euphoria. He was floating alone outside his little spaceship, at the end of a 10-foot tether, miles above the Earth's surface.

It was a spectacular achievement, but the mission went downhill from there. Leonov had trouble getting back into the spaceship because his suit had become unmanageably stiff from too much pressure. The automatic orientation system for the retrofire malfunctioned. (Retrofire is *the ignition of a retrorocket, a small rocket used to slow or change the course of a spaceship*.) And so the cosmonauts had to bring their ship back into Earth's atmosphere manually. They ended up landing in six feet of snow in a Siberian pine forest, hundreds of miles from their target area. Once rescuers had located them, it took them a whole day to cut through the trees and bring the crew out on skis.

When a Land Power Takes to the Heavens

The US-Soviet space race played out against the larger backdrop of the Cold War. The two countries' space programs reflected differences in the kind of power each was. Americans who watched the missions of the Mercury, Gemini, and Apollo astronauts took it for granted that they would "splash down" in the ocean. Then US Navy helicopters would pluck them from the water and ferry them to a waiting aircraft carrier. But the Soviets decided to have all their manned space missions "land" in the fullest sense of the word—on solid ground. Usually Soviet spacecraft landed in southern Kazakhstan. It's now an independent country, but back then it was part of the Soviet Union.

The United States was and is a great sea power, with bases around the globe. It has many small island bases across the Pacific. Russia is a great land power, and so was the Soviet Union before it, to an even greater degree. Spacecraft coming down in sparsely populated Kazakhstan were unlikely to hit anyone on the ground, and would be generally within reach of rescue crews coming overland. Soviet "land landings" were not always without incident, however, as *Voskhod 2*'s return to Earth showed.

Alexei Arkhipovich Leonov (1934–)

Alexei Arkhipovich Leonov attended the Kremenchug prep school for pilots and then the Chuguyev Higher Air Force School in Ukraine. After graduating in 1957, he served as a jet pilot in East Germany. He was a student at the Zhukovsky Air Force Engineering Academy when, on 5 March 1960, he got word that he was to be one of the first 12 Soviet cosmonauts.

Leonov is an artist as well as an aviator. His first stop after finishing secondary school was to enroll in the Academy of the Arts in Riga, Latvia, where he expected to train to become a professional artist. He withdrew soon after to begin his training as a pilot. But he continued his art studies in evening classes even as he was also qualifying to be a parachute instructor in the Soviet Air Force. The paintings he made of his spacewalk won him admission to the Soviet Artists' Union. He has displayed his paintings internationally and written a number of books.

A Soyuz rocket soars toward the International Space Station from the tarmac at the Baikonur Cosmodrome in Kazakhstan in March 2009 with the Expedition 19 crew.

Soyuz is also the name of the rocket used to launch the earliest *Soyuz* spacecraft.

Courtesy of NASA/Bill Ingalls

The History and Accomplishments of the Russian Soyuz Project

If the Vostok program was the Soviet counterpart to Project Mercury, and the Voskhod program the Soviet equivalent of Project Gemini, then the Soyuz Project was the Soviets' Project Apollo. *Soyuz* means *union*, as in the name of the former Communist state in Russia, the Soviet Union. *Soyuz* is also the name of the rocket used to launch the *Soyuz* spacecraft.

The History of the Russian Soyuz Project

In the year and a half between the last Voskhod mission and the first unmanned Soyuz flight, the Soviet space program lost three important advocates. Premier Nikita Khrushchev stepped down from his post in October 1964. And rocket designer Sergei Korolev and his chief assistant both died during this period. But their successors continued the work on Soyuz, under the new Soviet leadership.

Among the issues the designers faced was the need for a new upper stage for their basic launch vehicle. They needed a rocket with enough power to boost the *Soyuz* spacecraft, heavier than its predecessors, into orbit. They finally developed one. And then they made some test flights.

Soon they would be ready to launch *Soyuz 1*. The Soviet engineers designed their new spacecraft to take advantage of what the space program had learned from earlier flights. The Soviets also expected to learn more from *Soyuz 1* about how humans function in space and to investigate the problems of rendezvous and docking.

According to the Soviets, the purpose of the Soyuz missions was to develop a space station that would orbit Earth. Others speculated that the Soyuz program's real goal, like that of Apollo, was to put a man on the Moon.

And just as the United States lost its *Apollo 1* astronauts in a tragic fire, so, too, the first Soyuz mission ended in disaster. Vladimir M. Komarov, who had commanded the successful *Voskhod 1* mission a few years before, launched into space as the sole cosmonaut aboard the *Soyuz 1* on 23 April 1967.

After 24 hours and 18 orbits, he successfully accomplished retrofire. In this case, Komarov needed to use his retrorocket to slow down to reenter Earth's atmosphere and return to the ground. Communication proceeded normally. But the ship's parachute system failed. The main chute didn't deploy on schedule. *Soyuz 1* came crashing to Earth at high speed. The impact destroyed the ship and killed Komarov.

Star POINTS

At times people on both sides of the US-Soviet rivalry wondered whether cooperation might have served the two nations better than competition. But the competitive mode remained the most common.

The Docking Experiments Conducted During the Soyuz Project

Subsequent Soyuz missions were more successful. The next two represented an attempt to dock two craft in space. The next two after that succeeded in docking.

The Soviets launched the unmanned *Soyuz 2* on 25 October 1968. The next day *Soyuz 3*, with cosmonaut Georgy Timofeyevich Beregovoy aboard, went up. *Soyuz 3* went into co-orbit with *Soyuz 2*, and got within 200 meters of it. Based on reports in the Soviet newspaper *Pravda*, people in the US space program assumed the mission's goal was to dock the two spacecraft. But although Beregovoy made repeated attempts and did get very close, he didn't quite make the connection.

Then came *Soyuz 4*. On 14 January 1969 the Soviet Union made its first wintertime launch of a manned spacecraft: *Soyuz 4*, piloted by Vladimir A. Shatalov. The next day, *Soyuz 5* went up, and a day after that, *Soyuz 4* began a docking exercise with *Soyuz 5*.

Automatic systems brought the two ships within 100 meters of each other. The cosmonauts managed the final docking manually. During the 4½ hours the ships remained docked together, they completely interlocked controls, power, and telephones.

After *Soyuz 4* had made 51 orbits, two cosmonauts from *Soyuz 5* put on their spacesuits and opened their outer hatch. Floating and climbing hand over hand along the *Soyuz 5* handrails, they made their way through the open hatch of *Soyuz 4* and slipped inside. When *Soyuz 4* returned to Earth after three days, it carried, instead of the single man it went up with, a crew of three.

After their historic mission, *Soyuz 4* Commander Vladimir Shatalov illustrates how his spacecraft and *Soyuz 5* docked in Earth orbit on 16 January 1969.

Automatic systems brought the two ships within 100 meters of each other, but the cosmonauts managed the final docking manually.

Courtesy of NASA

The First Space Station

Salyut 1 ("salute") was the first space station of any kind. The Soviets put it into space—unmanned—on 19 April 1971. *Soyuz 10* was meant to deliver a crew to the space station. But it failed to dock properly. *Soyuz 11* did bring a crew to *Salyut 1*, and they had 23 productive days in space. The three cosmonauts perished on their return journey, however, because of an air leak on *Soyuz 11*. Three further space stations of the same class failed because they either never reached orbit or broke up before crews could get to them.

The Apollo-Soyuz Test Project

As the final mission of Project Apollo, the United States undertook a new joint project with the Soviet Union. The Apollo-Soyuz Test Project (ASTP) was the first human spaceflight mission managed jointly by two nations. It was also the first spaceflight in which ships from two different nations docked in space. Scientists designed the mission to test the compatibility of rendezvous and docking systems for the two countries' spacecraft. The test project was meant to prepare for future joint flights.

The Americans and Soviets had many challenges to work through, however. The two countries had developed their spaceships independently, so they weren't technically compatible for docking. The two space programs even had their astronauts and cosmonauts breathing different mixtures of air, at different pressures.

With the *Apollo* and *Soyuz* spacecraft docked in space, American astronauts (*upside-down from left*) Donald Slayton and Thomas Stafford visit with cosmonaut Alexei Leonov (*right side up*) in the *Soyuz Orbital Module*. The ASTP was a resounding success for both sides.

Courtesy of NASA

But the two sides met and overcame these issues. On 15 July 1975 the space agencies launched their two ships, *Soyuz 19* first and *Apollo* seven hours later. At 2:17 p.m. US Central Time on 17 July the two ships docked. Three cosmonauts and two astronauts carried out joint operations for two days. Then the ships separated. *Soyuz* landed in the Soviet Union on 21 July. *Apollo* splashed down near Hawaii on 24 July.

The mission was a resounding success for both sides. The two ships' crews had successfully docked. While docked, they had managed to move between spacecraft. And they carried out a series of scientific experiments.

Star POINTS

One of the most difficult problems the Apollo-Soyuz project faced was that of language differences. To bridge the language gap, the Americans studied Russian and the Russians studied English. The two groups found the best way for them to communicate was for the Russians to speak English and for the Americans to speak Russian.

The mission was important not only for its success as a space effort but for the mutual confidence and trust it engendered during the Cold War, when each country normally considered the other "the enemy." The repercussions of this mission continue today. Astronauts visited the Russian Space Station *Mir* numerous times over the years. And cosmonauts have flown on US space shuttle flights frequently as well. US and other astronauts regularly ride to and from the International Space Station aboard *Soyuz* spacecraft. Russian *Progress* cargo ships take supplies back and forth. This cooperation looks likely to last well into the future as both programs improve the means to get human beings into space, to the Moon, and maybe one day to planets in our Solar System.

CHECK POINTS

Lesson 2 Review

Using complete sentences, answer the following questions on a sheet of paper.

1. What two things did Colonel Yuri A. Gagarin accomplish on his one space flight?

2. How did Valentina Tereshkova's mission into space compare with that of L. Gordon Cooper Jr.?

3. What was Gherman Titov's accomplishment in space?

4. As Project Vostok and Project Mercury gave way to new space missions, what two goals did Soviet and American designers share?

5. For what accomplishment did the Soviets award Konstantin P. Feoktistov a place on the crew of *Voskhod 1*?

6. Who was Boris B. Yegorov and what was his particular work on *Voshkod 1*?

7. What "first" did Alexei Leonov achieve aboard *Voshkod 2*?

8. What, according to the Soviets, was the purpose of the Soyuz missions?

9. What was different about *Soyuz 4* when it returned to Earth?

10. What was the Apollo-Soyuz Test Project?

APPLYING YOUR LEARNING

11. What was the larger context or historical backdrop of the Apollo-Soyuz Test Project, and why did the project's success matter in that larger context?

LESSON 3 · Space Programs Around the World

Quick Write

Do you think it's important for astronauts on long space missions to have access to "comfort foods," familiar music, and other reminders of home? Why or why not? What food would you most miss on a long space voyage?

Learn About

- the history and accomplishments of the Chinese space program
- the history and accomplishments of the Indian space program
- the history and accomplishments of the European space program
- the history and accomplishments of the Japanese space program

So-yeon Yi, a 29-year-old biotech engineer, blasted off on 8 April 2008 to become the first South Korean in space. Among the things she carried in her kit was kimchi. Kimchi is a traditional dish in Korea, made of fermenting cabbage, with garlic and peppers. Koreans say they must eat it wherever they go, wrote *The New York Times* in an article before the mission. When South Korean mothers sent their sons off to war in Vietnam in the 1960s, for instance, they sent clay pots full of the stuff with them.

So when South Koreans decided it was time for one of their own to travel into outer space, they considered how to bring kimchi along. Three top government institutes got involved in the quest to create "space kimchi."

It was a quest on which South Korea spent years, and millions of dollars. The problem, the *Times* reported, was that microbes such as lactic acid bacteria thrive in kimchi. After all, these help ferment the cabbage. On Earth they're harmless. But would they cause problems in space? What if cosmic rays mutated the bacteria?

And how would fluctuating temperatures, common in space, affect kimchi? One nightmare scenario involved a bag of kimchi fermenting out of control, bursting, and then spilling all over essential aerospace electronics. These were serious questions, and some of the finest minds in the country sought answers.

"The key was how to make a bacteria-free kimchi while retaining its unique taste, color, and texture," Lee Ju Woon at the Korean Atomic Energy Research Institute told the *Times*. Eventually the scientists engineered a version of kimchi, plus nine other Korean recipes, that met with approval from the Russian space authorities. Their spaceship carried Yi to the International Space Station (ISS).

Scientists used radiation to kill the bacteria in the kimchi but were able to hang on to 90 percent of its flavor. The "space kimchi" had much less of an aroma, too. That probably made it more appealing to Yi's international crewmates on the ISS.

Vocabulary

- incendiary
- taikonaut
- monsoon
- geostationary

4 November 2004

SeaWiFS Project NASA GSFC ORBIMAGE

A NASA satellite captured this view of China in 2004. It shows a pollution-generated haze over much of the country. While the United States and Russia have dominated the field of space exploration over the decades, China is now counted as the world's No. 3 spacefaring nation.

Courtesy of SeaWiFS Project/NASA/Goddard Space Flight Center/ORBIMAGE

The History and Accomplishments of the Chinese Space Program

In the modern era, the United States and Russia have dominated the field of space exploration. But rockets were a Chinese invention—centuries ago. More recently China has made up for lost ground. It now is counted as the world's No. 3 spacefaring nation.

The Chinese Use of 'Fire Arrows' in the Thirteenth Century

China can claim to be the birthplace of the Space Age—or at least the Age of Rocketry. Historians say that Chinese warriors may have used "fire arrows" against their enemies as far back as 300 BC. The first fire arrows may simply have been flame-tipped arrows shot off by archers.

But by the thirteenth century AD, the fire arrows of the Sung Dynasty's army were rockets packed with gunpowder. These were the first solid-fuel rockets. The Chinese used them against the Mongol horde. In 1232 they repelled Mongol invaders at the battle of Kai-fung-fu.

Old records show that the biggest of these rockets carried iron shrapnel and incendiary material—material *able to set fire to something*. The noise from the blast of one of these early rockets could be heard for 15 miles.

China's National Space Administration

Seven centuries later, China developed a modern space program, which now includes satellites and manned spaceflight. The Asian giant has two major space agencies to organize these missions.

The People's Liberation Army, or PLA, is in charge of manned and military space missions. The China National Space Administration (CNSA) is in charge of civil and scientific projects. By early in the twenty-first century, China had become the world's No. 3 spacefaring nation, after the United States and Russia.

The Chinese Satellite and Human Spaceflight Programs

China's space program really took off in 1970 with the launch of its first satellite, the *Dong Fang Hong I*. It took many years of hard work and tests, plus lots of technical help from the United States and the Soviet Union, for China to get to this point.

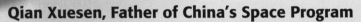

Qian Xuesen, Father of China's Space Program

China owes much of its modern success with rockets and missiles to a scientist who had a brilliant career in the United States but was later deported as a suspected communist. Qian Xuesen was born in China in 1911, as its imperial government was collapsing. He came to the United States as a young man and studied at the Massachusetts Institute of Technology. Later, he helped found the Jet Propulsion Laboratory, now one of NASA's premier research facilities.

His mentor and colleague, physicist Theodore von Karman, called Qian "an undisputed genius whose work was providing an enormous impetus to advances in high-speed aerodynamics and jet propulsion." In 1949 Qian wrote a proposal for a winged space plane. It would later be described as the inspiration for NASA's space shuttle.

But Qian's American career was over by 1950. This was the period known as "the Red Scare." Many Americans were worried that communists had gained influence over government and other institutions. And China had just gone through a communist revolution. When Qian tried to go there to visit his parents, FBI officials stripped him of his security clearance. They accused him of secretly being a communist.

And so in 1955 the United States sent him back to China. There the Chinese welcomed him as a hero. The government put him to work at once making rockets. He led the research that produced China's first ballistic missiles, its first satellite, and its Silkworm antiship missile.

Qian died at 98 years of age in January 2010.

According to a report by the US Department of Defense, China launched 78 satellites into space in addition to the *Dong Fang Hong I* through October 2003. In all, 67 succeeded in reaching orbit, but the other 12 failed—or at least didn't reach their correct orbit. These satellites included all kinds: communications, weather, remote sensing, navigation, and scientific.

Star POINTS

About three dozen countries have sent at least one astronaut into space since 1961. But except for Russia, the United States, and now China, those countries have all sent their people up in another country's spaceship.

China's space program has more recently expanded into manned spaceflight as well. On 15 October 2003 Lt Col Yang Liwei of the PLA took off into space aboard the *Shenzhou* 5 ("divine vessel"). His mission lasted some 21 hours and included 14 orbits around Earth. The spaceflight also brought China into the exclusive club of nations that have launched their own citizens into space.

Almost exactly two years after Col Yang's mission, China successfully launched a second manned spacecraft. The *Shenzhou* 6 sat atop a Long March-2F rocket. The launch occurred at a remote site on the edge of the Gobi desert. The two taikonauts (*Chinese astronauts*) were Fei Junlong and Nie Haiheng. They were both former pilots in the Chinese Air Force. Their mission lasted nearly five days. They orbited Earth 75 times.

China's third manned space mission, with three taikonauts, took place in September 2008. Mission commander Zhai Zhigang became the first taikonaut to walk in space. This mission was a sign of China's steady efforts to establish a permanent human presence in space.

Chinese Vice Premier Deng Xiaoping gets a briefing on NASA's manned space program from Johnson Space Center Director Christopher C. Kraft during a visit to the center in Houston, Texas, in February 1979. It took many years of hard work and tests, plus lots of technical help from both the United States and the Soviet Union, for China to join the space club.
Courtesy of NASA

China's Use of Communications and Weather Satellites

In 1986 China said it intended to enter the commercial space launch business. This is largely about lifting satellites into orbit. It gets less attention than do manned space missions. But it's an important part of the global aerospace industry.

And it's an area of Chinese industry that has close links to the United States. Almost all communications satellites needing commercial launch services are either built in the United States or include US components. This means that companies must obtain US export licenses before they can send satellites to China for launch. That is, the US government has to approve each move, generally for national security reasons.

In 1988 the Reagan administration gave conditional approval of the first export licenses to send three satellites to China. Early the following year, China met the conditions for the export to proceed. But then in June 1989 China's violent suppression of the Tiananmen Square uprising led to a souring of US-Chinese relations. Since then, China's satellite business has been subject to the changing political winds.

China has its own communication and weather satellites as well. These service customers in China and elsewhere.

China's Use of an Anti-Satellite Weapon

In fact, a Chinese weather satellite was recently involved in a very controversial space-news story. On 11 January 2007 the Chinese successfully "tested" an anti-satellite missile. They used it to shoot down one of their aging weather satellites. It was a stunning development—from the perspective of space science as well as foreign and military policy. The mobile missile demolished the satellite, the *Feng Yun 1C*, orbiting at 537 miles (859 km) above the Earth.

The missile's target was, after all, just a weather satellite, and one of China's own, at that. But the shootdown showed the Chinese had mastered key technologies important for advanced military operations in space. The missile was unmistakably a weapon. It was plain that China could someday direct such a weapon at satellites belonging to other countries.

The whole episode had outside analysts scratching their heads. For one thing, it occurred at a time when China's leaders were promoting the idea of "peaceful rising." By this they meant that China's advance as a world power did not have to threaten its neighbors.

The missile test was controversial for another reason, too. It looked like a deliberate decision by China to make the problem of space debris, or space junk, even worse. The destruction of the *Feng Yun 1C* created a cloud of hundreds of fragments. There had never been anything like it. Those fragments will clog the spaceways for many years to come.

The History and Accomplishments of the Indian Space Program

India's space program, in contrast with China's, has long been modest and low-key. The country has mainly focused on launching satellites that look back down on India itself, with its 1.2 billion people. It uses satellites to help Indians understand their own country, to put it another way. But India's space policy is changing. There's even talk of another race for the Moon—this time between India and China.

Vikram Sarabhai, the "father" of the Indian Space Research Organization, signs an agreement in 1969 with NASA Administrator Thomas Paine to work together to deliver TV shows via satellite to Indian villages. When India began its space program back in the 1960s, it concentrated on technology for economic development—particularly satellites.

Courtesy of NASA

The Indian Space Research Organization

When India began its space program back in the 1960s, it concentrated on technology for economic development—particularly satellites. The Indian Space Research Organization, the driving force of India's space program, has made the practical uses of space science, rather than prestige, a priority.

The government decided from the beginning to use space technologies to help ease poverty. India uses satellites to search for water and minerals, to map resources, and to monitor the health of its forests. Satellite technology also helps India monitor its monsoon season, *an annual period of heavy rainfall*. This is important to the health of India's huge agricultural economy.

But India is now moving beyond that traditional focus. It has planned its first manned space mission for 2015. India wants to join the club of spacefaring nations. If it succeeds in putting an astronaut up in 2015, India will join the club just a dozen years after China.

The *Chandrayaan-1* Moon Orbit Satellite

The first step along this new path for Indian space policy was *Chandrayaan-1*. This was an unmanned lunar probe launched in October 2008. Its two-year $83 million mission was to orbit the Moon and to map its surface in three dimensions. The probe carried payloads from NASA, the European Space Agency, and Bulgaria. Among the payloads were instruments used to detect water and minerals.

The mission experienced technical problems. Several months into its mission, *Chandrayaan-1* (the name is Hindi for "moon craft") overheated. And so the Indian space agency moved the probe into a new orbit 62 miles (100 km) farther from the Moon. Later a steering sensor failed. But even before a full year was out, India's space scientists declared the mission a success.

The Indian National Satellite System (INSAT)

Another of India's accomplishments is the Indian National Satellite System, known as INSAT. It's the largest domestic communication system in the Asia-Pacific region. Its satellites are geostationary. That term refers to *orbiting at a speed and altitude that keeps satellites in the same place above the Earth at all times*. The system began in 1983. INSAT meets India's telecommunications, broadcasting, meteorology, and emergency rescue needs.

The Mission of Indian Remote Sensing (IRS) Satellites for Earth Resources

The Indian Remote Sensing (IRS) satellite program supports India's national economy. It provides information on agricultural water resources, forestry, ecology, geology, watersheds, marine fisheries, and coastal management. This constellation of satellites is one of the largest in the world.

IRS can tell the healthy coconuts from the diseased ones hanging from the region's thick palms. The satellites can also spot swarms of mosquitoes in the jungle. Now high-resolution satellites are available. Indians are using them to study urban sprawl and plan infrastructure.

The Mission of the METSAT Weather Satellites

On 12 September 2002 India launched the first of a new series of meteorological (weather) satellites called METSAT. Scientists intended these to take a load off the rest of the INSAT system. Up to that point, INSAT's telecommunications and broadcasting satellites had also carried instruments for gathering weather data. But weather instruments impose certain design limits on a satellite. It's hard to design a satellite that "does everything." And as India and its economy grow, its telecommunications and broadcasting sectors make more demands than ever on India's satellite networks.

Most countries find weather data useful. But India has more reason than most countries to want the best possible weather data. Monsoons strike it annually. The country is also subject to cyclones (hurricanes) and floods.

The Moderate Resolution Imaging Spectroradiometer aboard NASA's Terra satellite took photos of flooding in India from a monsoon in 2007.

The black, or dark blue, portions show the Bay of Bengal, but the light blue scattered around the image is the actual flooding. Water mixed with mud shows up in the fainter shade. Due to annual monsoons, India has more reason than most countries to want the best possible weather data.

Courtesy of NASA/Jesse Allen/MODIS Rapid Response Team

CHAPTER 6 Space Programs

The History and Accomplishments of the European Space Program

Europe is a collection of wealthy but small countries. As in many other activities, so in space, too—Europeans have found strength in numbers. They have banded together to do more than any single country could do alone.

The Activities of the European Space Agency (ESA)

As its name suggests, the European Space Agency (ESA) is the organization through which Europe undertakes space research and missions. Eighteen European countries belong to the ESA. The agency has special arrangements with other countries as well. (Note that the ESA is not part of the European Union [EU]. The ESA's membership overlaps largely with the EU's, though.)

The ESA seeks to ensure that investment in space delivers benefits to the citizens of Europe and the world. ESA takes part in human spaceflight largely through the International Space Station program. ESA also carries out unmanned exploration missions to the Moon and planets in the Solar System. The agency maintains a spaceport at Kourou, French Guiana, and has a rocket program as well. It is a significant launcher of commercial satellites.

The ESA's headquarters are in Paris. The agency has several science centers around Europe. Cologne, Germany, is the home of the European astronaut training facility, for instance. The agency also includes liaison offices and tracking stations as needed around the world.

Space shuttle Discovery makes a brilliant arc in the sky as it heads toward the ISS for the station's 30th construction and maintenance mission on 28 August 2009. Mission STS-128's crew members included Christer Fuglesang of the European Space Agency. ESA takes part in human spaceflight largely through the International Space Station program.

Courtesy of NASA/Ben Cooper

The ESA's Member States

- Austria
- Belgium
- Czech Republic
- Denmark
- Finland
- France
- Germany
- Greece
- Ireland
- Italy
- Luxembourg
- Netherlands
- Norway
- Portugal
- Spain
- Sweden
- Switzerland
- United Kingdom

Canada has a cooperation agreement with the ESA. Hungary, Romania, and Poland are designated as European Cooperating States. Estonia and Slovenia have recently agreed to work with the ESA, too.

The ESA's Joint Activities With Russia and the United States

When ESA astronaut Frank De Winne, a Belgian, returned to Earth after six months in space on 1 December 2009, he landed at the Russian space facility in Kazakhstan. And he, along with Russian cosmonaut Roman Romanenko and Canadian astronaut Robert Thirsk, made the trip in a Russian reentry module. All this suggests how closely the ESA works with its Russian counterpart.

The ESA has also cooperated with Roscosmos on launch services, which have become a Russian specialty. In November 2009, for instance, the ESA launched two *Earth Explorer* satellites from a site in northern Russia. Russia possesses proven rocket technology. Sharing this with the Europeans gets the Russians access to the ESA launch site in French Guiana. The site's location near the equator makes it a better place from which to launch heavy payloads.

Saturn's icy moon Enceladus hovers in the foreground while Saturn's ring shadows fill the background. The *Cassini-Huygens* probe took this shot on 28 June 2007 from 181,000 miles (291,000 km) away. The Saturn probe is a joint venture of NASA, the ESA, and the Italian space agency, ASI.
Courtesy of NASA/JPL/Space Science Institute

The ESA also has an extensive program of joint activities with NASA. For instance, the two agencies will undertake a two-part exploration of Mars set for 2016 and 2018. The Hubble Space Telescope is a joint NASA-ESA project. And the *Cassini-Huygens* Saturn probe was a joint venture of NASA, the ESA, and the Italian space agency, ASI. The probe was the largest interplanetary spacecraft ever built.

The Mission of the *Venus Express*

You might call the ESA's *Venus Express* mission an example of space research on the economy plan. *Venus Express* is a follow-on from the *Mars Express* mission, which the ESA launched on 2 June 2003. The *Mars Express*—so called because it was developed on such a short, streamlined timetable—represented the ESA's first visit to another planet in the Solar System.

Many of the instruments aboard the *Venus Express* were simply upgraded versions of those sent off aboard the *Mars Express*. The ESA launched the *Venus Express* on 9 November 2005. After a 153-day cruise to Venus, the spacecraft went into orbit around the planet on 11 April 2006. Its mission is to run until the end of 2012. The *Venus Express* is looking for answers to a number of essential questions about Venus. The overarching question, however, is this: Why has Venus evolved so differently compared with the Earth, in spite of the similarities in terms of size, basic makeup, and distance to the Sun?

The Comet Intercept Mission of the Spacecraft *Rosetta*

Rosetta is the name of an ESA spacecraft with one of the most challenging missions ever: to rendezvous with a comet and study it up close for two years. *Rosetta* has two parts. One is a large orbiter, made to operate for a decade at great distance from the Sun. The other is a small lander called *Philae*.

Rosetta's target is Comet 67P Churyumov-Gerasimenko. *Rosetta* is to rendezvous with the comet in 2014. It will release its lander onto the comet. Then, as the comet heads for the Sun, the orbiter will continue to circle it. Scientists hope to learn from the *Rosetta* mission what conditions were like as the Solar System formed. They will also be looking for clues to the origins of life on Earth.

The ESA Role in the International Space Station

The ESA is also an important partner with the space agencies of the United States, Russia, Japan, and Canada in the International Space Station (ISS). The ESA describes it as "the greatest international project of all time." Once complete, it will provide more than 42,000 cubic feet (1,200 cubic meters) of pressurized living and working space. It will have room for a crew of seven, plus a wide array of scientific experiments.

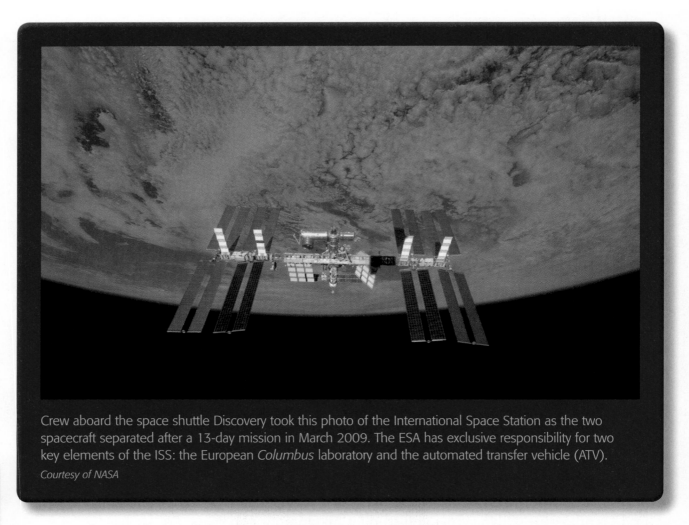

Crew aboard the space shuttle Discovery took this photo of the International Space Station as the two spacecraft separated after a 13-day mission in March 2009. The ESA has exclusive responsibility for two key elements of the ISS: the European *Columbus* laboratory and the automated transfer vehicle (ATV).
Courtesy of NASA

The ESA has exclusive responsibility for two key elements of the ISS: the European *Columbus* laboratory and the automated transfer vehicle (ATV). The *Columbus* lab represents a substantial part of the space station's research capability. It is specialized for research into fluid physics, materials science, and life sciences. The ATV is the space station's supply ship. It relies on the Ariane 5, the European rocket, to launch into space. The ATV can carry nearly eight tons of cargo.

The History and Accomplishments of the Japanese Space Program

After World War II, Japan rebuilt from postwar devastation to a position as one of the largest economies in the world. Under occupation by the United States, Japan essentially renounced war. These two factors mean that Japan's interests in space are more like the Europeans' than either those of the Cold War superpowers or the rising giants of Asia—China and India. Japan's interests are civilian and scientific, and pursued in cooperation with others.

The Japan Aerospace Exploration Agency (JAXA)

The Japan Aerospace Exploration Agency (JAXA) came into being on 1 October 2003. It was the result of a merger of three different bodies involved in Japanese efforts at space exploration. JAXA is an independent administrative institution like NASA in the United States. It houses all Japanese aerospace activities under one organizational roof.

The Work of Soichi Noguchi and Other Japanese Astronauts

The first Japanese in space was Toyohiro Akiyama. He was a TV journalist whose network paid $10 million for him to spend a week aboard the *Mir* space station in December 1990. He was also the first-ever paying passenger in space.

The first Japanese astronaut in space was Mamoru Mohri. He was on the crew of NASA's space shuttle mission STS-47 in September 1992. Takao Doi was the first Japanese astronaut to walk in space. On shuttle mission STS-87, in 1997, Doi made two spacewalks, in fact, logging 15 hours outside the ship.

Soyuz TMA-17 draws close to the International Space Station (while another *Soyuz* spacecraft already docked with the ISS is at top left in the photo). Among the crew members onboard the approaching *Soyuz* is JAXA astronaut Soichi Noguchi. Japan's interests in space are civilian and scientific.
Courtesy of NASA

More recently, on 20 December 2009 Soichi Noguchi launched in a Russian *Soyuz* spacecraft from Kazakhstan. He was off for six months aboard the ISS. Earlier, in 2005, he took part in the first flight of the US space shuttle after the 2003 *Columbia* accident. He has logged extensive periods of spacewalks, testing new procedures for shuttle inspection and repair.

Star POINTS

When Japanese astronaut Soichi Noguchi lifted off from Baikonur Space Center in Kazakhstan in December 2009, he carried with him the first sushi into space. He told a news agency, "We had a training in Japan and I was stupid enough to train [my fellow astronauts] to be sushi lovers."

Even astronauts have to let their hair down once in a while.

JAXA's Soichi Noguchi wears a fun and festive hat as he enters the ISS after docking via the *Soyuz* TMA-17 in December 2009.

Courtesy of NASA

The Mission of the Japanese Moon Probe *Kaguya*

JAXA launched its *Kaguya* probe of the Moon on 14 September 2007 from the Tanegashima Space Center. *Kaguya*'s mission objectives were to obtain data on the Moon's origin and to develop technology for future exploration there.

The probe consisted of a main orbiting satellite maneuvered into position about 60 miles (100 km) above the Moon's surface. It included two smaller satellites released by *Kaguya* into polar orbit around the Moon. Over the course of *Kaguya*'s mission, its controllers steered it into ever-lower orbits.

Kaguya, in Japanese folklore, is the name of a legendary princess who spurns earthly suitors and returns to the Moon, never to leave again. The probe lived up to its name. On 10 June 2009 it made a planned crash landing onto the Moon's surface.

Japan's Contribution to the ISS Program

Japan has spent at least $3 billion developing its *Kibo* lab for the International Space Station. *Kibo* means "hope." This major piece of hardware—and Japan's biggest contribution to the ISS—has two elements: a pressurized module and an "exposed facility." An indoor lab and an outdoor one, you might say. Astronauts will use the outdoor lab for long-term experiments in open space as well as for observations of Earth and the heavens. The control room at the Tsukuba Space Center, northeast of Tokyo, manages *Kibo's* operations 24 hours a day, 7 days a week.

At the height of the Cold War, the original TV series *Star Trek* featured an international (even interplanetary) crew aboard the starship *Enterprise*. Besides Americans, these included an African radio operator, a Russian ensign, a Japanese pilot, and a Scots engineer. Today more and more nations are launching into space or participating in the International Space Station. It makes far more sense for nations to pool their resources and cooperate in reaching out into space. It's a lot cheaper for everybody, too.

The next chapter will pick up the story of manned space missions after the Moon landings. It begins with the space shuttle and discusses the various American and Russian/Soviet space stations. You'll also learn about the many past, present, and future unmanned missions to different locations in the Solar System.

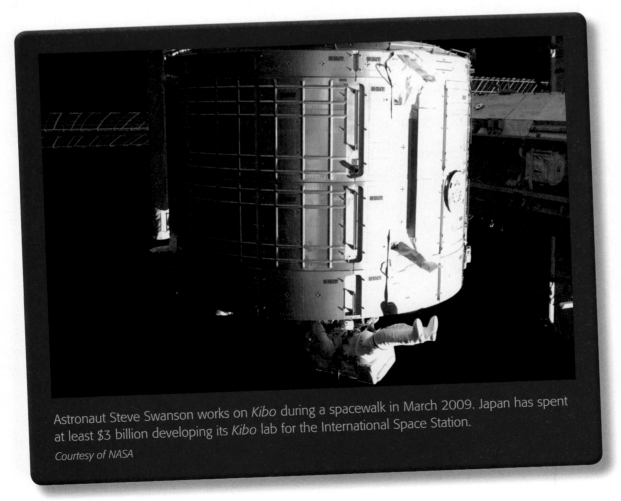

Astronaut Steve Swanson works on *Kibo* during a spacewalk in March 2009. Japan has spent at least $3 billion developing its *Kibo* lab for the International Space Station.

Courtesy of NASA

CHECK POINTS

Lesson 3 Review

Using complete sentences, answer the following questions on a sheet of paper.

1. Who had the first solid-fuel rockets, and against whom did they use them?

2. What are China's two major space agencies, and what does each do?

3. What did Lt Col Yang Liwei's October 2003 mission accomplish for China?

4. Why does China's commercial space launch business link it to the United States?

5. What does the Chinese satellite *Feng Yun 1C* have to do with the problem of space debris?

6. How is India moving beyond its traditional focus on the use of space technology for economic development?

7. Identify the mission of *Chandrayaan-1*.

8. How does the Indian National Satellite System rank within the Asia-Pacific region?

9. What does the Indian Remote Sensing satellite program do?

10. Why has India launched a new series of meteorological (weather) satellites?

11. Why is Kourou, French Guiana, important to the European Space Agency?

12. How does the December 2009 mission of astronaut Frank De Winne of Belgium show how closely the ESA works with the Russian space agency?

13. What overarching question are ESA scientists seeking to answer with their *Venus Express* probe?

14. Identify *Rosetta* and its mission.

15. For which two key elements of the ISS does the ESA have exclusive responsibility?

16. Identify JAXA.

17. What important space mission did Soichi Noguchi undertake in 2005?

18. How did the Japanese Moon probe *Kaguya* live up to its name?

19. Identify *Kibo*.

20. On 11 January 2007 the Chinese successfully tested an anti-satellite missile. What might this mean for the United States?

On 12 April 1981 *Columbia* blasts into space for the first time. The space shuttle made 36 orbits around Earth. It was the first of five orbiters to fly into space.

Courtesy of NASA

The Space Shuttle

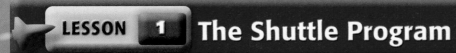

> " Whether outwardly or inwardly, whether in space or time, the farther we penetrate the unknown, the vaster and more marvelous it becomes. "
>
> *Charles Lindbergh*, Autobiography of Values

On 2 December 1993 space shuttle _Endeavour_ launched on an 11-day mission to repair the defective Hubble Space Telescope. Faulty optics meant that the scope's performance fell far short of what astronomers had hoped for. The telescope's primary mirror was shaped incorrectly, and so it could not focus all the light from an object to a single sharp point. Instead, Hubble "saw" everything with a halo around it.

Although NASA developed computer software that significantly improved Hubble's focus, a very difficult and complex repair mission seemed the only way to fully correct the situation. NASA had designed Hubble in such a way that astronauts could service it while it remained in orbit. But not everyone was convinced that NASA was up to the job. NASA's reputation was on the line.

Endeavour's mission, known by its NASA designation as STS-61, was one of the most sophisticated in the shuttle's history. This was the first mission ever having five spacewalks. One of these lasted nearly eight hours, the second-longest spacewalk in history.

On flight day No. 2 _Endeavour_ began a series of engine burns to let the shuttle close in on Hubble, in orbit high above the Earth. With each 95-minute orbit, _Endeavour_ came about 70 miles closer to Hubble. Finally _Endeavour_ got close enough that commander Richard O. Covey could maneuver it manually to within 30 feet of the Hubble. Then Claude Nicollier, a mission specialist, used _Endeavour_'s robot arm to grab the telescope and pull it into the shuttle's cargo bay.

Endeavour's crew spent five days tuning Hubble up. They installed two devices that compensated for the incorrect shape of Hubble's primary mirror.

The astronauts also replaced the telescope's solar arrays, which collect energy from the Sun to power Hubble. The old arrays wobbled too much during the stressful transitions from cold darkness to bright warmth and back again as Hubble orbits. The new arrays are steadier. The crew also replaced defective gyroscopes—*a device with a spinning wheel at its center that helps objects retain their balance*—and other electronic equipment.

Hubble's improved vision soon led to a string of remarkable discoveries. And STS-61 also proved that "on-orbit servicing" could be an effective way to keep a valuable scientific instrument in shape.

Vocabulary

- gyroscope
- aft
- foreign national

Three engine test stands pop up above a thick fog at NASA's Stennis Space Center in Mississippi. In many ways these test stands symbolize where space exploration—old and new—begins. They are not only where the space agency tested the *Saturn V* rockets for the Apollo program, but they are where engineers have tested the space shuttle program's main engines as well. The space shuttle began as part of a larger vision at NASA for what to do after the country had reached the Moon.

Courtesy of NASA/SSC/Allen Forsman, Pratt and Whitney Rocketdyne

Why the Space Shuttle Was Developed

Putting astronauts on the Moon was a dramatic achievement. But it wasn't clear what NASA would do for an encore. The *Apollo 11* lunar landing had come only a few brief years after President Kennedy's ringing call to make the race for the Moon.

America's political and economic environment changed vividly over those years. Kennedy himself was assassinated in 1963. His successor, Lyndon Johnson, soon had his hands full with his ambitious War on Poverty. Not long after, the costs and domestic political turmoil of the Vietnam War would consume his presidency. When Richard Nixon took office as the next president, he had little appetite for anything like another Project Apollo. He called for cutting NASA's budget as sharply as was politically feasible.

The Original Concept of the Shuttle

The space shuttle began as part of a larger vision at NASA for what to do after the country had reached the Moon. After the lunar landings, according to this vision, the next step would be the construction of a series of ever-larger space stations circling Earth, holding up to 100 people at one time. Then would come space stations circling the Moon, and then lunar bases to be used as a staging ground for the exploration of Mars.

The idea of a space shuttle, a vehicle that would take crews and supplies to low-Earth orbit, grew out of this vision. And to save money, NASA planned to develop a fully reusable spacecraft.

But the concept of a shuttle cruising to space stations ran afoul of political realities. With budget cuts hitting the agency, NASA placed its space station program on hold. It canceled further production of the *Saturn V* rockets, which had lifted the *Apollo* modules into space. This left NASA with the space shuttle as the only manned spaceflight program it could hope to undertake. But what would the shuttle do if it weren't ferrying people and parts to orbiting construction sites?

NASA needed a new rationale for the shuttle. That rationale emerged from three intense years of technical study and budget negotiations. These discussions tried to reconcile the conflicting interests of NASA, the Department of Defense, and the White House.

NASA managed to win approval from the White House by attempting to meet certain terms. The shuttle had to be reusable. It had to be capable of delivering private and government satellites to space (the Department of Defense was particularly interested in this ability). And it also had to carry out repair missions on satellites. NASA ingenuity met many of these demands.

The Original Six Orbiters

"Orbiter" is NASA's term for what most people think of as "the space shuttle." It's the space plane with the distinctive wings that rides into space with the help of a couple of rockets and an enormous fuel tank.

The first orbiter NASA and its contractors built was the *Enterprise*. It was a test aircraft that flew but lacked systems to go into space. NASA used it, however, for approach and landing tests and several launch pad studies in the late 1970s.

Enterprise, a test model for the space shuttles to follow, takes a test run atop the 747 shuttle carrier aircraft (SCA). During early trials in the shuttle program, the SCA would take off and land with the shuttle on its back. During later checks, such as this one in 1977, the shuttle would release from the SCA's back around 19,000 feet to 26,000 feet altitude to glide without power back to Edwards Air Force Base. *Enterprise* never went into space.

Courtesy of NASA/DRFC

Columbia was the first actual space shuttle orbiter delivered to NASA's Kennedy Space Center in Florida in March 1979. It was the heaviest of the orbiters. Its role was somewhat limited during its years of service. It weighed too much and lacked the equipment needed to help assemble the International Space Station. But *Columbia* achieved a number of notable "firsts" in spaceflight, including flying the first orbiter mission in space.

Challenger started out as just a test vehicle for the space program. But it eventually made it into space. NASA wanted a lighter-weight orbiter than *Columbia*. Once the agency had a lighter airframe design, it needed a test vehicle to make sure the new design could handle the stress of spaceflight. Computer modeling software wasn't advanced enough at the time to be of much help in this. And so the best test NASA engineers could come up with was to submit the vehicle to a year of intense vibration and heat exposure.

By early 1979 NASA decided that the lighter airframe had passed the test. Making *Challenger* spaceworthy was a big job. It involved taking apart and replacing many parts and components. But in July 1982 *Challenger* arrived at the Kennedy Space Center, as the second space-rated orbiter in the fleet.

The third spaceworthy orbiter was *Discovery*. It arrived for the first time in Florida in November 1983. NASA launched it on its first mission, to deploy three communications satellites, on 30 August 1984. *Discovery* can claim the distinction of flying more than 30 successful missions. That's more than any other orbiter. *Discovery*'s notable missions have included the launch of the Hubble Space Telescope in April 1990 and the second and third missions to service Hubble, in February 1997 and December 1999.

Next came *Atlantis*. Its construction began on 3 March 1980. Thanks to lessons learned from building and testing the earlier orbiters, construction of *Atlantis* took only about half as many person-hours as were spent on *Columbia*. This was largely because *Atlantis* had large thermal protection blankets wrapped around its upper body rather than individual tiles. Those tiles had caused problems from the start.

Atlantis was also nearly 3.5 tons lighter than *Columbia*. The fourth orbiter arrived at the Kennedy Space Center on 9 April 1985 to prepare for its first flight. On that first spaceflight, launched 3 October 1985, it carried a classified payload for the Department of Defense. *Atlantis* later carried four more Defense payloads.

Endeavour was authorized by Congress to replace *Challenger*, lost in an accident in 1986. (You will read about that in the next lesson.) *Endeavour* arrived at the Kennedy Space Center on 7 May 1991, piggybacked atop NASA's new space shuttle carrier. It lifted off exactly a year later, on 7 May 1992. One of its primary objectives on that mission was to retrieve and repair a communications satellite that was still orbiting but no longer functioning.

The shuttle wasn't expressly designed for this mission. But *Endeavour* succeeded anyway, after four spacewalks, including the longest one in history up to that time, lasting more than eight hours.

The Shuttles' 'Enterprising' Names

Enterprise, the first space shuttle orbiter, was originally going to take the name *Constitution*. But fans of the popular show *Star Trek* campaigned to have the orbiter named for the spaceship that figures in the television series. NASA named all its other space shuttles after famous sailing ships, each with a role in exploration.

- *Columbia* got its name from a sailing frigate—one of the first that the US Navy sent around the globe—that launched in 1836.

- *Challenger* took its inspiration from a Navy ship of the same name that spent four years—from 1872 to 1876—exploring the Atlantic and Pacific oceans.

- *Discovery* had two famous namesakes. One was Henry Hudson's ship in which the famous explorer looked for a northwest passage from 1610–1611. He ended up finding the bay now named for him, Hudson Bay. The other famous early *Discovery* was the ship in which the eighteenth-century British explorer Captain James Cook discovered the Hawaiian Islands.

- *Atlantis*'s namesake was a twentieth-century exploring ship, a two-masted ketch, sailed by the Woods Hole Oceanographic Institution of Massachusetts from 1930 to 1966. It covered more than half a million miles conducting ocean research.

- *Endeavour* took its name from another of Captain Cook's ships.

The Shuttle's First Mission

Columbia was the first of the space shuttles to go into outer space. NASA launched that first mission, designated as STS-1, on 12 April 1981. Captain John W. Young, a veteran of the Gemini and Apollo programs, was the commander. Robert L. Crippen was the pilot.

On this maiden voyage the goals were to check out the overall system, ascend safely into orbit, and return safely to Earth. NASA was confident that the shuttle met these objectives. Officials decided that the shuttle was spaceworthy. The spacecraft had only one payload—a package of sensors and measuring devices to track the orbiter's performance and record stresses that occurred at each step of the flight.

Post-flight inspection of the *Columbia* revealed that the ship had lost 16 heat-shield tiles and that an additional 148 had been damaged. In all other respects, however, *Columbia* came through with flying colors.

After orbiting Earth 36 times, *Columbia* landed on a dry lakebed runway at Edwards Air Force Base in California. The ship made three more research missions to test its performance. Its first operational mission, with a four-man crew, was STS-5, launched on 11 November 1982.

Star POINTS

NASA launched *Columbia* on its first flight on the 20th anniversary of Yuri Gagarin's first flight into space in 1961.

<u>Columbia</u> blasts into space for the first time on 12 April 1981. The shuttle was the first of five orbiters to go into outer space.

Courtesy of NASA

Numbering Shuttle Flights

NASA designates individual shuttle missions (flights) with numbers beginning with the initials STS, short for "space transportation system." *Columbia*'s first mission, for instance, was STS-1.

Following STS-9, NASA changed the flight numbering system for space shuttle missions. Thus, the next flight, instead of being designated STS-10, became STS 41-B. The new numbering system was designed to be more specific in that the first numeral stood for the fiscal year in which the launch was to take place, the "4" being 1984. The second numeral represented the launch site "1" for the Kennedy Space Center and "2" for Vandenberg AFB, California. The letter represented the order of launch assignment—"B" was the second launch scheduled in that fiscal year. After the *Challenger* accident, NASA reestablished the original numerical numbering system. Thus the first flight following STS-51-L is STS-26.

CHAPTER 7   The Space Shuttle

The Space Shuttle's Main Features

The space shuttle is not only the world's first reusable spacecraft. It's also the first spacecraft that can carry large satellites into orbit, and then retrieve them. The shuttle is launched like a rocket, maneuvers in orbit like a spacecraft, and lands like an airplane.

The Orbiter

As noted earlier, the orbiter is what many people think of as "the space shuttle." Actually, the orbiter is but one element in the space shuttle or "space transportation system." But it's both the brains and the heart of the shuttle. It contains a pressurized crew compartment, which can carry up to eight people, a huge cargo bay, and three engines mounted aft— *the rear of a spacecraft or any other ship*.

The orbiter is boosted into space partly on the strength of those engines—but mostly on the power of the solid rocket boosters (SRBs). You'll read more about them shortly.

NASA likes to describe the orbiter as comparable in size and weight to the DC-9— a workhorse civilian jetliner long in use for short- to mid-length passenger flights. In other words, the orbiter is a good-sized aircraft, much bigger than the *Mercury* capsules that Navy helicopters plucked from the sea. On the other hand, the orbiter is much smaller than big long-haul aircraft such as the Boeing 747 "jumbo jet."

Discovery lifts off on 12 September 1993 for STS-51 from the Kennedy Space Center in Florida. This image offers great views of all the elements working together: the orbiter with main engines firing, the external tank (the tallest element in the photo painted a rust color), and the solid rocket boosters giving off the brightest glow at their base. But the orbiter is what many people think of as "the space shuttle."
Courtesy of NASA

The Shuttle's Main Engines

The shuttle has three main engines. These are known in NASA-speak as Space Shuttle Main Engines, or SSMEs. These engines combine with the boosters to get the shuttle off the ground in the initial ascent. They operate for 8.5 minutes after launch.

After the shuttle jettisons its boosters, the engines provide thrust—that is, they push the shuttle forward. This accelerates the shuttle from 3,000 mph (4,828 km per hour) to 17,000 mph (more than 27,358 km per hour) in just six minutes to reach orbit. The engines create a combined maximum thrust of more than 1.2 million pounds.

During this acceleration, the engines burn through half a million gallons of liquid propellant. The propellant consists of liquid oxygen and liquid hydrogen.

Space shuttle Atlantis takes off from Kennedy Space Center in Florida on 2 December 1988.
While the rocket boosters give off the most dramatic flames, the space shuttle's three main engines (in a triangle of fainter glows at the shuttle's base) are nonetheless giving off 375,000 pounds of thrust. The main engines operate for 8.5 minutes after launch.
Courtesy of NASA

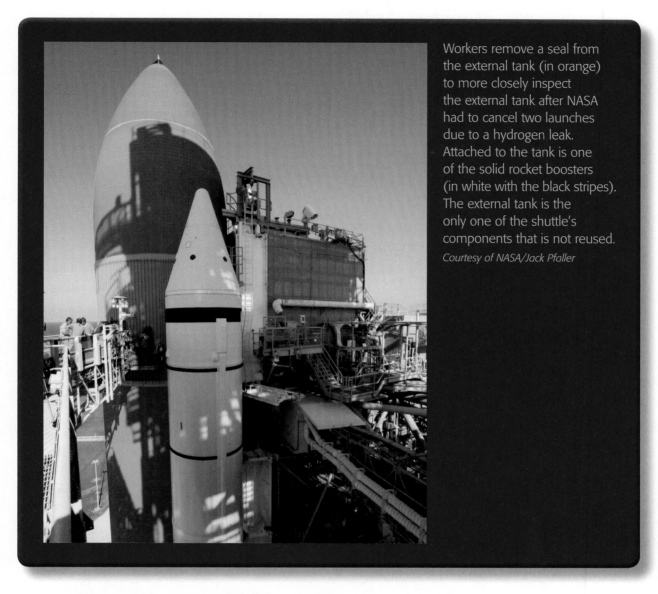

Workers remove a seal from the external tank (in orange) to more closely inspect the external tank after NASA had to cancel two launches due to a hydrogen leak. Attached to the tank is one of the solid rocket boosters (in white with the black stripes). The external tank is the only one of the shuttle's components that is not reused.

Courtesy of NASA/Jack Pfaller

The External Tank and the Solid Rocket Boosters

The shuttle's three main engines get their propellants from the huge rust-colored external tank (ET). It's the only one of the shuttle's components that is not reused. Rather, at each shuttle launch, the fuel tank burns up in the Earth's atmosphere.

The solid rocket boosters (SRBs), on the other hand, are reusable. They are the largest solid-propellant motors ever flown. And they are the first designed for reuse. Each is nearly 150 feet long and a little more than 12 feet in diameter.

The SRBs provide most of the power for the first two minutes of a shuttle flight. At launch, their weight is about 1.3 million pounds apiece. Most of that is fuel. Once they have spent their fuel, the empty boosters fall away into the Atlantic Ocean, about 140 miles off Florida's east coast.

Enormous parachutes slow their journey, but the spent boosters typically hit the water a little less than five minutes after separation. They are equipped with radio beacons and flashing lights. Once the recovery crew locates the boosters, they tow them back to the launch site for cleaning up and refurbishing for a future launch.

The Shuttle Crew Positions

In Chapter 5 Lesson 2, you read about the different types of astronauts. The crews of space shuttle missions range from five to seven people. They vary according to different mission objectives. But the crew members fall into the same classifications that you read about earlier.

Mission commanders come from the ranks of the pilot astronauts. The commander is the ship's captain, so to speak, with overall responsibility for the orbiter, the crew, mission success, and safety. A *mission pilot* helps control and fly the ship, and may also help with tasks such as deploying or retrieving satellites. Astronauts typically accomplish these tasks with the shuttle's robotic arm.

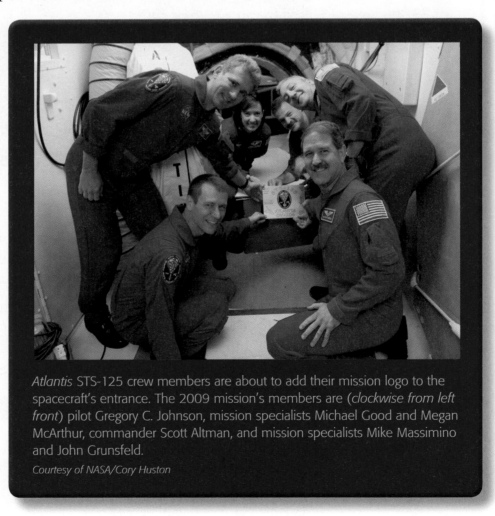

Atlantis STS-125 crew members are about to add their mission logo to the spacecraft's entrance. The 2009 mission's members are (*clockwise from left front*) pilot Gregory C. Johnson, mission specialists Michael Good and Megan McArthur, commander Scott Altman, and mission specialists Mike Massimino and John Grunsfeld.

Courtesy of NASA/Cory Huston

Mission specialists are the astronauts who plan, coordinate, and manage the mission's activities. They don't actually fly the ship. But NASA fully trains them in the details of all the onboard systems. They have the lead role in EVAs, or spacewalks, and in operating the robotic manipulator system. They are also responsible for payloads and specific experiments.

If the demands of a given mission require, its crew may also include one or more *payload specialists*. As the name suggests, the focus of these crew members is on specific payloads—a scientific experiment, for instance. Foreign nationals fly aboard the shuttle as payload specialists, but not as mission specialists or pilot astronauts. (A foreign national is *someone who owes allegiance to a foreign country*—someone not a US citizen, in other words.)

The Shuttle's Legacy

As of this writing in early 2010, NASA plans to shutter its space shuttle program with five final flights the same year. It all depends on whether the numerous space agencies working on the International Space Station can complete construction of the orbiting lab by that time. All five final shuttle flights support this goal. Only three shuttles are still in use: *Discovery*, *Atlantis*, and *Endeavour*. NASA has scheduled *Discovery* to make the final flight in September 2010.

For nearly 30 years, NASA's space shuttles have served as the foundation for the human-spaceflight program in the United States. The shuttles represented a leap in thinking about rockets—reusing significant parts.

True, the shuttles couldn't lift as much mass into space as NASA's old Saturn V rocket. But their spacious cargo bays allowed scientists to envision, build, and use space-based observatories such as the Hubble Space Telescope and the Spitzer Space Telescope. These are giving humanity new views of the vast cosmos it inhabits.

The shuttles also paved the way for a new, higher level of global cooperation in space. This was first hinted at by the *Apollo-Soyuz* mission. It resumed modestly with the first flight in 1983 of *Spacelab*, a space-station-like laboratory built in Germany. *Spacelab* fit into *Columbia*'s cargo bay. It provided astronaut-scientists inside with a comfortable environment for conducting experiments in microgravity. It was a foretaste of what would come with the International Space Station, now on orbit. Major components of the station have come from Europe, Japan, Canada, and Russia, as well as the US. Astronauts who serve on the space station come from these and other countries. (You'll read more about space stations in Chapter 8.)

But perhaps the shuttle program's most important legacy is its reminder that human spaceflight must always be treated with respect. One successful launch after another made shuttle flights look routine, almost easy. The loss of *Challenger* and its crew in 1986 and the loss of *Columbia* and its crew in 2003 emphasized while space exploration is exciting, it also is risky. The next lesson will review what NASA learned from those accidents.

Xenon lights and lightning dramatically draw attention to Launch Pad 39A at NASA's Kennedy Space Center in Florida.

The bad weather delayed a *Discovery* launch in 2009 for a few days. NASA has many missions ahead for its astronauts that include the Moon and, perhaps one day, Mars.

Courtesy of NASA/Ben Cooper

CHAPTER 7   The Space Shuttle

Shuttle Firsts

Guion S. Bluford Jr. has the distinction of being the first African-American in space. He was a mission specialist aboard STS-8 in August-September 1983. It was *Challenger*'s third mission and the shuttle program's first nighttime launch and landing.

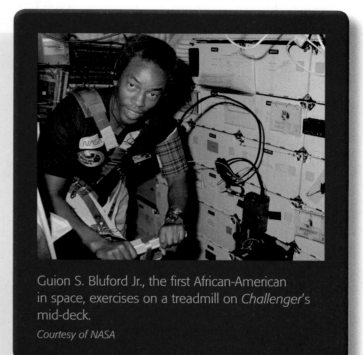

Guion S. Bluford Jr., the first African-American in space, exercises on a treadmill on *Challenger*'s mid-deck.

Courtesy of NASA

Franklin Chiang-Diaz was the first Hispanic American to fly in space.

Courtesy of NASA

Franklin Chiang-Diaz was the first Hispanic American to fly in space. He first flew on STS-61-C and flew on seven flights in all, logging more than 1,601 hours in space. This included 19 hours and 31 minutes in three space walks. Holder of a doctorate in applied plasma physics, he served from 1993 to 2005 as director of the Advanced Space Propulsion Laboratory at the Johnson Space Center in Houston, Texas.

US Air Force Col **Eileen M. Collins** was the first woman shuttle pilot. On her first mission, STS-63, in February 1995, she piloted the *Discovery* on the first flight of the new joint Russian-American space program. She flew three more missions after that. On her third, in July 1999, she made history again as the shuttle's first woman commander. On her fourth and final shuttle mission in July 2005, she commanded STS-114 as well. This was celebrated as the "return to flight" mission after the loss of the orbiter *Columbia*. (You'll read more about that in the next lesson.)

Col Eileen M. Collins was the shuttle program's first woman pilot and commander.

Courtesy of NASA/Robert Markowitz

Col **Sidney Gutierrez** was the first Hispanic shuttle pilot. A native of New Mexico and a US Air Force Academy graduate, he made two shuttle flights. On the first, STS-40, in June 1991, he was the pilot. On the second, STS-59, he was the spacecraft commander. After his first flight, he served as spacecraft communicator, or CAPCOM, for several shuttle missions. This is a crucial role—the voice link between the flight crew and Mission Control during a mission.

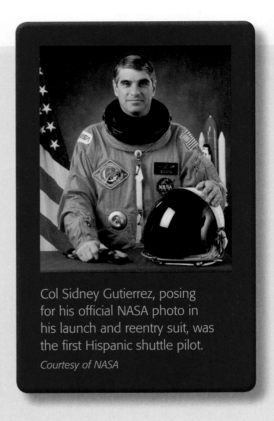

Col Sidney Gutierrez, posing for his official NASA photo in his launch and reentry suit, was the first Hispanic shuttle pilot.
Courtesy of NASA

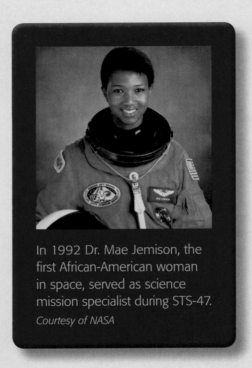

In 1992 Dr. Mae Jemison, the first African-American woman in space, served as science mission specialist during STS-47.
Courtesy of NASA

Dr. Mae Jemison was the first African-American woman in space. A physician who also has a background in engineering, she was the science mission specialist on STS-47 in September 1992. This was a joint mission with the Japanese space agency.

Dr. Ellen Ochoa was the first Hispanic woman in space. A native Californian, she is a veteran of four spaceflights, starting with STS-56 in 1993, on which she was a mission specialist. She holds a doctorate in electrical engineering from Stanford University and is as of this writing deputy director of the Johnson Space Center in Houston, Texas.

Dr. Ellen Ochoa was the first Hispanic woman in space.
Courtesy of NASA

The first Asian-American in space was Col **Ellison Onizuka**. A native of Hawaii, he was also an Air Force officer with two degrees in aerospace engineering from the University of Colorado. In January 1985 he was a mission specialist on STS-51C, the space shuttle's first Defense Department mission. The next year he was a mission specialist aboard *Challenger*. He died along with his crewmates when their ship blew up shortly after launch. Onizuka was posthumously promoted to colonel and awarded the Congressional Space Medal of Honor.

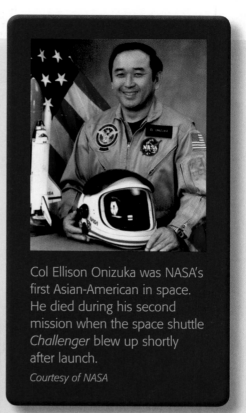

Col Ellison Onizuka was NASA's first Asian-American in space. He died during his second mission when the space shuttle *Challenger* blew up shortly after launch.
Courtesy of NASA

Dr. Sally K. Ride floats on *Challenger*'s mid-deck during STS-7.
Courtesy of NASA

Dr. Sally K. Ride was the first American woman in space. A physicist by training, she was part of the STS-7 crew in June 1983. This was *Challenger*'s second mission.

Commander John W. Young (left) and pilot Robert L. Crippen pose with a model of *Columbia*. Each marked a first in spaceflight: Young was the first space shuttle commander, and Crippen was the first shuttle pilot.

Courtesy of NASA

John W. Young was the first space shuttle commander. He was already an experienced astronaut when he took off on STS-1, in the orbiter *Columbia*, in April 1981. It was his fifth spaceflight. **Robert L. Crippen** was the first shuttle pilot. His 36-orbit mission with Young was a shakedown cruise, meant to test out the new shuttle's performance. *Columbia* was the first winged reentry vehicle to return from space to a runway landing.

CHECK POINTS

Lesson 1 Review

Using complete sentences, answer the following questions on a sheet of paper.

1. The space shuttle began as part of what larger vision?

2. How did *Atlantis* benefit from lessons learned in construction of earlier orbiters?

3. What problem did post-flight inspection of *Columbia* identify?

4. The orbiter is comparable to what familiar civilian aircraft? In what way?

5. How much fuel do the shuttle's engines burn during acceleration? What kind of fuel is it?

6. What is unique about the shuttle's external tank?

7. When foreign nationals fly on shuttle missions, what is their role?

8. Who was the first shuttle commander? The first American woman in space? The first African-American in space?

APPLYING YOUR LEARNING

9. Do you think it is important for the space program to have a specific goal, such as going to the Moon or building a space station? Why or why not?

Lessons Learned: *Challenger* and *Columbia*

Quick Write

_____

_____

How do Richard Feynman's efforts to understand the *Challenger* disaster show the value of bringing an outsider's fresh thinking into an organization facing a big problem?

Learn About

- the *Challenger* accident
- the *Columbia* accident

World-renowned physicist Richard P. Feynman tended to avoid the committees and commissions that scientists of his stature are usually called to serve on, wrote *The New York Times* upon his death in 1988. But he made an exception for the inquiry into the *Challenger* space shuttle disaster of 1986.

He wasn't much of a team player, however. He asked tough questions of witnesses. And he sometimes skipped regular meetings of the commission to perform his own research. He questioned engineers and examined rocket parts in storage at NASA facilities, according to the *Times*. When he found out something, he wasn't shy about going before television cameras to share it.

None of this pleased the chair of the commission, William P. Rogers. He was a Washington lawyer who had once served as his country's top diplomat. He wanted an "orderly investigation."

One day at the commission's hearings, attention focused on the O-ring seals. (An O-ring is *a flat ring of rubber or plastic, used as a gasket or seal.*) These were supposed to seal the joints between parts of the booster rockets. As members of the commission passed around a piece of the O-ring material, Feynman asked for a glass of ice water. He dipped the ring into the ice water and then squeezed it briefly with a clamp. When he released the ring from the clamp, the rubbery material failed to spring back.

He then confronted the former chief of the solid rocket booster program: "I took this stuff that I got out of your seal and I put it in ice water, and I discovered that when you put some pressure on it for a while and then undo it, it doesn't stretch back. It stays the same dimension. In other words, for a few seconds at least and more seconds than that, there is no resilience in this particular material when it is at a temperature of 32 degrees," he said.

But the whole point of the O-rings was to spring back under pressure. The inquiry concluded that if NASA had conducted the same experiment and discovered that the O-rings did not seal at low temperatures, the disaster could have been avoided.

Vocabulary

- O-ring
- resilient
- posthumously
- spar
- normalization of deviance
- organizational culture

NASA took this photo only seconds after *Challenger*'s accident on 28 January 1986. It shows exhaust plumes from the space shuttle main engines and solid rocket boosters twisting around a fiery ball from the external tank. Mission STS-51L was NASA's first fatal accident in almost 20 years.

Courtesy of NASA

The *Challenger* Accident

On 28 January 1986 the space shuttle *Challenger*, on a mission designated STS-51L, blew up just a little more than a minute into its flight. All seven crew members perished.

It was NASA's first fatal accident in almost 20 years. And it occurred with the whole world watching. *Challenger*'s crew included Christa McAuliffe, selected as NASA's "Teacher in Space." She was part of an effort to help the public identify with the shuttle program and rekindle Americans' romance with space exploration. Her presence aboard the shuttle meant more people were watching—including children—than would have been otherwise.

How the *Challenger* and Its Crew Were Lost

After the accident, President Reagan named a commission to investigate and recommend steps to prevent such a thing from happening again. He put former Secretary of State William P. Rogers in charge.

The investigators faced a grim task. But at least they could draw on the vast amount of data NASA routinely collects in connection with all space missions. And from this, they were able to get a very clear idea of what had gone wrong.

The photo record of *Challenger's* brief flight showed that in literally the first second, a puff of gray smoke spurted from a particular joint on the right solid rocket booster. That joint, known as the "aft field joint," was supposed to seal two parts of the rocket together. But the smoke indicated a break in the seal.

Over the next couple of seconds, eight more distinctive puffs of smoke—each blacker than the one before—emerged from the damaged seal. The smoke was dark and thick. From this, investigators concluded that the solid rocket booster's hot propellant gases were burning the grease, insulation, and the rubber O-rings in the joint seal. The joint simply wasn't strong enough to contain the hot propellant gases.

This weak spot on the solid-fuel rocket booster faced the external tank. At 58.788 seconds into the flight, the first flicker of a flame appeared. It grew into a large plume and spread to the external tank. Soon after, the external tank ruptured and leaked liquid hydrogen fuel. This liquid propellant mixed with flame from the solid rocket booster. At 73.124 seconds, the external tank's liquid hydrogen tank shot upward into its liquid oxygen tank. The solid-fuel rocket booster also collided with the liquid oxygen tank.

At 73.137 seconds, *Challenger* began to break up. At 78 seconds, *Challenger* was an enormous fireball in the sky. People on the ground in Florida, and before their television screens around the world, could only look on in horror.

The Rogers Commission released its report on 6 June 1986. The consensus of the commission and the other agencies that took part in the investigation was that the failure of the joint between the two lower parts of the right solid-fuel rocket caused the accident. No other part of the space shuttle was a factor.

Search and recovery teams found this joint from the right solid-fuel rocket booster, which had traces of O-ring seal tracks. The consensus of the Rogers Commission and the other agencies that took part in the *Challenger* investigation was that the failure of the joint between the two lower parts of the right solid-fuel rocket booster caused the accident.
Courtesy of NASA

How the Weather's Effect on the Solid Rocket Boosters Caused the Accident

When the space shuttle was first in development, its solid-fuel rocket boosters were something new for NASA. The agency had used solid-fuel rockets for some small unmanned spacecraft. But the astronauts of the Mercury, Gemini, and Apollo missions had all been boosted into space by liquid-fuel rockets.

And even as NASA weighed the possibility of solid-fuel rockets for the shuttle, price estimates indicated that liquid-fuel rockets offered potentially lower operating costs. But the solid-fuel rockets offered lower development costs—they would cost less to make, in other words. So that's what the agency opted for. NASA awarded the contract for the rockets to a company called Morton Thiokol.

The Rogers Commission sharply criticized the solid-fuel rocket booster and particularly the faulty design of its joint. The problem got worse, the report noted sternly, as both NASA and its contractors "first failed to recognize it as a problem, then failed to fix it, and finally treated it as an acceptable flight risk."

At the heart of the controversy over the joint on the solid-fuel rocket boosters were the O-rings. That little washer that seals the connection between your garden hose and the spigot is an O-ring. You can find O-rings in industrial settings as well as backyards. For instance, engineers may seal together sections of pipe with O-rings. But at whatever scale, an O-ring works because it's made of a resilient material— one *capable of bouncing back to its original shape after being compressed*. An O-ring is pressed between two lengths of pipe, as one example, and then bounces back to fill the space between them, creating a perfect air- or gas- or water-tight seal.

NASA used the rockets repeatedly. Many of them took some knocks after being launched into space several times. The solid fuel was a rubbery material that didn't always fit just right inside the rocket. And the O-ring material wasn't as resilient as it should have been, especially at low temperatures. All these factors made it hard to form a perfect seal.

Star POINTS

The Rogers Commission included some of the most famous names in American aerospace exploration: Neil Armstrong, the first man to walk on the Moon; Charles (Chuck) Yeager, the first pilot to fly faster than the speed of sound; and Sally Ride, the first American woman in space. Richard P. Feynman, a Nobel Prize-winning physicist, also served.

Although engineers conducted many tests on the shuttle, they really had no test data to predict the safety of a launch if the temperature were below 53 degrees F (12 degrees C). In January, even Florida can get cold. As scientists prepared *Challenger* for what would turn out to be its final flight, the temperature hovered around the freezing mark: 32 degrees F (0 degrees C).

While pieces of the right solid-fuel rocket booster are falling to Earth, the left booster can be clearly seen still thrusting. The reddish-brown plume surrounds the disintegrating orbiter. The Rogers Commission sharply criticized the solid-fuel rocket booster and particularly the faulty design of its joint.

Courtesy of NASA

How NASA Management Contributed to the Accident

In its inquiry, the Rogers Commission found failures in communication. These failures led to a launch decision made on the basis of incomplete and sometimes misleading information. The commission found conflict between engineering data and management judgments. The commissioners also found that NASA's management structure allowed flight-safety questions to bypass key shuttle managers.

As early as 1977 a test of the Thiokol rocket had identified a defect in the seal. This defect meant that elements tended to come apart, rather than become more tightly sealed together, under the pressures of launch. But NASA managers never addressed this problem.

On the eve of the launch, Thiokol engineers thought they had warned their managers that NASA should not launch the shuttle in cold weather because of doubts about the joint seal. But the managers did not interpret the engineers' remarks as advice to hold off on a launch.

Star POINTS

The *Challenger* accident has often been used as a case study in courses and seminars on decisionmaking and workplace ethics.

The commissioners also noted that they heard nothing from, or about, NASA's safety team during the investigation. No witness mentioned safety engineers, and no one said whether safety engineers had done a good job or not. And no one thought to include safety staff in the meeting where NASA made the final decision to go ahead with the launch.

Changes NASA Made to Reduce the Possibility of Another Accident

The Rogers Commission recommended nine steps NASA should take to reduce the possibility of another *Challenger* disaster. The steps included big changes. Almost all of them had many parts. They addressed technical issues (redesign of the troublesome O-ring seals) as well as "human factors" issues such as communication. The commission found that too many managers tended to get stuck in their cubicles and not see the bigger picture.

President Reagan asked NASA to provide, within 30 days of the issue of the Rogers Commission Report, a plan to carry out the report's recommendations.

In response to these recommendations:

- NASA had the solid rocket booster redesigned. Engineers made changes in the segment joints and case-to-nozzle joints, the nozzle, propellant grain shape, ignition system, and ground support equipment. The O-rings were replaced by new rings made of a better-performing material called nitrile rubber.

- NASA added an orbiter to the fleet to lighten the burden of a heavy flight schedule on too few spacecraft; the agency also reassigned some tasks to unmanned spacecraft.

- NASA reorganized the shuttle program's management structure to ensure that dissenting voices got a say in launch decisions. It also strengthened its support for its safety staff.

- The space agency ordered improved communication among managers, and an end to the isolation of managers from one department to the next.

- NASA strengthened the flight readiness review—the pre-launch process that had given *Challenger* its green light in 1986. Staff members now record reviews and take minutes (a formal kind of note-taking).

- NASA committed to "criticality review and hazard analysis." This involved looking over every shuttle component to see which ones needed upgrades to make them reliable.

- The agency's scientists developed new systems to allow astronauts to escape in the case of another faulty liftoff. NASA also improved the orbiters' landing systems—tires, wheels, and the like—so that in the case of an aborted mission, the shuttle crews would have further options for landing.

On 29 September 1988 NASA celebrated the shuttle's "return to flight" as the orbiter *Discovery* blasted off from the Kennedy Space Center. On this mission, designated STS-26, *Discovery* carried a tracking and data relay satellite as one of its payloads. It was identical to the one that had been lost two and a half years before. The shuttle had returned to the skies, much safer than before.

Not everyone was convinced that NASA had taken all possible steps to avoid another accident, however. Some critics felt the changes had not been thorough enough, and that sooner or later, there would be another disastrous accident. Tragically, they would be proved right less than 15 years later.

Discovery's STS-26 launch on 29 September 1988
Courtesy of NASA

The *Challenger* Crew

Francis R. (Dick) Scobee, the commander of *Challenger*'s last mission, STS-51L, was a native of Washington state who enlisted in the Air Force as a teenager. After training as an engine mechanic, he got a university education through the Air Force. Once he became part of the astronaut corps, he served as an instructor pilot on the NASA/Boeing 747 shuttle carrier airplane. STS-51L was his second spaceflight.

Michael J. Smith was the pilot of STS-51L. Born in North Carolina and educated at the US Naval Academy, he served in the Navy for many years before NASA selected him as a candidate astronaut in 1980. STS-51L was his only spaceflight.

Judith A. Resnik, an Ohio native, was an electrical engineer by training and a classical pianist by avocation (her hobby). NASA accepted her as an astronaut candidate in January 1978. On her first spaceflight, as a mission specialist aboard STS-41D in August–September 1984, she and her crewmates earned the nickname "Icebusters" for the skill they showed in using the shuttle's robotic arm to dislodge dangerous bits of ice from the orbiter. STS-51L was her second spaceflight.

Ronald E. McNair was a native South Carolinian, a jazz saxophonist, and a fifth-degree black belt in karate. He also had a doctorate in physics from MIT. His first spaceflight was aboard STS-41B, in February 1984. STS-51L was his second spaceflight. He was a mission specialist.

Ellison Onizuka, a Japanese-American born in Hawaii, studied aerospace engineering at the University of Colorado and spent many years as a flight test engineer, working on several different kinds of aircraft. He was aboard STS-51L as a mission specialist.

Gregory B. Jarvis was a native of Detroit. He was an electrical engineer by training, and he specialized in missiles and satellite design. He was a civilian payload specialist aboard STS-51L. It was his only mission in space.

S. Christa Corrigan McAuliffe was aboard *Challenger* as part of NASA's Teacher in Space Program. A Boston native who had taught English and social studies in middle and high schools, she was teaching American history at Concord High School in New Hampshire when NASA selected her as a teacher in space. Unlike their *Challenger* crewmates, McAuliffe and Gregory Jarvis were not federal employees.

All seven of the *Challenger* crew were awarded the Congressional Space Medal of Honor posthumously (*after death*).

The crew of *Challenger*'s last mission, STS-51L. *Left to right*: Christa McAuliffe, Gregory Jarvis, Judith Resnik, Dick Scobee, Ronald McNair, Michael Smith, and Ellison Onizuka.

Courtesy of NASA

The *Columbia* Accident

On 1 February 2003 the shuttle *Columbia* was on its way back home at the end of an intense 16-day science mission, designated STS-107. Sixteen minutes before its scheduled touchdown in Florida, *Columbia* broke up on reentry into Earth's atmosphere. All seven of the crew perished. The accident left debris scattered across much of the Southwestern United States.

To find out what had gone wrong and prevent another occurrence, NASA convened the Columbia Accident Investigation Board, or CAIB. Adm Hal Gehman was its chair.

How the *Columbia* and Its Crew Were Lost

The physical cause of the *Columbia* disaster was a breach in the thermal protection system—a kind of protective plating made of several materials—heat-resistant tiles, thermal blankets, and reinforced carbon-carbon. The system was supposed to shield the orbiter on its way back to Earth. A chunk of insulating foam broke off at launch and struck the orbiter's left wing within the first two minutes of flight. This caused the breach.

Columbia STS-107 lifts off on its fateful flight on 16 January 2003.
Courtesy of NASA

The chunk of foam, about the size of a small briefcase, made a chink in the reinforced carbon-carbon protecting the wing's leading edge. The chink let hot air penetrate the wing's interior. This superheated air was more than 5,072 degrees F (2,800 degrees C) hot. That ultimately melted the wing's thin aluminum spar, or *structural support*. That, in turn, weakened the whole orbiter's structure. As *Columbia* hurtled to Earth with a broken wing, aerodynamic forces acting on the shuttle caused loss of control, wing failure, and the orbiter's breakup.

How Damage to the Thermal Protection System Caused the Accident

By the end of the investigation into the *Challenger* disaster, almost everyone in America had heard about O-rings. After the *Columbia* accident, everyone had heard about heat-shield tiles. These made up the shuttle's thermal protection system, or TPS. *Columbia*, as you may recall, was the first orbiter to go into space. As you read in Chapter 7, Lesson 1, it returned from its maiden flight with 16 of its tiles missing and another 148 damaged.

It wouldn't be the last time such damage would occur to an orbiter. Many times chunks of insulating foam around the shuttle's external tank broke off at launch and dinged the TPS. As the CAIB report later made clear, this phenomenon became so common that NASA officials developed their own standard term for it—"foam shedding." And since they had observed it so many times on orbiters that did return safely, they didn't consider it a serious problem.

Normalization of deviance is the term for this *process of reclassifying defects as acceptable*. "Lowering the bar" is a common phrase that captures much the same meaning.

Space shuttle *Columbia* debris sits in a hangar at Barksdale Air Force Base in Shreveport, Louisiana. The debris was catalogued before being shipped to the Kennedy Space Center in Florida.

Courtesy of NASA

The foam strike in the case of STS-107 was first noticed the day after launch, as NASA officials on the ground reviewed high-resolution photography of the launch. Some engineers suspected real trouble lay ahead. But the strike seemed to have no effect on *Columbia*'s mission.

Some weak signals appeared early on that the orbiter was in trouble, the CAIB concluded afterward. But mission management failed to detect them and take corrective action. Clear signs of trouble emerged only after reentry had begun. And by then it was too late.

How NASA Management Contributed to the Accident

Like the Rogers Commission, the *Columbia* inquiry looked not only at hardware that failed but also at management systems. The CAIB report used forceful and direct language. "We are convinced that the management practices overseeing the Space Shuttle Program were as much a cause of the accident as the foam that struck the left wing." The board's conversations with Congress also suggested that the nation needed a "broad examination" of NASA's Human Space Flight Program, rather than just an investigation into the specifics of the *Columbia* accident.

The CAIB faulted NASA for its overly ambitious flight schedule. The board also faulted the agency for not putting into place a truly independent office for safety oversight. Too many people at NASA had responsibility for both sticking to the flight schedule and maintaining safety. As an organization, NASA clearly needed to do both. But as items on any individual's to-do list, those two responsibilities conflict with each other. The safety function needed to be separated from the responsibility for sticking to the flight schedule.

Star POINTS

The search for debris from *Columbia* was the largest land search ever conducted.

Star POINTS

Can you identify aspects of the organizational culture of any group you're part of, at school or elsewhere?

The CAIB report also touched on NASA's organizational culture as a factor in the *Columbia* accident. Organizational culture refers to *the values, norms, and shared experiences of an organization.* It's "the way we do things around here."

The CAIB noted that the Apollo program's dramatic achievements had given NASA staff a sense of their organization as a "perfect place." NASA failed to adapt from the high drama of historic firsts in space to the bureaucratic routine of the space shuttle program. And as NASA budgets got smaller, more and more work was done by outside contractors, rather than people who worked for NASA. That should have prompted more effort to develop effective communications and safety oversight processes. But it did not.

In 1992 *Columbia* rockets to the skies on its 12th flight. The Columbia Accident Investigation Board faulted NASA for the shuttle program's overly ambitious flight schedule.
Courtesy of STS-50, NASA

Changes NASA Made to Reduce the Possibility of Another Accident

The CAIB report acknowledged that the changes it recommended would be hard to make, and resisted by NASA. But NASA did put in place the changes the board called for. These included:

- Efforts to reduce "foam shedding" and also to strengthen the orbiter's heat shield
- Improved inspection routines before launches
- Improved imaging—video and photos—of the shuttle, both at launch and during orbit
- Establishment of a Technical Engineering Authority responsible for all technical requirements for the shuttle system—to identify, analyze, and control hazards within the system.

NASA now has contingency plans to launch a rescue mission, should an orbiter get into trouble in space. And in late 2008 NASA released a report outlining what it had learned from the *Columbia* accident with regard to crew safety and survivability for future spaceflight. It called for changes in the harnesses that hold astronauts in place during reentry on a space mission, for example.

Challenger on mission STS-7 in 1983 as photographed by the Space Pallet Satellite (SPAS). The shuttle orbits with its payload bay packed with satellites, its robot arm, and more. Even with the shuttle program now coming to a close, lessons learned from the *Challenger* and *Columbia* accidents remain critical.

Courtesy of NASA

It also called for automatic parachutes that could bring even unconscious astronauts safely back to Earth in case of accident. The report said that with current technology, another accident like that of *Columbia* would not be survivable. But providing for crew escape from a damaged spacecraft would widen the margin of human safety.

Almost everything people do involves some risk, whether it's spaceflight, driving a car, or flying in a commercial airplane. The question for NASA and the American public is how much risk is acceptable. The federal and state governments, reacting to public pressure, have enacted many laws and regulations to make cars and planes safer—from requiring seatbelts and airbags in cars to limits on how many hours a day commercial pilots can fly. To retain public support for the space program—and protect the lives of its astronauts—NASA must do all it can to reduce the risk of spaceflight as much as possible.

Even with the shuttle program now coming to a close, lessons learned from the *Challenger* and *Columbia* accidents remain critical. Scientists, engineers, and managers are taking these lessons and applying them to future missions and spacecraft design. Through hard work and bright ideas—as well as never forgetting the shuttle crews' sacrifices—the highly skilled men and women on the ground hope to create safer spaceflight for astronauts in years to come.

The *Columbia* Crew

Rick Douglas Husband was an Air Force colonel who commanded *Columbia* for STS-107. He was a Texan by birth and a mechanical engineer by training. He flew more than 40 different types of aircraft. Chosen as an astronaut candidate in December 1994, Husband made two flights on the shuttle.

Willie McCool was the pilot on mission STS-107. He was an Eagle Scout and a 1983 graduate of the US Naval Academy, where he was second in his class of 1,083. NASA selected him as an astronaut in 1996. STS-107 was his only spaceflight.

David M. Brown, a native of Virginia, was a physician by training. But he was also an accomplished Navy pilot. NASA chose him to be an astronaut in 1996. As part of the crew for STS-107, his only spaceflight, he logged nearly 16 days in space.

Laurel Blair Salton Clark, of Racine, Wisconsin, was a Navy captain and a physician. Her education included Navy undersea medical officer training. In the course of her military service she performed many medical evacuations from US submarines. STS-107 was her only spaceflight.

Michael Anderson was born in upstate New York but considered Spokane, Washington, home. He had a bachelor's degree in physics/astronomy from the University of Washington and a master's in physics from Creighton University. He was a lieutenant colonel in the US Air Force. STS-107 was his second spaceflight.

The crew for *Columbia*'s final mission, STS-107. *Left to right*: David Brown, Rick Husband, Laura Blair Salton Clark, Kalpana Chawla, Michael Anderson, William McCool, and Ilan Ramon.
Courtesy of NASA

A native of India and a naturalized American, **Kalpana Chawla** had aeronautical and aerospace engineering degrees from schools in India, Texas, and Colorado. She was also a highly trained pilot, licensed for airplanes and gliders. She worked for NASA for several years and then in private industry. She was selected as an astronaut in December 1994. STS-107 was her second mission.

Ilan Ramon was a colonel in the Israeli air force and a payload specialist on STS-107. He had a degree in electronics and computer engineering. While still in his teens, he fought in the Yom Kippur War of 1973.

Like the members of the *Challenger* crew, all members of the *Columbia* crew, including Ramon, who was not a US citizen, were awarded the Congressional Space Medal of Honor posthumously.

Lesson 2 Review

Using complete sentences, answer the following questions on a sheet of paper.

1. What caused the *Challenger* accident?

2. What was the problem with *Challenger*'s O-rings at low temperatures?

3. What specific problem with NASA's management structure did the Rogers Commission identify?

4. NASA made many changes in response to the Rogers Commission report; list three of them.

5. What caused the *Columbia* accident?

6. What is "foam shedding," and why didn't NASA see it as a serious problem?

7. The Columbia Accident Investigation Board faulted NASA for allowing which conflict in responsibility to continue with regard to safety?

8. NASA made many changes in response to the Columbia Accident Investigation Board's report; list three of them.

APPLYING YOUR LEARNING

9. How did faulty leadership and management contribute to the two shuttle disasters? How might you apply the lessons to a problem at your school or in your community?

CHAPTER 8

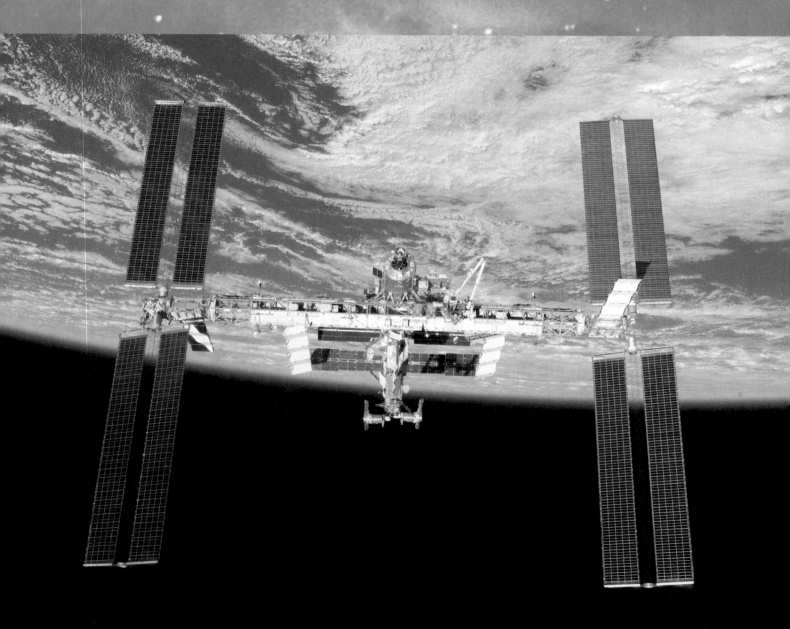

Earth's horizon provides a dramatic backdrop for the International Space Station. Crew aboard the space shuttle *Atlantis* snapped this shot after undocking from the station for STS-117, the 21st shuttle journey to the space station.

Courtesy of STS-117 Crew/NASA

Space Stations and Beyond

Chapter Outline

LESSON 1 From *Salyut* to the International Space Station

LESSON 2 The Future in Space

> Oh, I have slipped the surly bonds of Earth
> And danced the skies on laughter-silvered wings;
> Sunward I've climbed, and joined the tumbling mirth
> Of sun-split clouds—and done a hundred things
> You have not dreamed of—wheeled and soared and swung
> High in the sunlit silence....
> And, while with silent lifting mind I've trod
> The high untrespassed sanctity of space,
> Put out my hand, and touched the face of God.

John Gillespie Magee Jr., High Flight

LESSON 1 | From *Salyut* to the International Space Station

Quick Write

Why was it important for both astronaut and cosmonaut to enter the International Space Station at the same time? What does that mean to you?

Learn About

- the *Salyut* space station
- the *Skylab* space station
- the *Mir* space station
- the International Space Station

On 4 December 1998 NASA and its partners began building a dream. That was the day space shuttle *Endeavour* lifted off on a 12-day mission to deliver NASA's *Unity* module to orbit.

Up in space, *Endeavour* and its crew would meet up with the Russians, whose *Zarya* control module was already on orbit. Just getting *Zarya* aloft was an accomplishment. But *Zarya* plus *Unity* added up to a truly international space station—even though only in its first stages.

Astronaut Robert Cabana was *Endeavour*'s commander. He remembers that trip vividly. He recalls how he and Russian cosmonaut Sergei Krikalev prepared to be the first to enter the newly joined modules.

As he said afterward, "We finally got all the hatches open and we're up to the main hatch going into Node 1 [*Unity*]. We open the hatch and Sergei Krikalev was with me. I just waved my hand toward the hatch and the two of us entered together." International cooperation was the whole point of the space station, he said. And the two men's action symbolized that. "You know, there wasn't a first person in. It was, we went in together."

Cabana, now Kennedy Space Center director, found it remarkable that people from so many different countries could all work together on the International Space Station. They produced components at separate locations around the globe to be bolted together more than 200 miles above Earth's surface.

"You take all those different cultures, people, and hardware built around the world and it comes together for the first time on orbit and it works flawlessly—that's phenomenal," he said.

The *Salyut* Space Station

Vocabulary

- stellar spectrum
- orbital decay
- attitude
- deorbit

In 1903 a Russian schoolteacher named Konstantin Tsiolkovsky wrote a book called *Beyond Planet Earth*. It was a work of fiction, but the author based it on sound science. It described space stations in orbit around the Earth, where humans would learn to live in space. Tsiolkovsky believed space stations would lead to explorations of the Moon, Mars, and even the asteroids. He wrote his science fiction for another three decades, inspiring generations of Russian space scientists.

As Americans pursued manned missions to the Moon, the Soviet Union (as it was then) built several space stations. Later space stations, including the International Space Station, which you'll read more about in this lesson, built on the foundation of these early efforts.

Salyut 1, the First Space Station Put Into Orbit

The first-generation Soviet space stations had one docking port. The Russians launched these space stations "unmanned"; crews followed later. They could not resupply or refuel the space stations, however, because crews used the single docking port to visit and enter. This left no port for a resupply vessel.

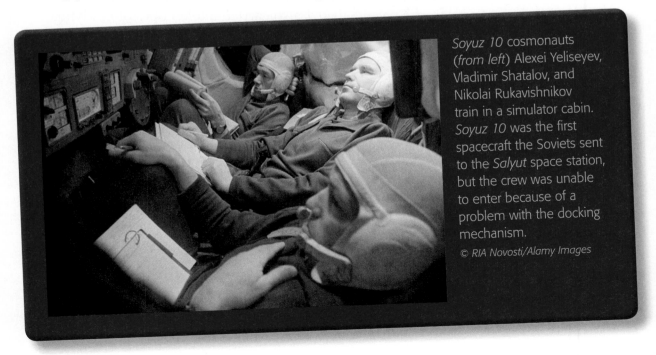

Soyuz 10 cosmonauts *(from left)* Alexei Yeliseyev, Vladimir Shatalov, and Nikolai Rukavishnikov train in a simulator cabin. *Soyuz 10* was the first spacecraft the Soviets sent to the *Salyut* space station, but the crew was unable to enter because of a problem with the docking mechanism.

© RIA Novosti/Alamy Images

Salyut (which means "salute" in Russian) was the name for this first series of space stations. The Soviets were actually running two space-station programs at the same time. There was a civilian one, and a military one known internally as *Almaz* ("diamond"). To confuse Westerners, however, the Soviets referred to both types of stations as *Salyut*.

The first *Salyut* stations included some notable failures. On 19 April 1971 the Soviets launched *Salyut 1* (which actually *was* a civilian space station) atop a *Proton* rocket. It was the first space station in history. It reached orbit unmanned. But the crew of *Soyuz 10*, the first spacecraft sent to the station, was unable to enter because of a problem with the docking mechanism. The second crew to attempt entry, that of *Soyuz 11*, succeeded. The crew members lived aboard the station for three weeks. But on their return, an air leak developed in their spacecraft. It proved fatal. Tragically, all three cosmonauts aboard died.

Three more space station failures followed. The stations either couldn't reach orbit or broke up in orbit before crews could get to them.

But then, during the years 1974–77, came a string of successes, two military space stations and a civilian one. They supported five crews among them.

Salyut's Main Purpose

When the Soviets launched *Salyut 1*, they said the mission's purpose was to test the space station's systems and to conduct scientific research and experiments. But in another sense, the *Salyut 1*'s principal mission was to study the effect on the human body of long trips into space.

The Experiments and Research Conducted From *Salyut*

Another important mission was astronomical observation. *Salyut 1* was to take photographs of Earth from Space. *Salyut 1* also carried a telescope for studying stellar spectra. A stellar spectrum is *the characteristic pattern of light that a star emits*. A star's spectrum is like its "fingerprint," and studying it helps scientists understand what materials it's made of.

An example of stellar spectra
Courtesy of NOAO/AURA/NSF

Salyut 1 included a greenhouse as well to analyze plant growth, and a camera and film plates to study cosmic rays. The cosmonauts had also planned to study gamma rays from the Sun. But their special telescope proved unusable. The cosmonauts performed engineering tests during their stay at the space station, too. These tests helped develop the next generation of space stations.

Skylab orbits over Brazil's Amazon River Valley, seen below.
Courtesy of NASA

The *Skylab* Space Station

When America entered the space station era a couple years after the Soviet Union, it named its first attempt *Skylab*. Mechanical troubles beset *Skylab* in the beginning. And its end came in a little different fashion from what NASA planners originally had in mind for it. But in between, it was a successful space station.

The First US Space Station

NASA launched *Skylab* into orbit around the Earth with a *Saturn V* rocket on 14 May 1973. The space station was unmanned.

It was not a picture-perfect launch. The pressures of the launch tore off a shield meant to protect against tiny meteoroids and to shade *Skylab* from the Sun. The takeoff also ripped off one of the space station's main solar panels. These were to generate electric power for *Skylab*. And it got worse: Debris from the damaged meteoroid shield pinned the remaining solar panel to the station's side. In this position, it couldn't capture solar energy. So *Skylab* was left seriously short of the power it needed.

As NASA officials scrambled to devise a way to save *Skylab*, ground controllers maneuvered it into a "Goldilocks" position ("just right") that let it collect just enough solar energy to function but not so much that it could overheat. This was an acceptable "holding" position until a crew of astronauts could get there to make repairs. Among the concerns NASA faced: If the Sun melted *Skylab*'s plastic insulation, the gases this would release would render it uninhabitable.

Skylab's first crew—Pete Conrad, Paul Weitz, and Joe Kerwin—launched on 25 May 1973. Like later *Skylab* crews, they traveled to and from the space station aboard *Apollo* spacecraft. They went to work at once, making repairs that provided some protection against the Sun while also allowing *Skylab* to generate solar electricity. They spent 28 days in orbit.

Skylab crew members Joseph Kerwin (*left*) and Paul Weitz get into their suits.

Courtesy of NASA

The second crew—Alan Bean, Jack Lousma, and Owen Garriott—made additional repairs to *Skylab* during their 59-day stint in orbit. This crew faced other major technical problems: a thruster leak that made it harder than expected to rendezvous with the space station, and then, once the crew were on board, a second thruster leak. The situation was serious enough that NASA began drawing up plans for a rescue mission. (This would come up as the *Columbia* disaster was under investigation years later. Some critics suggested that NASA should have attempted a rescue of the *Columbia* crew once the foam-strike damage had come to light.)

In the end, though, Bean, Lousma, and Garriott completed their mission as planned. *Skylab*'s final crew, Jerry Carr, Bill Pogue, and Edward Gibson, set another record for long-duration spaceflight. When they returned to Earth on 8 February 1974, they had been gone 84 days. No American would beat that record for more than 20 years.

Skylab's Two Main Purposes

Like the Soviet space stations, *Skylab* was meant to prove that humans could live and work in space in reasonable comfort over extended periods. *Skylab* also had a mission to expand solar astronomy beyond the observations that could be made from Earth. Despite the mechanical troubles at the start, NASA deemed *Skylab* a success in all respects.

The Scientific and Technical Experiments Conducted From *Skylab*

You may not think of something as simple as brushing your teeth as a scientific experiment. But in fact, aboard *Skylab*, all kinds of everyday activities were part of the scientific research program. NASA launched *Skylab* to help study the feasibility of long-duration space missions. And though they were free-falling in Earth orbit, and moving at 16,000 miles (25,600 km) per hour, astronauts aboard *Skylab* said everyday life there was really pretty normal.

Their day would begin at 6 a.m. (This was 6 a.m. Houston time; they didn't call Houston "Mission Control" for nothing.) Astronauts would check their Teletype machine to see what their orders for the day were. (Teletypes are devices similar to typewriters, except that someone from far away can send an electronic message to the Teletype where it prints out on paper as if someone were sitting right at the machine.) They would then use the restroom, weigh themselves, and eat breakfast.

The astronauts would rotate their daily science assignments every day. They took turns at solar observation and at serving as the "guinea pig"— the test subject—for various medical evaluations.

Free time was between 8 p.m. and 10 p.m. One of the astronauts' favorite activities was simply looking out the window. They also devised scientific experiments of their own—watching blobs of water react to microgravity, for instance.

Star POINTS

In one famous prank that astronauts played on Mission Control, they shocked the ground crew by (apparently) having Owen Garriott's wife, Helen, call down from the station. The crew in Houston sat stunned and confused until the *Skylab* crew burst into laughter and explained the joke. Garriott had recorded his wife's voice before liftoff, and enlisted the help of his voice link on the ground, Robert Crippen, to play along.

Some fun and games aside, *Skylab* was about serious science. It was the greatest solar observatory of its time, a lab for the study of microgravity, a medical lab, and an observatory for studying Earth. Most important, though, it was a home away from home for its astronauts—a first step on that long journey to other worlds.

In practical terms, *Skylab* led to new technologies. NASA had special showers, toilets, sleeping bags, exercise equipment, and kitchen facilities designed for the low-gravity environment of Earth orbit.

Star POINTS

Local officials in the Shire of Esperance, in Western Australia, fined NASA $400 for littering after *Skylab* deposited debris across their community, among other places, as it fell to Earth. On the 30th anniversary of *Skylab*'s return, a radio host in California raised money to pay the fine for NASA.

NASA originally planned that *Skylab* would stay in orbit for another eight to 10 years and even receive a visit from the space shuttle, then in development. But high solar activity at the time foiled that plan. Such activity tends to increase the atmospheric drag on satellites such as *Skylab*. That, in turn, leads to orbital decay—*a gradual reduction in the height of a satellite's orbit*. Two years before the first space shuttle flight could arrive and boost *Skylab* into a higher orbit, the empty spacecraft reentered Earth's atmosphere and broke apart on 11 July 1979. It scattered debris across a sparsely populated section of Western Australia and the Indian Ocean.

Astronaut Charles Conrad rides an exercise bike in *Skylab*'s crew quarters.
Courtesy of NASA

Showering on *Skylab*

NASA engineers built a compact shower assembly for use on *Skylab*. The shower remained stored on the floor when not in use. To take a shower, an astronaut would step inside a ring on the floor and raise a fireproof beta cloth curtain strung on a hoop. Then he would attach it to the ceiling. A flexible hose with a push-button shower nozzle could spray three quarts (2.8 liters) of water from the personal hygiene tank during each shower. The astronaut would then vacuum used water from the shower enclosure into a disposable bag and deposit it in the waste tank.

Crew on *Atlantis* took this photo on their approach to *Mir* during STS-74.
Courtesy of NASA

The *Mir* Space Station

After America's *Skylab*, the Soviet Union took the next notable step in space station technology. The *Mir* was Soviet Russia's third-generation space station, after the two groups of *Salyut* stations. When it was fully assembled, *Mir* was second only to the Moon as the heaviest object orbiting Earth.

Mir's 10-Year Assembly

Cosmonauts put *Mir* together piece by piece over 10 years, starting in 1986. It was under construction, in fact, for most of its time in space. The Soviets launched *Mir*'s core in 1986. Russian engineers based its construction on the last *Salyut* space station, but *Mir* had six ports instead of two. It weighed 20.4 tons.

The second piece of *Mir* to go up was the *Kvant 1* module. (*Kvant* means "quantum" in Russian.) The Soviet Union launched it on 31 March 1987 and attached the module to *Mir* on 12 April 1987. *Kvant 1* was small—only 11 tons—and carried astrophysics instruments along with life-support and attitude-control equipment. (Attitude refers to *a spacecraft's position, or angle, relative to the direction in which it is traveling.*)

Kvant 2 went up in 1989, with an airlock for spacewalks, solar arrays, and life-support equipment. The Soviets based this larger module (19.6 tons) on the spacecraft originally intended for the *Almaz* military space station.

The Russians added *Kristall* to *Mir* in 1990. It carried scientific equipment, retractable solar arrays, and a docking node that let it receive spacecraft weighing up to 100 tons.

On 20 May 1995 the Soviet Union launched *Spektr* ("spectrum" in Russian) on a *Proton* rocket from Central Asia. It carried four solar arrays plus some scientific equipment, more than 1,600 pounds of it from the United States. *Spektr's* special scientific mission was observation of Earth's natural resources and atmosphere. *Spektr* also served as *Mir's* "guest room," the living quarters for visiting American astronauts.

Priroda ("nature" in Russian) was the last science module the Soviets added to *Mir*. *Priroda's* primary purpose was to provide *Mir* with Earth remote-sensing capability. *Priroda* also contained hardware and supplies for several US-Russian science experiments. In November 1995 a US space shuttle (STS-74) delivered a docking module to *Mir*, to make it easier for the orbiters to dock there.

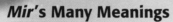

Mir's Many Meanings

Mir is commonly translated "peace"—as in the great Russian novel, *War and Peace*. *Mir* also means "world." But as Frank L. Culbertson, a NASA official who worked on the Shuttle-*Mir* Program once wrote, *mir* originally "meant what I think we would call a village, or even a commune, in the countryside, where all the local people lived in close or communal proximity to better share the limited resources of building supplies, food, child care, and, most critically in the harsh Russian winters, heat."

The Shuttle-*Mir* Program

The first major effort by Americans and Russians to work together in space was the *Apollo-Soyuz* Test Project in 1975. You read about this in Chapter 6. US-Soviet relations went through a rough patch after that and got much worse after the Soviets invaded Afghanistan in 1979. They remained poor until President Ronald Reagan met with Soviet President Mikhail Gorbachev in Geneva, Switzerland in 1985. A steady improvement began, and in 1987, the two countries signed an agreement to cooperate in space. They renewed this in 1992, after the Soviet Union's breakup.

The two countries fulfilled their pledge to work together in space with the Shuttle-*Mir* Program. As its name suggests, it involved the US space shuttle flying missions to the Russian space station *Mir*. The shuttle carried Russian cosmonauts to *Mir*, and an American astronaut flew there aboard a *Soyuz* spacecraft. As with other programs involving space stations, the focus was long-duration missions.

In February 1994 Sergei Krikalev became the first Russian to fly aboard the space shuttle. By then, plans for US-Russian cooperation in space included the International Space Station.

Many historic "firsts" took place as part of the Shuttle-*Mir* Program. In 1995 Norman Thagard became the first American aboard *Mir*, where he spent 115 days in orbit with his Russian colleagues. Later that year, on mission STS-71, *Atlantis* became the first shuttle to dock with *Mir*. From February 1994 to June 1998, space shuttles made 11 flights to the space station. American astronauts spent seven residencies aboard *Mir*. Shuttles also conducted crew exchanges and delivered supplies and equipment.

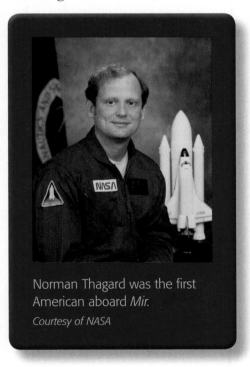

Norman Thagard was the first American aboard *Mir*.

Courtesy of NASA

The Experiments Conducted From *Mir*

Mir served as a floating laboratory. It's estimated that astronauts carried out some 23,000 scientific and medical experiments there during the space station's time in orbit. Some of these were purely Russian experiments. Others were collaborations with Americans, within the framework of the Shuttle-*Mir* Program.

The records of shuttle missions to *Mir* include references to hundreds of pounds of equipment and data samples being transferred from the shuttle to *Mir* and back again.

Between March 1995 and May 1998 NASA and Russian scientists carried out experiments meant to answer questions about how plants and animals (including humans) function in space and how the Solar System developed. Their work also addressed questions of how to build better technology in space and how to build future space stations.

Kvant 2, for instance, contained an airlock. It also had equipment for taking pictures of Earth. It allowed cosmonauts and astronauts to step outside for spacewalks, and let them gather biotechnology research data. The airlock also made it possible to research the effects of space exposure on electronics and construction materials.

Precautions Taken in Controlling *Mir*'s Deorbit

A spacecraft's launch is magnificent. But, as with *Skylab*, bringing an empty worn-out space station back to Earth is considerably less grand. It has to be done, though. Anything in low Earth orbit needs continual boosting up higher, or else gravity will eventually drag it back down to Earth.

Star POINTS

Over time *Mir* got a reputation as somewhat accident prone, even as it racked up years of service. At one point in February 1997, a 15-minute equipment fire imperiled the space station. Then came problems with attitude and environmental controls, computer malfunctions, and power outages. In June 1997 a visiting supply vehicle breached *Spektr*'s hull and rendered the module uninhabitable. But *Mir*, and its crews, endured.

So it was with *Mir* after 15 years or so. Russia's space agency was looking ahead to the International Space Station, which would require most of its attention and money. Members of the Russian parliament mounted a rear-guard action to keep *Mir* in orbit. But on 30 December 2000 Russian Prime Minister Mikhail Kasyanov signed a resolution to sink *Mir* in the ocean by early 2001. The Russian space agency guided *Mir* flawlessly back to Earth on 23 March 2001.

By then the Russians had built up considerable experience bringing space stations back to Earth. They had already managed to deorbit—*to cause to go out of orbit*—80 *Progress* spacecraft and five *Salyut* space stations since 1978. *Mir* was bigger than any of these—but the technique needed was the same.

Russian authorities were fairly certain any *Mir* debris would land in the ocean. But one official was quoted saying, "We don't have a 100 percent safety guarantee."

Mir's path took it over areas of Earth as far north as Alaska's Aleutian Islands and as far south as the southern Andes of southern South America. Pieces of Russian space litter had already crashed down upon Canada, Australia, and southern South America—fortunately with no casualties or damages. But Russia took out insurance in case *Mir* caused some damage. The Japanese kept close watch, too, because *Mir* would sweep over their country during its final orbit.

In the end, though, after more than 86,000 orbits, *Mir* reentered Earth's atmosphere on 23 March 2001. Its larger pieces sizzled harmlessly into the South Pacific, about 1,800 miles (2,897 km) east of New Zealand. Observers on Fiji reported spectacular gold and white streaming lights. A highly successful space program had ended safely, going down in a blaze of glory.

STS-79 crew aboard <u>Atlantis</u> recorded this shot of <u>Mir</u>.
Courtesy of NASA

Astronaut Dave Williams continues construction on the ISS during STS-118's first spacewalk.
Courtesy of NASA

The International Space Station

The different individual Russian and US space stations eventually led to a true joint project, a step beyond the Shuttle-*Mir* Program: the International Space Station. It is a venture born of budget cuts and a new spirit of cooperation between Russia and the United States since the end of the Cold War. Russia had intended to follow up its highly successful *Mir* station with *Mir 2* in the 1990s. The United States had planned to build a station called *Freedom*, in cooperation with space agencies in Europe, Canada, and Japan.

But by the early 1990s space planners were having trouble finding funds for either project. And the idea of cooperating in space had taken root in both the United States and Russia. So in 1993 the two countries agreed to build one big station together: the International Space Station.

Construction of the International Space Station

Like many a real estate project, the International Space Station (ISS) has experienced some delays. The United States and Russia plan for it to remain in orbit until at least 2016. NASA's 2011 budget request, were Congress to approve it, would extend the ISS to 2020. It will have been under construction for most of its useful life.

At this writing, the ISS is still under construction, over budget, and behind its original schedule, but expected to be finished in 2011. Estimates are that by the time it's complete, the project will have taken 80 flights of US space shuttles and Russian rockets to assemble. The Japanese and Europeans will contribute with supply vehicles launched on the ESA's *Ariane 5* and Japan's H-2A booster rockets. These partners expect their contributions to require another 20 flights.

NASA and its teammates calculate that wrapping up ISS construction will take more than 140 spacewalks, totaling nearly 800 hours. This is more spacewalking time than in all of US space history before station construction began.

The first piece of the ISS to go up was a Russian-built, US-funded module called *Zarya*, or "sunrise." (A less poetic name is "functional cargo block," known by the Russian initials FGB.) The Russians launched *Zarya* on 20 November 1998. Two weeks later, the *Unity* module lifted off from Florida. The former Cold War rivals had reached new levels of cooperation in space when they put those first two modules of the ISS together.

Nations Working Together

The United States, through NASA, has the lead in the ISS project. But it has 15 other countries as partners. As noted, Russia has a major role. Eleven members of the European Space Agency (Belgium, Britain, Denmark, France, Germany, Italy, the Netherlands, Norway, Spain, Sweden, and Switzerland) are involved. So are Japan, Canada, and Brazil.

Canada's special expertise is in building robotic arms (the so-called Canadarm systems) that have been a mainstay of space exploration since the early 1980s. Brazil has a special contract with NASA. In exchange for providing equipment to the United States, Brazil will have access to US equipment, as well as permission to send a Brazilian astronaut to the station. Some other countries have sent up experiments to the ISS, too.

Stephen Robinson rides the Canadarm2 at the International Space Station during mission STS-114.
Courtesy of STS-114 Crew/ISS Expedition 11 Crew/NASA

The International Space Station's Elements, Systems, and Facilities

NASA has likened the construction of the ISS to building an enormous ship while the ship is already at sea. It has been one of the most extraordinary construction projects in human history. When complete, the ISS will be larger than a five-bedroom house. Its cabin volume will be 33,023 cubic feet (935 cubic meters), and its mass will be 1 million pounds. It will be four times as big as *Mir.* The ISS will measure 354 feet (108 meters) by 290 feet (88 meters). It will have almost an acre of solar panels to produce electrical power.

The ISS will include six state-of-the-art laboratories and living quarters as well. It will also have systems for data processing, communication, navigation, and guidance, plus utilities such as electricity supply, heating and cooling, and life support.

The ISS orbits Earth from about 221 miles (356 km) up. Its orbit is angled at 51.6 degrees to Earth's equator. This orbit gives excellent coverage for observations of Earth. The ISS flies over 85 percent of the globe and 95 percent of Earth's population.

While space science has warmed up relations between the Americans and Russians over the decades, political differences remain to this day. In fact, tensions have increased between the two countries in recent years. The issues range from conflicting views on missile defense systems to an expanding European Union and NATO that push ever harder against Russia's borders. Whether any of this will get in the way of continuing cooperation on the International Space Station—and whatever may one day replace it—is yet to be seen. But one thing is certain: Both countries have deep scientific communities that will press—either alone or together—to venture ever farther into space.

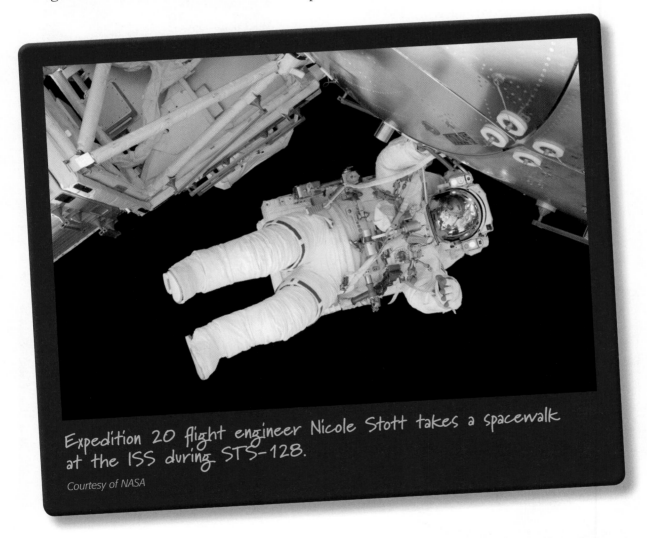

Expedition 20 flight engineer Nicole Stott takes a spacewalk at the ISS during STS-128.

Courtesy of NASA

Some ISS Astronauts

Many American astronauts, men and women, have spent time aboard the ISS. The following short biographies are just a sampling of the astronauts showing their different backgrounds and duties.

Edward Michael Fincke

Edward Michael Fincke is a Pittsburgh native. He graduated from the Massachusetts Institute of Technology, which he attended on an Air Force ROTC scholarship. After MIT, he studied cosmonautics at the Moscow Aviation Institute in the former Soviet Union. He continued his education at Stanford University, and then went on to pick up two additional degrees. An Air Force colonel, he has flown on 30 different types of aircraft. Selected as an astronaut in April 1996, he has served as part of two ISS expeditions. At this writing NASA has him scheduled for a third. As of August 2009 he had logged an entire year in space.

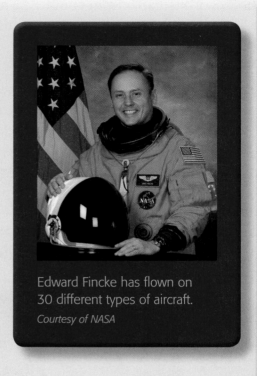

Edward Fincke has flown on 30 different types of aircraft.

Courtesy of NASA

But the wider public probably knows Fincke best as the astronaut who roots for the Pittsburgh Steelers. In February 2009, on the eve of Super Bowl XLIII, Fincke encouraged his team by waving his yellow "Terrible Towel," just like the ones Pittsburgh fans wave at Heinz Field, in front of a television camera aboard the ISS. Fincke, who later gave the "space towel" to the Steelers' team president, was part of NASA's Hometown Heroes program. Its purpose is to return astronauts to their hometowns to build support for space exploration and foster young people's interest in career opportunities in science.

PS: Pittsburgh won the Super Bowl, defeating the Arizona Cardinals 27–23.

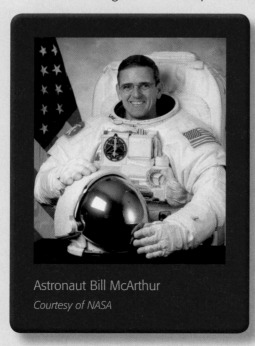

Astronaut Bill McArthur

Courtesy of NASA

Bill McArthur

A North Carolinian by birth, William Surles McArthur graduated from West Point in 1973. He became an Army flier in 1976. He served in South Korea and in Savannah, Georgia. After additional study at Georgia Tech, he returned to West Point as a professor. Following more specialized training, he was first assigned to NASA in 1987.

McArthur became an astronaut in July 1991. He is a veteran of four space missions, three on the space shuttle and the fourth aboard the International Space Station. As the commander and science officer of the space station's *Expedition 12* crew, he was in space from 30 September 2005 to 8 April 2006.

CHAPTER 8 Space Stations and Beyond

Unit 3: Manned and Unmanned Spaceflight

Sandra Magnus

Sandra H. Magnus is a native of Belleville, Illinois. She studied at the University of Missouri-Rolla and Georgia Tech. She spent several years working for the McDonnell Douglas Aircraft Company as a stealth engineer, working on ways to keep US military aircraft hidden from enemies. Chosen an astronaut in April 1996, she began two years of training and qualified as a mission specialist.

Her first spaceflight was a space station assembly mission. She flew as part of the crew of STS-112 in 2002. Shuttle *Atlantis* delivered, and crews installed, the third piece of the station's 11-piece integrated truss structure—its main framework. Magnus operated the space station's robotic arm during three spacewalks required to outfit and activate the new component.

Later she flew on STS-126 to the ISS, where she had a 4 1/2-month stay. She served as flight engineer No. 2 and science officer. She helped upgrade the ISS so that it could support a six-member crew.

Sandra Magnus once worked as a stealth engineer.
Courtesy of NASA

Like Mike Fincke, Magnus has been involved in the Hometown Heroes program. Her participation involved going back to one of her alma maters, Georgia Tech, to speak to women engineering students on campus. She told them how her return to higher education after the phaseout of her program at McDonnell Douglas led to her fulfillment of a childhood dream—acceptance into the astronaut corps.

Astronaut Tracy Caldwell
Courtesy of NASA

Tracy Caldwell

Tracy Caldwell Dyson is a Californian, with chemistry degrees from Cal State University, Fullerton, and the University of California at Davis. After NASA picked her to be an astronaut in June 1998, she reported for training in August. The following year the space agency assigned Caldwell to the Astronaut Office ISS Operations Branch as a "Russian Crusader." This meant she took part in the testing and integration of Russian hardware and software products developed for the ISS.

In August 2007 she flew aboard *Endeavour* on a 13-day mission to the ISS. On the mission, STS-118, Caldwell operated the robotic arm to help maneuver new components of the ISS into position. She also was the "IV" or "intravehicular" crew member, the one who stayed inside to direct other members' spacewalks.

Clay Anderson

Although NASA receives more applications for the astronaut corps than it can ever accept, the agency encourages those who aren't accepted at first to reapply. Clayton Anderson followed this advice—more than once, in fact.

Anderson, a Nebraska native, had wanted to be an astronaut since he was 4 or 5 years old. He did wait until 1983, when he was an adult, with a master's degree and a year of work experience, before he applied. And when that application wasn't accepted, he applied again—15 times in all before getting an interview. Then he waited two more years before NASA selected him in 1998.

Anderson had established contact with NASA early on. A guidance counselor at his college introduced him to an alumnus who worked for NASA. That led to Anderson's getting a summer internship at NASA that resulted in a permanent job. That was the second-best thing to being an astronaut, Anderson said.

But he kept updating his resume, looking for ways to shine in his job at the Johnson Space Center in Houston and be picked as an astronaut. Then finally something changed. He got an interview and NASA eventually accepted him into the program.

And after nine more years, Anderson actually flew. On 8 June 2007 he launched aboard *Atlantis* with the crew of STS-117. Arriving at the ISS, he assumed the roles of flight engineer and science officer. He was aboard the ISS for 152 days, during which he made three spacewalks.

Astronaut Clay Anderson
Courtesy of NASA

✔ CHECK POINTS

Lesson 1 Review

Using complete sentences, answer the following questions on a sheet of paper.

1. What was the significance of *Salyut 1*?

2. What was *Salyut 1*'s principal mission, and what did the Soviets say its main purpose was?

3. Why did the *Salyut 1* crew want to study stellar spectra?

4. What job did *Skylab*'s first crew tackle right away?

5. What were *Skylab*'s two main purposes?

6. What new technologies did *Skylab* lead to?

7. How did cosmonauts construct *Mir* and over what time frame?

8. Name a couple of "firsts" associated with the Shuttle-*Mir* Program.

9. How did *Kvant 2* figure into *Mir*'s research program?

10. Why were the Japanese particularly worried about *Mir* in March 2001?

11. How much spacewalking time is the International Space Station expected to require to complete? How does this compare with the amount of time American astronauts had spent walking in space before ISS construction began?

12. What special expertise does Canada bring to the ISS project? What other nations are working on the ISS project?

13. How big will the ISS be when it's complete?

💡 APPLYING YOUR LEARNING

14. When the International Space Station has outlived its usefulness, do you think the participants should replace it with another station? Why or why not?

Quick Write

_____

_____

How do you think Percival Lowell's promotion of the idea of "canals" on Mars might have influenced space exploration efforts?

Learn About

- the planned return trip to the Moon
- the plans for a Moon outpost
- the plans for a manned mission to Mars

Years before there was NASA, or the Apollo program, or the space shuttle, there was Percival Lowell. In Chapter 3, you read about his connection with the effort to find Pluto. But Mars was his first love. A headline in the *Boston Globe* would one day call him "The Man Who Invented Mars."

Lowell lived from 1855 to 1916. According to the *Globe*, he was a Boston Brahmin—a wealthy, well-educated businessman. He devoted much of his time, money, and energy to proving that Mars had intelligent life. He looked through his telescopes, among the best of their day, and thought he saw canals. If there were canals, there had to be canal builders, didn't there?

His fascination with astronomy began when he received a telescope as a gift when he was boy. But he was really "hooked" after he discovered the writings of the Italian astronomer Giovanni Schiaparelli. The director of the Milan Observatory, Schiaparelli described the *canali* he saw on Mars. The Italian word simply meant "channels," and referred to naturally occurring cuts on the planet's surface.

But Lowell understood them as canals, perhaps naturally so. After all, the *Globe* pointed out, he lived in a canal-building age. His family had made much of their money in the textile business in Lowell, Massachusetts, a city crisscrossed with industrial canals. The Suez Canal in Egypt had opened in 1869, reenergizing efforts to build a canal across Panama.

Lowell built an observatory in Flagstaff, Arizona. It was far from the East Coast's bright lights and often cloudy weather. This made it all the better to get a good look at Mars during the winter of 1894–95, when Mars was closer to Earth than usual.

Lowell was a tireless popularizer of his ideas. He published books and articles in prestigious magazines. And he lectured. He was a commanding presence, handsome and intense.

Lowell was never able to prove the existence of canal-building Martians. But his idea that "we are not alone in the Solar System" captured the popular imagination. Some say it has influenced space exploration efforts ever since.

Lowell died of a stroke in 1916. He is buried in Flagstaff in a domed mausoleum built to look like an observatory.

In the early years of the twenty-first century, the Mars *Opportunity* rover found evidence that at some point in the past, Mars had been "soaking wet." Where there was water, there was life, or at least potential for life.

Soon after NASA announced the *Opportunity* findings, someone left a glass of champagne, with a note, at Lowell's tomb. The *Globe* reported that the note contained a quotation from a play by the ancient Greek poet Euripides: "Far away, hidden from the eyes of daylight, there are watchers in the sky."

Vocabulary

- geodetic
- contour map
- albedo
- regolith
- volatile
- fission

The Planned Return Trip to the Moon

On 14 January 2004 President George W. Bush gave a speech in which he announced a new "Vision for Space Exploration." He said the United States would "finish what it started," by completing the International Space Station. Next, the country would develop a new manned space vehicle, a successor to the *Apollo* spacecraft. And third, America would "return to the Moon by 2020, as the launching point for missions beyond."

The speech raised far more questions than it answered. But the president was firm: "We do not know where this journey will end, yet we know this: Human beings are headed into the cosmos."

As this book was written, it was far from certain that the United States could return an astronaut to the Moon by 2020. President Barak Obama has recommended canceling the *Constellation* program—including the *Ares* launch vehicle and the *Orion* crew capsule under development. Whether Congress would go along with this or maintain the original program was also not clear. But whenever missions to the Moon and Mars take place, they will involve the issues and technologies this lesson discusses.

The Development of the 2006 Global Exploration Strategy (GES)

Sometimes when you decide to do something big—to plan a big trip, for instance—it's a good idea to sit down and think deeply about it. Your decision may be intuitive. In other words, it may be the kind of decision that comes to you all at once and just feels right. But still it can be good to flesh out your thinking about it. In the case of a trip, you may think about what you really want to get out of it, how to prepare for it, what to bring, what other destinations to include on the itinerary, and so forth.

Returning to the Moon is a big trip that's worth careful thought. To prepare for this mission, NASA developed a program of lunar exploration themes and objectives. The agency was trying to figure out its goals for a trip to the Moon. NASA refers to this planning program as the Global Exploration Strategy. And to put together this strategy, NASA gathered the ideas of people from all around the world and from many different stakeholder groups. (*Stakeholder* originally meant someone who owned part of a business. Nowadays people use stakeholder to mean anyone who is involved in or affected by a certain course of action.)

NASA's process for developing its Global Exploration Strategy (GES) began in April 2006 with a workshop lasting several days in Washington, D.C. As the workshop was under way, NASA sent out a formal request for information to other parties whose views the agency wanted. These included, for instance, all the NASA centers, such as the Goddard Space Flight Center and the Kennedy Space Center, as well as the Space Commerce Roundtable, a private organization seeking to develop business opportunities in space.

CHAPTER 8   Space Stations and Beyond

Space Agencies Taking Part in the GES

These bodies all contributed ideas to NASA's Global Exploration Strategy:

- The British National Space Centre
- The Canadian Space Agency
- The Chinese National Space Agency
- The Commonwealth Scientific and Industrial Research Organization (Australia)
- The European Space Agency
- The French National Space Agency
- The German Aerospace Agency
- The Indian Space Research Organization
- The Italian Space Agency
- The Japanese Aerospace Exploration Agency
- The Korean Aerospace Research Institute
- The National Space Agency of Ukraine
- Roscosmos (the Russian space agency).

After completing a rough draft of its themes and objectives, NASA sought ideas from 13 other space agencies around the world. These agencies had expressed interest in taking part in a trip to the Moon.

As it brought everyone's ideas together, NASA chose not to identify the source of each idea. Nor did the agency edit the contributions to bring them into line with its own policies or plans. The GES isn't a fixed program, either. As it moves forward, NASA will keep asking others in the global space community for their input.

The strategy, by the way, doesn't represent US or NASA policy. It doesn't commit the United States to any particular activities. But the GES does represent a global consensus as to the value of lunar exploration.

The Moon and Earth from space
Courtesy of NASA

Unit 3: Manned and Unmanned Spaceflight

Figure 2.1 An illustration of a futuristic human base camp on the Moon
Courtesy of NASA

Benefits Expected From Lunar Exploration

Nearly 200 lunar exploration objectives emerged from this idea-gathering process. They represent answers to the question, "Why should people return to the Moon?" NASA grouped the answers into six broad themes. These areas of pursuit define, in the world's eyes, the value of going to the Moon. Here they are, with a little explanation of the benefits lunar exploration could provide (Figure 2.1):

- *Human civilization:* The human presence should be extended to the Moon to make eventual settlement there possible.

- *Scientific knowledge:* A return to the Moon will help answer fundamental questions about the history of the Earth, the Solar System, and the universe, and humanity's place there.

- *Exploration preparation:* The Moon is a logical "base camp" for further manned spaceflight throughout the Solar System, notably to Mars. Trips to the Moon will offer opportunities to test new techniques and technologies close to home, too.

- *Global partnerships:* Space exploration can provide a challenging, but peaceful, shared activity among different nations.

- *Economic expansion:* The exploration of the Moon should lead to economic growth at home, including quality-of-life improvements.

- *Public engagement:* A vibrant space exploration program can capture the public imagination. It can also help develop the high-tech workforce needed to face the challenges of tomorrow. Not everyone will be an astronaut. But a strong space program can draw bright students into the study of math, physics and other sciences, engineering, and technology.

CHAPTER 8   Space Stations and Beyond

Suni Williams

People take many different routes to NASA's astronaut corps. Sunita Williams, you might say, arrived by helicopter.

Born in Ohio to parents who had emigrated from India, Williams considers Needham, Massachusetts, her hometown. She went to the US Naval Academy and graduated in the middle of her class. But she didn't do well enough to train as a diver, as she wanted. So she went for her second choice: flight school. Again, she did well, but not quite well enough to get one of the spots for a female jet pilot. So she opted for a third choice: becoming a helicopter pilot instead.

Once Williams tried it, she loved it. She became the test pilot for her squadron, and then went on to test pilot school. That's how she got her first glimpse of NASA. She visited the Johnson Space Center. There she heard former astronaut John Young speak.

"He talked to us about landing on the Moon and told us that they had to learn how to fly helicopters to do the lunar landing," Williams said in an interview. Up to that point, she didn't know much more about space travel than what she'd picked up from TV shows she'd watched as a child. "So that was the first time I was really like, 'Wow— maybe if I really want to do this, I should apply. Maybe we'll go back to the Moon.' "

NASA picked her for its astronaut corps in 1998. She has served as a flight engineer on the International Space Station (2006–2007), setting records for the longest time in space for a woman (195 days), and greatest number of spacewalks by a woman (four—a record broken in 2008).

Expedition 15 flight engineer Sunita Williams exercises aboard the International Space Station in 2007. She attended the US Naval Academy and eventually learned how to pilot a helicopter.

Courtesy of NASA

The Plans for a Moon Outpost

Important parts of any plan to return to the Moon are the *Lunar Reconnaissance Orbiter* (LRO) and the work of the Lunar Architecture Team.

Figure 2.2 A contour map of the Moon used for *Apollo 16*

Courtesy of NASA

The *Lunar Reconnaissance Orbiter* (LRO)

The LRO was the first mission in the Vision for Space Exploration. NASA launched it on 18 June 2009 from Florida aboard an *Atlas V* rocket for a yearlong exploration of the Moon. The LRO's objective is to conduct the preliminary surveys that will make a human return to the Moon possible.

The LRO is to provide data for day-night temperature maps of the Moon. It will capture data for a global geodetic grid. (Something that's geodetic is *related to measuring the Earth's shape, or the shape of another celestial body*.) That is, the LRO will take the measurements needed to produce a contour map—*a map that shows an area's different elevations*—of the entire lunar surface (Figure 2.2). This map would include a grid for navigating on the Moon. The LRO will also gather data for high-resolution color imaging and information on the Moon's ultraviolet albedo—*its reflecting power, expressed as a ratio of reflected light to the total amount falling on the surface.*

The LRO's particular emphasis, though, will be on the Moon's polar regions. NASA scientists suggest that sunlight shines on these areas all the time. Water may also be present.

The Work of the Lunar Architecture Team

The next logical phase in NASA's return to the Moon is for engineers to come up with ways to make extended stays possible once the new space fleet gets there. "Architecture" usually suggests buildings. But the Lunar Architecture Team isn't charged with designing just a Moon dwelling for space travelers. To this team, architecture includes designs for rovers—the wheeled vehicles that you first read about in Chapter 3 that can travel around the surface of a moon or planet. The team's job involves spacesuits as well. There's plenty of engineering that has to go into those suits. They're not what the salesclerks at your local clothing store would call "unstructured" garments!

NASA planners have their eye on a site known as Shackleton Crater, near the Moon's south pole, as a possible location for their lunar outpost. The rim of this crater is near both permanently shadowed regions and peaks that are in sunlight for much of the year. This would be a good location for solar power arrays to provide electricity to the outpost. The crater may also harbor water ice, which could be an important source of water for a base there. Planners don't intend to limit their options, though. And so the architecture team is designing hardware that would work at any of a number of locations on the Moon.

The original idea for building an outpost on the Moon was to send up smaller elements and have the astronauts assemble them on site. But the Lunar Architecture Team has found that sending up larger elements—more preassembled modules— would help get the outpost up and running more quickly. The architecture team is also considering a mobile habitat module—a sort of lunar RV. This would let a couple of astronauts move away from base camp, for instance, to explore other places on the Moon as mission needs dictate.

Small pressurized rovers are another idea NASA is considering. These would go out in pairs, two astronauts in each, and could venture nearly 125 miles (200 km) from the main outpost for research or other purposes. If one rover broke down, astronauts could ride "home"—back to the outpost—in the other.

NASA says that the rovers should provide a "shirtsleeve environment." That means that astronauts don't have to wear spacesuits inside them. Rather, they would "step in" to spacesuits attached to the rovers' exterior. Do you see why NASA calls this "architecture"?

The Lunar Architecture Team's lunar rover

Courtesy of NASA/Sean Smith

The Plans for a Manned Mission to Mars

Current thinking is that once NASA sets up its lunar outpost, it will turn its attention to Mars. At a workshop entitled "Why Mars," held in August 1992 in Houston, Texas, a team of consultants laid out six arguments in favor of human exploration of Mars. You may find them similar to the arguments for a return to the Moon:

- *Human evolution:* Mars is the next step toward expanding the human race into the stars.

- *Comparative planetology:* Understanding Mars will help us understand Earth better.

- *International cooperation:* A joint effort to explore Mars could help the cause of global unity.

- *Technological advancement:* New technologies would inevitably come out of a mission to Mars. These will improve the lives of people on Earth and encourage high-tech industry.

- *Inspiration:* A mission to Mars will test human capacities to the limit. A population mobilized in service to such a mission would be an inspiration to later generations.

- *Investment:* The cost of a mission to Mars is reasonable in comparison with other social spending.

Star POINTS

One of the arguments for sending people to Mars, and not just more robots, is that the red planet is a very complex environment—much more so than the Moon, for instance. And so it would be worth it to bring the greater intelligence of human explorers to bear. What NASA's robotic rovers take all day to do, a human explorer could accomplish in about a minute.

Building a New Fleet of Spacecraft

NASA's Web sites and other public communications are full of excitement about future missions to Mars. But there's not much discussion about building a new fleet of spacecraft to get there. It's likely that whatever spacecraft eventually goes to Mars will be adapted from whatever program NASA uses to return to the Moon. NASA compares these much shorter missions to using a brand new car to run errands in town before taking it out on a long road trip.

The Greatest Challenges of a Manned Mission to Mars

The challenges of a trip to Mars would include the concerns you read about in Chapter 5. These include the physical effects of long-term exposure to high-energy cosmic rays and other forms of radiation. The low-gravity, low-light environment of a long space mission would also pose some problems for astronauts.

Some of the trials would be psychological: a sense of isolation from Earth and from loved ones back home during the two and one-half years it would take to get there and back. On the other hand, astronauts bound for Mars would also face the test of too much company of people they couldn't get away from: their crewmates. (Even your best friend can get on your nerves sometimes.) A Mars-bound crew would surely include at least one physician, but the astronauts would have to face the reality of having only limited medical facilities.

Other challenges to a mission to Mars would be political and economic. A journey to another planet would take a long time just in terms of actual space travel. It would also take time in terms of planning and budgeting. But the US Congress—which ultimately has to approve NASA's budget every year—can change course every two years. The space program has always been at the mercy of the changing political winds. A mission to Mars will require sustained political and financial support.

Star POINTS

The orbits of Earth and Mars produce a 15-year cycle divided into seven launch windows. About every 26 months, when the two planets get closer to each other, a launch window opens up. A roundtrip mission to Mars would take about two and one-half years—six months to travel to Mars, about 500 days on Mars, and then six months to return home.

Endeavour's crew is hard at work in the onboard Spacelab-J module. This photo shows the types of crowded conditions astronauts would face on a long journey to Mars.

Courtesy of NASA

New Technologies to Support the Mission

NASA says that sending humans to Mars "currently lies on the very edge of our technological ability." A successful Mars mission would be a milestone achievement, and a testament to the possibilities that technology presents to civilization. Here are some of the technologies under development at NASA to support missions to the Moon and ultimately to Mars:

Structures, Materials, and Mechanisms

This NASA project develops technologies to build lightweight vehicles and dwellings. The project team is also working on low-temperature devices—equipment that works where it's very cold. Technology for Mars will rely on lightweight composites and inflatable structures to reduce the amount of mass that has to be launched into space.

Protection Systems

This team focuses on heat and dust. It's developing a heat shield to protect the next crew vehicle, as well as protection systems for equipment that will land on the Moon and Mars. Both these bodies are full of dust—dust that is essentially tiny bits of broken glass. Regolith is the term for this *top layer of silt-fine dust.* NASA is working to figure out how to protect astronauts and their equipment from this threat.

Nontoxic Propulsion

Fuels are usually dirty or even poisonous. But NASA has a team looking for nontoxic—nonpoisonous—propulsion systems. The idea is that nontoxic propellants may take some effort to develop but will be much easier to handle.

Energy Storage and Power Systems

If you've ever had a flashlight fail on a camping trip, you have some flicker of an idea how important batteries will be on future space missions. NASA has a team working on advanced lithium-ion batteries and regenerative fuel cells for energy storage. These systems will make it possible to store solar energy during the day for use at night, and provide power for mobile systems such as rovers, too.

Thermal Control for Surface Systems

This project is developing heat pumps, evaporators, and radiators to make sure the capsule stays the right temperature at all times. Scientists are also working on thermal control for systems for use on the lunar surface, and eventually on Mars, too: habitats, power systems, and spacesuits.

Avionics and Software

In Chapter 5 you read about the danger that radiation poses to electronics, including the computers that would guide spacecraft to their destinations. This project concentrates on "radiation hardened" electronics and reliable software. It's also working on systems that will function in extreme cold.

Environmental Control and Life Support

This project works on ensuring that astronauts have enough air to breathe and water to drink, as well as safe ways of handling their wastes—including the carbon dioxide they breathe out. The team also designs systems to ensure safe and comfortable "room temperatures" for astronauts. It's developing special monitoring instruments for testing on the International Space Station.

Crew Support and Accommodations

This project is at work on an advanced extravehicular activity (EVA) suit. Astronauts will need such a suit for walking around on the Moon or Mars. The current suit used on the space shuttle and the International Space Station is too heavy and constrictive to allow astronauts to move freely.

ISS Research and Operations

This project conducts basic microgravity research in biology, materials, fluid physics, and combustion aboard the International Space Station. It takes advantage of the ISS as a specialized lab in space.

In-Situ Resource Utilization

"In-situ resource utilization" means, essentially, "making the most of what you've got on the scene." "In situ" is a Latin phrase that means "in place." Up to now, space travel has required astronauts to bring everything with them—even the bacteria-free "space kimchi" that you read about in Chapter 6. But now NASA is beginning to work on technologies that would let astronauts draw on resources available on the Moon or on Mars. These include producing oxygen from regolith, and collecting and processing lunar ice and other volatiles. A volatile, or volatile chemical, is *a substance that readily changes from solid or liquid to a vapor.* Scientists expect use of in-situ resources to help cut the amount of water, oxygen, and rocket fuel that NASA must ship in from Earth to the Moon or to Mars.

Robotics, Operations, and Supportability

Whatever the advantages of human over robotic exploration, robots will certainly be part of the program. Robots can move over rough terrain and can help astronauts build and maintain their lunar outpost.

Fission Surface Power Systems

Solar power can be an effective source of energy on the Moon and on the way to Mars. As pointed out earlier, that's what makes the Shackleton Crater such a prime location. But Mars is a different story. The farther you get from the Sun, the less effective solar arrays are. Mars is 1.5 times farther from the Sun than Earth, and the intensity of sunlight there is only about half that on Earth. This means you need bigger solar arrays to generate the same amount of power you get near Earth. Solar power works on Mars, but nights there are very long and require storing energy in batteries. In addition, dust in the atmosphere can degrade solar cells' operation.

By the time you get to Jupiter, the intensity of sunlight is only $^{1}/_{25}$th that in Earth orbit. For this reason, spacecraft traveling away from the Sun to the Jovian planets and beyond—such as the *Voyagers*—need nuclear power.

Star POINTS

Fission is the opposite process from what you read about in Chapter 3, which explained that the Sun generates massive amounts of energy from nuclear *fusion*, the combining of atoms.

Likewise, space scientists have long thought in terms of nuclear fission to meet the energy needs of long stays on the Moon or Mars. Fission is *the splitting of an atom's nucleus*. It creates huge amounts of energy in the form of heat. This heat is then converted to electricity, much as happens in a nuclear power plant here on Earth. NASA is working with the Department of Energy on what the space agency calls "affordable nuclear fission surface power systems" for its Moon and Mars missions.

For Mars in particular, fission would be a better power source—if engineers can find a way to transport the equipment there in a cost-effective manner.

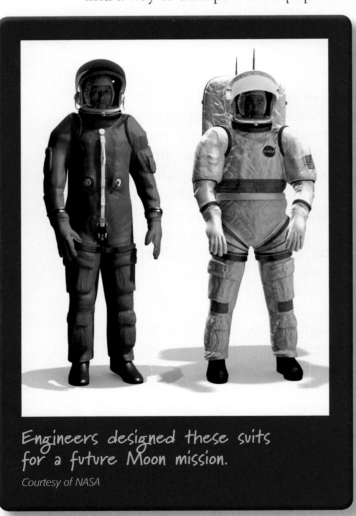

Engineers designed these suits for a future Moon mission.

Courtesy of NASA

While human beings have already achieved tremendous technological leaps that nineteenth- and twentieth-century science fiction writers only dreamed about, scientists have even greater steps ahead—some of them as yet unimagined by even novelists and filmmakers. NASA takes the process a step at a time. That's why the US space agency is first calling for a return to the Moon. Then a lunar outpost. And only then a journey to Mars. Just as you must master the subjects in one grade before you may advance to the next, so science gathers knowledge in increments. This is how to build a solid foundation—or a launch pad—to the next phase of exploration.

CHECK POINTS

Lesson 2 Review

Using complete sentences, answer the following questions on a sheet of paper.

1. From whom did NASA seek ideas after it had completed a rough draft of its Global Exploration Strategy?

2. Why is a return to the Moon a logical first step toward further exploration of the Solar System?

3. NASA launched the *Lunar Reconnaissance Orbiter* on 18 June 2009 with what mission objective?

4. How has the original idea for building an outpost on the Moon changed?

5. From what program will the rocket to go to Mars likely be developed?

6. What two kinds of psychological challenges would astronauts on a mission to Mars probably face?

7. What is in-situ resource utilization and how does it represent a change in space travel?

APPLYING YOUR LEARNING

8. Do you find the arguments for the exploration of Mars presented in this lesson persuasive? Why or why not?

Unit 3: Manned and Unmanned Spaceflight

The Lunar Reconnaissance Orbiter witnesses an "Earthrise" in this painting of a satellite orbit low over the Moon's surface. In 2009 NASA launched LRO to investigate the Moon's atmosphere, environment, and terrain.

Courtesy of Chris Meaney, NASA's Conceptual Image Lab

The Unmanned Missions of Space Probes

Chapter Outline

> Exploration is in our nature. We began as wanderers, and we are wanderers still. We have lingered long enough on the shores of the cosmic ocean. We are ready at last to set sail for the stars.
>
> *Carl Sagan*

_____

_____

Based on the experiences of Russia's cosmonaut dogs, do you think it is right to use animals as test subjects in space rather than send humans in untested spacecraft?

Learn About

- spacecraft that have studied the Sun
- unmanned exploration of the Moon
- unmanned exploration of Venus
- unmanned exploration of Mars

Before humans could go and explore the Moon, or anywhere else in space, scientists tested space travel's effects on other living organisms. The United States tested using monkeys and chimpanzees; the Soviets used dogs. Some of these early space explorers successfully returned to Earth, while others did not.

The Russian stray dog, Laika, is perhaps the best known of these early explorers. She died in space because the spacecraft the Russians sent her in, *Sputnik 2*, was not equipped to sustain life for an extended period. The purpose of her flight in 1957 was to see if humans could withstand takeoff and the various pressures of spaceflight.

At the time the Soviets said that she lived for a few days. But in 2002 the Russians released reports that indicated she lived for five to seven hours, and died in part because of the stress of takeoff. In any event, her spacecraft was not designed for safe reentry.

Laika has lived on in the hearts of many, though, with a statue dedicated to her honor in Moscow, as well as numerous stamps issued with her likeness from around the world.

In 1960 the Soviet Union sent two more dogs into space, Belka and Strelka. They became the first living things to survive orbital flight. A day after takeoff, they returned safely to Earth. Strelka later gave birth to six puppies. As a peace offering, Soviet Premier Nikita Khrushchev gave one of the puppies to Caroline Kennedy, the daughter of President Kennedy.

Spacecraft That Have Studied the Sun

Over the past few decades, scientists have launched more than a dozen unmanned spacecraft to study the Sun. This section focuses extensively on those explorations.

The First Explorations

NASA's *Pioneer 7*, along with its three siblings *Pioneers 6, 8,* and *9*, blasted into space in the mid-1960s to explore the solar wind, interplanetary relations, and cosmic rays—among other things. The four spacecraft formed a ring of solar weather stations allowing scientists on Earth to monitor solar storms. Commercial airlines and communication companies, among countless other interested parties, were able to use the data for guidance and planning purposes. The information helped them predict solar storms, which can interfere with communications and expose airline passengers to increased radiation.

One of the highlights of *Pioneer 7*'s tenure came in 1986, when it flew within 7.6 million miles (12.2 million km) of Halley's Comet. During this time it monitored the relationship between the comet's tail and the solar wind.

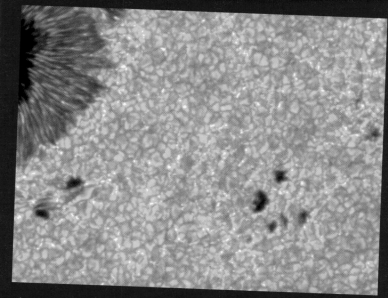

The solar probe *Hinode* took these photos of solar granulation on the Sun's surface in 2007. Each granule is about 620 miles (1,000 km) across and represents hot fluids rising to the surface only to cool and sink. The never-ending process takes about 20 minutes for each granule. Scientists have launched more than a dozen unmanned spacecraft to study the Sun over the past few decades.

Courtesy of NASA/Goddard Space Flight Center Scientific Visualization Studio

If You Can't Stand the Heat …

An old saying advises: If you can't stand the heat, get out of the kitchen. *Helios 1* and *Helios 2* however *had* to stand the heat when they studied the Sun. They endured temperatures up to 698 degrees F (370 degrees C).

Helios 1 got as close as 29 million miles (47 million km) from the Sun. *Helios 2* flew 27 million miles (43.5 million km) from the fiery orb. Since *Helios 2* passed closer to the Sun than the first probe, it had to endure 10 percent more heat.

Each spacecraft survived the heat by spinning once every second. Furthermore, scientists equipped each with mirrors to bounce the powerful solar rays away from the probes (this took care of 90 percent of the heat).

In the 1970s the United States and Germany worked together to send *Helios 1* and *Helios 2* into space. This joint venture allowed astronomers on Earth to study the relationship between the Sun and the planets. *Helios 1*, which continued to send information back to Earth until 1982, told scientists that the Sun has 15 times more micrometeorites—*meteorites that have a diameter of less than a meter*—passing near it than the Earth does.

Star POINTS

Helios was the Greek sun god.

Helios 2 was only in orbit four months in 1976, but during that time traveled closer to the Sun than had any previous manmade object. It also provided important information on solar plasma, the solar wind, cosmic rays, and cosmic dust.

Ulysses

Ulysses was one of the most successful space probes ever launched. The United States and the European Space Agency (ESA) worked together on the mission. This study had a significant effect on the way scientists view the Sun and its relationship with the planets.

The space shuttle *Discovery* transported *Ulysses* into outer space. Once outside the Earth's atmosphere, *Ulysses* and *Discovery* parted ways. The space probe traveled on its own toward Jupiter and studied the largest planet for more than two weeks. Once it was finished studying Jupiter, *Ulysses* used Jupiter's gravitational pull like a slingshot to launch toward the Sun.

Star POINTS

Ulysses had a life expectancy of five years. It lasted more than 17. During its mission, *Ulysses* encountered three comet tails and gave scientists enough information to write more than 1,000 scientific articles.

Why go to Jupiter first if the aim was to study the Sun? By using the momentum gained from Jupiter's gravity, *Ulysses* was able to pick up enough speed to leave the ecliptic plane—*the plane of a planet's orbit around the Sun*—and fly over the Sun's north and south poles. Without the gravity assist, the probe would have needed far more thrust to carry it out of the plane—thrust the launch vehicle carried aboard the shuttle could not provide.

Ulysses was important because it allowed scientists to study the Sun's north and south poles for the first time. It examined the poles several times in six-year increments. Because the satellites were there for extended periods of time, and on numerous occasions, scientists were able to observe the Sun's poles during both calm and stormy periods.

Ulysses also made the first direct measurements of interstellar dust particles and neutral-charged helium atoms in the Solar System. It discovered that the magnetic field that leaves the Sun is balanced across latitudes. The observations redefined the way scientists think about space weather.

The space shuttle *Discovery* heads to the heavens carrying five crew members and the *Ulysses* solar explorer on 6 October 1990. Once outside the Earth's atmosphere, *Ulysses* and *Discovery* parted ways.
Courtesy of NASA

When the two space agencies launched *Ulysses* in 1990, the probe achieved the fastest velocity ever achieved by a manmade object. Ten years later, the ESA announced that *Ulysses* had discovered the most distant gamma-ray burst yet recorded, about 11 billion light years from Earth. A gamma-ray burst is *a short-lived burst of gamma-ray photons, the most energetic form of light.*

Hinode

Hinode is a multinational space operation the United States works on with Japan (the project's leader), Britain, and the ESA. *Hinode* means "sunrise" in Japanese. The probe launched in 2006 and has excited scientists with its photos and information-gathering skills ever since.

Hinode's primary goal is a basic one—to give scientists a better understanding of the Sun's physics. The probe orbits the Earth in a pattern that keeps it permanently facing the Sun. One of the many things *Hinode* studies is the solar wind.

In late 2009 the space probe sent information back to Earth telling astronomers about a new type of magnetic field. *Hinode* has a high-powered telescope that came across sunspots completely different from those previously discovered. Astronomers refer to these new sunspots as THMFs—transient horizontal magnetic fields.

Current Missions to the Sun

In addition to *Hinode*, several other missions are exploring the Sun at this writing. Many are international projects.

In 1995 NASA and ESA launched the Solar and Heliospheric Observatory, or SOHO. The mission's goal is to determine where the solar wind comes from specifically, and what the Sun's interior structure is really like.

SOHO has alerted astronomers to hundreds of previously unknown comets. In early 2010 SOHO helped scientists capture images of a comet as it disintegrated in the Sun's heat, while at the same time showing images of Venus and Mercury in the background.

● ← **Approx. size of Earth**

SOHO grabbed this photo in 2002 of a solar prominence—dense gas thrown into space. Illustrators added an image of Earth to the bottom of the picture to show how small the blue planet is compared with the Sun. SOHO's mission is to determine where the solar wind comes from and what the Sun's interior is really like.

Courtesy of SOHO-EIT Consortium/ESA/NASA

Unit 3: Manned and Unmanned Spaceflight

Working in conjunction with SOHO is NASA's Transition Region and Coronal Explorer (TRACE). The American space agency launched this space probe in 1998 with the goal of determining the predictability of solar activity.

TRACE monitors various aspects of the Sun's plasma. The probe should also help scientists better understand why solar flares take place and how solar heating occurs. A big question for astronomers is why the Sun's temperature gets warmer moving away from its center. Think about a pond in winter. The last part to freeze over is the middle because it's the warmest. Why is this not true for the Sun?

NASA sent the Advanced Composition Explorer (ACE), into space in 1997. This spacecraft has enough power to keep running until 2024. Right now it works much like the Doppler Radar your local TV station displays on the evening news. It provides real-time space-weather information and advanced warnings of geomagnetic storms.

ACE has nine instruments. Together they have power that is anywhere from 10 to 1,000 times greater than anything previously launched. With ACE, astronomers should be able to better predict space storms to allow for safer manned space travel in the future.

Unmanned Exploration of the Moon

The space race to the Moon between the United States and the former Soviet Union resulted in dozens of missions beginning in 1959. While exploration of the Moon died down after the manned landings took place in the 1960s and 1970s, recent unmanned missions are paving the way for more manned trips in the future, perhaps for extended periods.

The First Explorations

The Russians first examined the Moon in January 1959 with the *Luna 1*, which soared about 3,700 miles (5,995 km) from its surface. Two months later the United States launched its probe *Pioneer 4* on a fly-by mission. Ten years later on 20 July 1969, US astronaut Neil Armstrong was first to set foot on the Moon.

Landing men on the Moon was a monumental feat and a victory for the United States during the decades-long Cold War. But scientists had earlier learned crucial facts about the Moon from the many more numerous unmanned spacecraft that had studied the body.

Both the United States and the Soviet Union had numerous failed missions before they were successful. What these missions taught both countries was just as important as what they eventually would learn when they did land on the Moon—on manned or unmanned missions. With each failure, scientists learned more about the dynamics of outer space.

The streak of light heading toward the Moon is the *Lunar Prospector* catching a ride on Lockheed Martin's *Athena II* launch vehicle. In 1998 America boosted the *Lunar Prospector* into space to look for water on the Moon, but it found none.

Courtesy of NASA

The first American spacecraft to return pictures of the Moon was the *Ranger 7* in 1964. It was one of nine Ranger missions costing the United States nearly $170 million in research, and one of only three that successfully landed on the Moon. The Russians sent more than 15 unmanned probes to study the Moon, including one in 1968 that carried turtles, worms, and flies to examine the effects of Moon travel on living organisms.

Once both countries hit their strides in the mid-1960s, space missions returned thousands of pictures and numerous material samples to Earth. In 1969 the *Apollo 12* crew brought back the camera portion of *Surveyor 3*, which the United States had launched in 1967. That camera is on display at the Smithsonian Institution's National Air and Space Museum in Washington, D.C. With it, *Surveyor 3* took more than 6,000 photos of the area of the Moon around its landing site. *Surveyor 3* also conducted experiments to ensure that the Moon's soil would support an Apollo lunar module.

The last time an American was on the Moon was in 1972. Several unmanned probes have launched since then, however. In 1998 America boosted the *Lunar Prospector* into space to look for water on the Moon and found none. Perhaps most notably though, new partners in Moon exploration have emerged. Japan sent a probe in 1990 and again in 2007, the ESA in 2003, China in 2007, and India in 2008.

CHAPTER 9 The Unmanned Missions of Space Probes

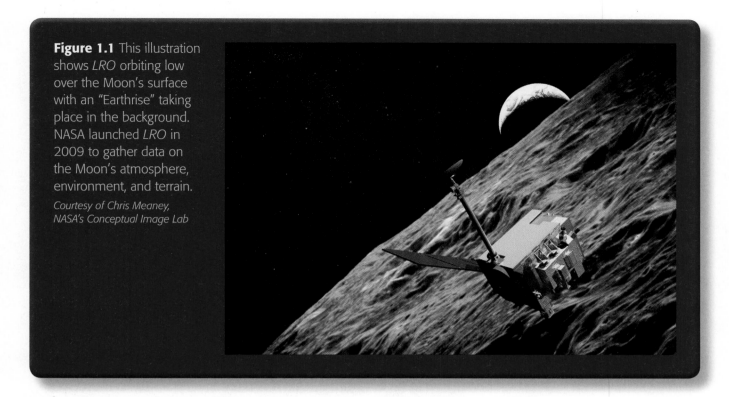

Figure 1.1 This illustration shows *LRO* orbiting low over the Moon's surface with an "Earthrise" taking place in the background. NASA launched *LRO* in 2009 to gather data on the Moon's atmosphere, environment, and terrain.

Courtesy of Chris Meaney, NASA's Conceptual Image Lab

The Lunar Reconnaissance Orbiter (LRO)

Before either a return to the Moon or a mission to Mars can become a reality, scientists need to learn much more about the topography, resources, and varying temperatures on the Moon.

NASA launched the Lunar Reconnaissance Orbiter (LRO) in 2009 with the purpose of studying everything it could about the Moon. It is orbiting the Moon approximately 31 miles (50 km) above the surface to gather data on its atmosphere, environment, and terrain (Figure 1.1). NASA planned for the LRO to spend more than a year conducting these observations.

While the ultimate goal is extended human visits to the Moon, scientists still want to learn more about the rock structure and water on the Moon. They also want to better understand the Moon's shape—ever so slightly like an egg—and what this means about the Moon's formation millions of years ago.

During the six manned American visits to the Moon, scientists collected rocks and other particles. But with the LRO they should learn more about the Moon's regions that are always in shadow—such as the floors of craters at the pole. The LRO will look for potential ice deposits, as well as rough terrain, rock abundance, and other landing hazards. It will also study how radiation in the lunar environment will affect humans.

In late 2009 NASA announced that with the help of LRO scientists were able to record the coldest temperature ever measured in the Solar System. The orbiter registered −416 degrees F (−249 degrees C) at the base of one of the lunar craters. This is important because these cold temperatures trap numerous particles, gases, and compounds, thus opening new areas for discovery.

Getting a ride to the launch pad is the *Atlas V/Centaur* rocket with the payloads Lunar Reconnaissance Orbiter and Lunar Crater Observation and Sensing Satellite on top. NASA launched LCROSS as a companion probe to LRO in 2009.
Courtesy of NASA/Jack Pfaller

Unit 3: Manned and Unmanned Spaceflight

The Lunar Crater Observation and Sensing Satellite (LCROSS)

In late 2009, scientists made an exciting discovery—they finally confirmed that there is water on the Moon.

NASA launched the Lunar Crater Observation and Sensing Satellite, (LCROSS), as a companion probe to the LRO in 2009. LCROSS revealed the presence of water after impacting in a crater near the Moon's south pole. The water researchers found in the debris plume was not just a tiny amount, but about 24 gallons. Scientists believe this water has been present on the Moon for billions of years.

NASA deliberately sent LCROSS into the Cabeus crater with the hope of locating water. The impact and subsequent pictures and materials collected with LCROSS gave scientists more information than they had hoped for. Using a spectrometer, *a special instrument equipped with devices for measuring the wavelengths of the radiation it observes*, scientists detected numerous compounds in this region, including water and carbon dioxide. Now that scientists know that water does exist on the Moon, they have endless questions to answer. Where did the water come from? How did it get there? How long has it been there? Several theories speculate that it came from the solar wind or from comets.

The Role of the National Space Science Data Center (NSSDC)

Just as the National Archives preserves important documents such as the Constitution and Declaration of Independence, scientists needed a place to house all the artifacts and pictures taken during space missions. In 1966 NASA created the National Space Science Data Center (NSSDC) at the Goddard Space Flight Center in Greenbelt, Maryland. The NSSDC is responsible for the long-term archiving and preservation of all space science information.

"Space science" includes astronomy, astrophysics, as well as solar and space plasma physics, and planetary and lunar science. The center works with scientists, educators, and in some cases the general public to help preserve and learn more from the vast amounts of research NASA has conducted. The center gives scientists examining different aspects of space access to a number of data and information services, including a large digital archive.

Unmanned Exploration of Venus

The Moon may be Earth's closest neighbor, but Venus has sometimes been called Earth's sister planet because the two are in many ways very similar. As you have learned in previous chapters, however, the atmospheres of the two are very different. Venus's gases have made taking pictures of the planet's surface a challenge. But scientists have gotten a somewhat better glimpse at Earth's neighbor through numerous unmanned missions.

The very first successful interplanetary spacecraft was sent to explore Venus. *Mariner 2*, launched in 1962, flew within 21,000 miles (34,000 km) of the planet. It measured the planet's hot temperature for the first time and was the first craft to measure the solar wind.

The Missions and Findings of the Soviet *Venera* Spacecraft

The Soviets also sent spacecraft to Venus in the 1960s. Much like the American *Ranger* program to study the Moon, the Soviet *Venera* program had many failures before it was successful. As a result of these failures, though, scientists were able to learn more about the planet named after the Roman goddess of love and beauty.

The Soviets lost the first *Venera* after it left Earth's orbit in 1961. Over the next 22 years, the Russians sent 15 more *Venera* spacecraft to study Venus. *Venera 4* in 1967 found that the temperature on Venus was about 932 degrees F (500 degrees C), and that Venus's atmospheric pressure is roughly 75 percent greater than on Earth. This probe was also important because it told scientists the composition of Venus's atmosphere was about 90 percent to 95 percent carbon dioxide with no nitrogen. Scientists considered this mission a partial failure, however, because the air pressure destroyed the probe before it was able to land.

In 1970 the first successful transmission of data from another planet back to Earth took place when *Venera 7* landed on Venus. Two years later, *Venera 8* gave scientists enough information to determine that sunlight on Venus is similar to sunlight on Earth during the dawn hours.

Venera 9 and *Venera 10* were the first of several twin missions the Soviets sent to study Venus. They launched *Venera 9* in 1975, and it transmitted information for more than a year. This was the first spaceship to orbit Venus. It landed on the planet on 22 October 1975 and sent back the first close-up photo of the planet's surface. Three days later *Venera 10* landed on the planet. It sent panoramic pictures back to Earth, measured the atmospheric pressure, and determined the makeup of rocks on the surface. It was able to send information back for more than an hour—which at that point was a record.

Venera 11 and *Venera 12* also worked well together. These probes, launched in 1978, helped determine that the beautiful planet could also be volatile because for the first time scientists detected thunder and lightning.

In 1981 the Soviets launched their next set of twin satellites—*Venera 13* and *Venera 14*—which sent back to Earth the first color pictures of Venus. These images showed a surface filled with orange and brown rocks. The mission also detected a wind speed of less than 1 mile (1–2 km) per hour. *Venera 13* helped scientists test and analyze the first soil samples from Venus. The Russians used *Venera 14* in coordination with other space probes examining Halley's Comet.

In 1983 the Soviets launched their final two missions to Venus, *Venera 15* and *Venera 16*. These missions helped scientists for the first time map the top half of the planet's northern hemisphere. When the mission was complete a year later, researchers had more than 400 miles (600 km) worth of tape to analyze.

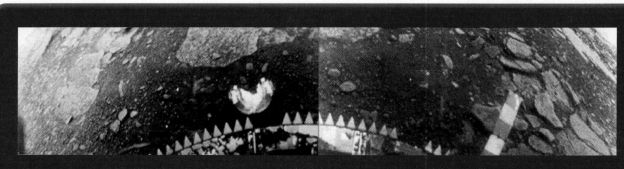

Part of *Venera 13* shows at the bottom of this photo, which the spacecraft took of Venus's surface. In 1981 the Soviets launched the twin probes *Venera 13* and *Venera 14*. They sent back to Earth the first color pictures of Venus.
Courtesy of NASA

The Missions and Findings of *Pioneer Venus*

While the *Ranger* and *Venera* missions were filled with trial and error, the American *Pioneer* probes that studied Venus were like the "little engine that could." US scientists expected the orbiter mission alone to last eight months. It kept going for 14 years.

NASA launched *Pioneer Venus* in 1978. The orbiter's main objective was to make a radar map of Venus's surface. It was also to study the solar wind's effects on the atmosphere (Figure 1.2).

When all was said and done, the orbiter collected enough material to construct a topographical map of a large portion of the planet. At the end of its mission in 1992, astronomers had determined that Venus had a much flatter surface than previously believed. But they also learned the planet had a mountain larger than Mt. Everest and a gorge deeper than the Grand Canyon.

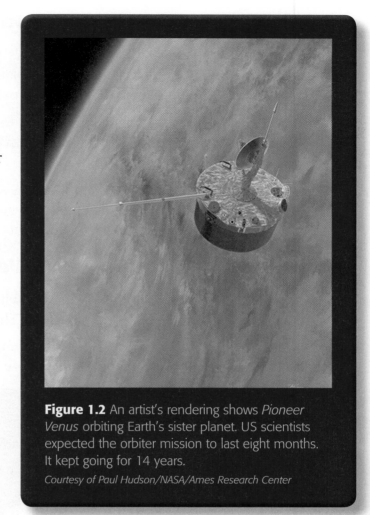

Figure 1.2 An artist's rendering shows *Pioneer Venus* orbiting Earth's sister planet. US scientists expected the orbiter mission to last eight months. It kept going for 14 years.
Courtesy of Paul Hudson/NASA/Ames Research Center

The *Pioneer* probe was also able to determine that the planet has a negligible magnetic field and that its clouds are made up mostly of sulfuric acid. Over time the amount of sulfuric acid decreased, however, leading scientists to guess that the probe may have arrived shortly after a volcanic explosion. Scientists were also able to detect a near-continuous pattern of lightning strikes.

NASA turned off the probe in 1981 but it remained in Venus's orbit. Ten years later the agency turned it back on. It delivered images and data to Earth until 1992, when scientists believe it disintegrated due to age and burned as it fell closer to Venus's surface.

Other *Venus Pioneer* probes had varying success. *Venus Pioneer 2* crashed when it landed on the planet in 1978, but was able to relay important information back to Earth during its descent. Notably, it appears that below a thick haze, from about 18 miles (30 km) high, the atmosphere is relatively clear.

Star POINTS

After the success of the *Venus Pioneer* mission, international scientists decided that, from that time forward, they would name any discoveries on the planet after historical or mythical women.

The Missions and Findings of the ESA *Venus Express*

Like the US *Venus Pioneer* probe, the ESA's *Venus Express* has outlived its predicted functional period. The European agency launched it in 2005 expecting it would last roughly two years. In late 2009 ESA extended the mission through 2012.

The *Venus Express* is so named because it took only three years from the mission planning stages to blast off. This is a relatively quick time in the world of space exploration.

The probe's main goal was to map Venus's southern hemisphere. In this regard it has been very successful. Scientists have released images that show highlands they believe are the remnants of continents once surrounded by oceans.

The Venus Radar Mapping Mission of *Magellan*

In 1989 NASA sent *Magellan* on its way with the goal of mapping Venus's surface to a degree that hadn't been done before. The probe mapped 98 percent of the planet.

Magellan enjoyed instant success. In the mission's first year, it sent more data back to NASA than all previous NASA planetary missions combined.

Magellan's early orbit around Venus varied in distance from the surface, anywhere from 182 miles (294 km) to 5,296 miles (8,543 km). The probe orbited from north to south and south to north. During this time it determined, among other things, that volcanic flows cover 85 percent of the planet's surface.

Magellan was the first mission to practice aerobraking, *the use of the planet's atmosphere to slow an orbit and thereby lower a satellite closer to a planet.* To give you an idea of how this works, think of a duck landing on a pond. It doesn't just plop into the water. It drags its feet and skims the surface to help slow down. That is very similar to aerobraking. Scientists like this because it costs considerably less than the more traditional methods of slowing orbiting modules.

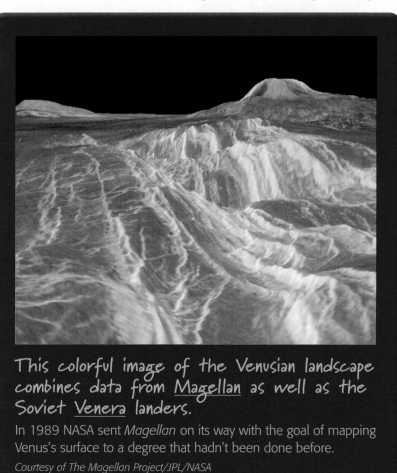

This colorful image of the Venusian landscape combines data from Magellan as well as the Soviet Venera landers.

In 1989 NASA sent *Magellan* on its way with the goal of mapping Venus's surface to a degree that hadn't been done before.

Courtesy of The Magellan Project/JPL/NASA

Unmanned Exploration of Mars

As you have seen with the unmanned explorations of the Sun, Moon, and Venus, studying the various parts of the Solar System involves a lot of trial and error. Such was the case with Mars as well.

The First Explorations

Unmanned missions to Mars are the only type of exploration that scientists have undertaken of the red planet so far. While there is talk of sending humans to Mars, it is something that is still years in the future. When that time comes, however, scientists will use much of what they have acquired over the years during various unmanned space missions to help them prepare.

Mariner 9 took this photo of Olympus Mons—the biggest volcano in the Solar System—as a dust storm rages on the Martian landscape around it. Olympus Mons is 40 miles (64 km) across. *Mariner 9*'s 1971 mission represented the first time a spacecraft orbited another planet.
Courtesy of NASA/JPL

As with the race to the Moon, the United States and the former Soviet Union battled it out in the 1960s and 1970s to see who would get to Mars first. Oftentimes their missions would occur within days of each other, and sometimes both failed miserably. *Mars 1*, a Russian spacecraft, was the first to fly by the planet in 1962.

Mariner 4, an American space probe, was one of NASA's early success stories. It flew by Mars on 14–15 July 1965 and sent back 21 black and white pictures. The probe determined that Mars had a very cold, battered, and barren surface with daytime temperatures of about −212 degrees F (−135 degrees C). It also discovered a very weak radiation belt around the planet. NASA turned off *Mariner 4* in late 1965 but reactivated it two years later to support the *Mariner 5* mission to Venus.

The United States had three other Mariner spacecraft of note. In 1969 the space agency launched the twin probes *Mariner 6* and *Mariner 7*. They sent nearly 200 pictures back to Earth. But their images gave a very different impression from those that *Mariner 4* sent back and left scientists with more unanswered questions. The *Mariner 9* mission in 1971 was the first time a spacecraft orbited another planet. While a dust storm interfered with its mission of mapping 70 percent of Mars' surface, it eventually sent nearly 7,000 images back to Earth. It identified about 20 volcanoes, including Olympus Mons, the biggest volcano in the Solar System. *Mariner 9* completely revolutionized scientists' view of the red planet and paved the way for the Viking program. The now-inactive satellite will remain in Mars orbit for several more decades.

The Viking Missions

The Soviet Union was so determined to beat the United States in Martian research that it sent up several missions in the early 1970s, even though scientists determined they had at best a 50 percent chance of success. These missions failed.

Shortly after these failed Soviet missions, NASA sent up the twin probes *Viking 1* and *Viking 2* within a couple of weeks of each other. Both had a lander and an orbiter. The project's combined cost was $1 billion.

When the *Vikings* took off, they both traveled to Mars and orbited until scientists could find a soft landing site. Once they found a spot, the lander separated from the orbiter, deployed parachutes, and landed.

Viking 1 and *Viking 2* conducted a series of tests, including water-vapor studies. They determined the soil of Mars was rich in iron, but they detected no life forms. The point of the missions was to study the planet's biology (if any), chemical composition, and physical properties.

Viking 1 took several 360-degree pictures, which gave scientists the best images of Mars until that time. *Viking 2* took some 16,000 pictures and orbited the planet 706 times. The missions themselves only lasted a few years, but were considered a big success.

The Findings of the Mars Scout Lander, the *Phoenix*

The *Phoenix* is the first in a new series of orbiter robots to combine the resources and research abilities of government, universities, and the space industry. The primary goal of the mission, which began in 2007, was to further search for water on Mars.

During a five-month period in 2008, the *Phoenix* took ground samples of Martian dirt. NASA scientists equipped it with an arm that could dig nearly two feet into the ground for specimens, which it returned to the craft for analysis. The exploration took place primarily on the icy plains of Mars' North Pole region.

What scientists discovered astounded them. First, the presence of perchlorate, *a chemical that attracts water*, indicated the potential exists to sustain life forms. Some living things on Earth use perchlorate for food. In the long term, scientists believe perchlorate could be used to help humans on Mars develop rocket fuel or even generate oxygen. Scientists also found calcium carbonate deposits, which is an indication liquid water is or has been present.

Scientists observed snow falling from the clouds on Mars, which surprised them. Up until this time, astronomers were not aware of any precipitation falling onto the planet.

Phoenix took more than 25,000 pictures and sent them back to Earth, giving scientists clearer images of the red planet.

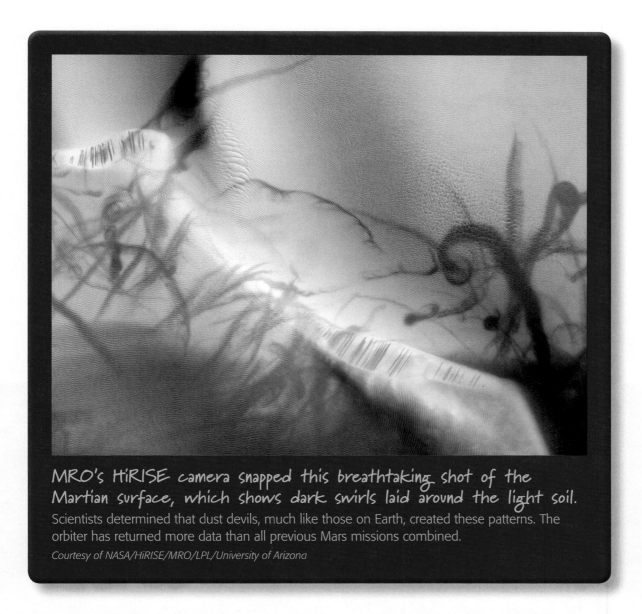

MRO's HiRISE camera snapped this breathtaking shot of the Martian surface, which shows dark swirls laid around the light soil. Scientists determined that dust devils, much like those on Earth, created these patterns. The orbiter has returned more data than all previous Mars missions combined.

Courtesy of NASA/HiRISE/MRO/LPL/University of Arizona

The mission officially ended in November 2008, when colder climate conditions affected *Phoenix's* ability to function. The original mission was only scheduled to last three months, or 92 Earth days—so like many other probes it outlived expectations and provided valuable research for scientists for years to come.

The Mission of the Mars Reconnaissance Orbiter

The Mars Reconnaissance Orbiter had four goals when it reached the planet in 2006. First, NASA scientists planned to see if it were possible that life ever existed on Mars, and if so, whether some form of life might still exist there. Their second aim was to learn more about the climate on Mars and what it could tell them about the past. Third, they intended to find out more about the rocks and minerals on Mars and how they are different from those here on Earth. Finally, the scientists designed the MRO to help them plan for future explorations of the red planet.

The MRO flies in Mars' atmosphere, orbiting anywhere from 160 miles (255 km) to 200 miles (320 km) above the surface. Since it began its mission, it has returned more data to Earth than all previous Mars missions combined. Some of the most interesting photos demonstrate the presence of gullies and dunes on the surface that no one had observed before.

NASA's engineers equipped the MRO with many different instruments. It has a powerful camera that can identify objects on the surface as small as a dinner plate. These detailed pictures illustrate areas where water once flowed, or any possible "hot spots" where geothermal pools like those found at Yellowstone National Park may exist.

Other equipment on the MRO can identify minerals on the planet's surface and see how dust and water move in Mars' atmosphere. Besides taking detailed pictures of smaller objects on Mars, the MRO also has a second camera to take larger, medium-resolution photos that provide a broader picture of conditions on Mars. As of this writing, MRO's mission is supposed to last through 2010.

Spirit, stuck in sand since 2009, took hundreds of photos to compile this view of the Martian landscape.
The rise in the far distance scientists dubbed Husband Hill. Since landing in 2004, *Spirit* has traveled roughly five miles.
Courtesy of NASA

CHAPTER 9   The Unmanned Missions of Space Probes

The Findings of the Mars Exploration Rovers *Spirit* and *Opportunity*

The Mars rover *Spirit* first landed on Mars in January 2004. A second rover, *Opportunity*, landed halfway across the planet three weeks later. As with MRO, these two probes were still operating as of early 2010.

Scientists designed the rovers with special tools needed to study Mars. The rovers have cameras that serve as eyes, and robot arms and wheels to move around the Martian surface. The rovers operate on batteries charged by solar panels, and are equipped with scientific instruments to study rocks and determine which minerals and chemicals are present. They also have a grinding tool to help dig into the surface.

Opportunity made the mission's first major discovery when it found mineralogical evidence that Mars had liquid water in the past. Since its landing, *Opportunity* has conducted a two-year investigation of a half-mile-wide crater called Victoria. The rover has traveled more than 11 miles and sent more than 132,000 images from its cameras back to Earth.

Spirit has had a tougher time on the red planet. Its right front wheel stopped working in 2006, and the right rear wheel stalled in late 2009. The rover also became stuck in a sand trap nicknamed "Troy" in mid-2009 from which scientists were unable to free it. Since landing, Spirit traveled roughly five miles. In early 2010 NASA declared it a stationary science platform and gave up trying to free it.

The US space agency planned for the rover missions to last just three months. Instead, they continued for more than six years. During its time on the red planet, *Spirit* has found evidence of a very warm and violent environment during Mars' ancient past. *Opportunity*, however, saw evidence of very different conditions, showing a wet and acidic past.

Additionally, the two rovers have given scientists panoramic views of different parts of the red planet. They have also adjusted these pictures to provide "natural color" to show Mars as it would appear if you were standing on it.

Scientists haven't limited their explorations of the Solar System to the Sun, Moon, Venus, and Mars. Their gaze also extends to speeding asteroids as well as the outer planets—Jupiter, Saturn, Uranus, and Neptune. Even though these planets aren't friendly to human habitation, they can still shed light on the universe's composition with possible applications on Earth. You will read more about these missions in the next lesson.

Lesson 1 Review

Using complete sentences, answer the following questions on a sheet of paper.

1. What did scientists monitor with the aid of the four *Pioneer* spacecraft?

2. How did *Ulysses* use Jupiter's gravitational pull to its advantage?

3. What does THMF stand for?

4. What is the SOHO mission's goal?

5. Which country first sent a spacecraft to the Moon?

6. What was the coldest temperature recorded in the Solar System and where was it recorded?

7. Why did NASA send LCROSS into the Cabeus crater, and what did they discover?

8. What is the National Space Science Data Center responsible for?

9. What happened to *Venera 4* when it attempted to land on Venus?

10. What was the *Pioneer Venus* probe's main objective?

11. What do scientists believe Venus's highlands are remnants of?

12. What is aerobraking?

13. Where is the biggest volcano in the Solar System?

14. What makes scientists believe Mars could potentially sustain life forms?

15. How small an object can the camera on the Mars Reconnaissance Orbiter detect?

16. What power sources allow *Spirit* and *Opportunity* to operate on Mars?

APPLYING YOUR LEARNING

17. Since the rocks and minerals the rovers *Spirit* and *Opportunity* dig up will never be sent back to Earth, how do you think scientists analyze the materials from such a great distance? What must the scientific instruments on the rovers be able to do?

Unit 3: Manned and Unmanned Spaceflight

Quick Write

What does this story tell you about the many different people who contribute to the space program? What skills do you have that agencies like NASA might be able to use?

Learn About

- how the Hubble Space Telescope aids the exploration of space
- scientific discoveries among the outer planets
- scientific investigations of comets and asteroids

Tucked away in a basement building at NASA's Goddard Space Flight Center in Greenbelt, Maryland is a unique facility. Workers here precisely measure, cut, and carefully sew custom-made thermal blankets for the Hubble Space Telescope and other space missions. The telescope already sports several that astronauts installed on previous servicing missions.

According to Shirley Adams, group leader for blanket fabrication, her employees come from very diverse backgrounds. "Some have designing backgrounds in upholstery work, costume designing, and one even has a background in ice skating costume-making," Adams said.

Such talents have proven very beneficial since sewing, stitching, and custom-fitting the different thermal blankets for the telescope is accomplished in-house at Goddard. Coupled with experts in materials and mechanical engineering, the expertise at Goddard makes the center the logical home for the development and production of the blankets. Goddard also performs the analysis of blankets the astronauts have brought back on previous servicing missions.

How the Hubble Space Telescope (HST) Aids the Exploration of Space

Vocabulary

- spectrograph
- Cepheid variable stars
- heliosheath
- hot spot
- magnetosphere

Scientists in the 1920s first tossed around the idea of a spacecraft that could study the universe in detail. But it wasn't until 1990 that NASA launched the Hubble Space Telescope (HST). It sends back to Earth the clearest images ever captured of outer space. The much-anticipated telescope has forever changed the way scientists view the universe.

The History of the Hubble Space Telescope

Hubble's great advantage over most other telescopes is that it circles Earth from hundreds of miles up in space beyond interference from the planet's atmosphere. Its images therefore offer clear views of the universe in all directions. HST has also returned such amazing photos because of its extraordinarily precise and advanced cameras.

Hubble's ride hasn't been easy, though. Its journey has experienced numerous bumps along the way.

The Hubble Space Telescope was born when NASA decided to fund the mission in the 1970s. The agency planned to launch it in 1983, but several delays, including the *Challenger* tragedy, pushed the launch back to 1990. The space shuttle *Discovery* delivered it into orbit around Earth on 25 April.

NASA built HST to be a serviceable spacecraft, meaning astronauts could repair it in space. It looks like a giant school bus made from the same material as stovetop instant popcorn containers. It is made up of many drawer-like compartments so new parts can easily replace those that are older and out-of-date.

Star **POINTS**

Hubble orbits 353 miles (569 km) above the Earth. On 11 August 2008 it circled Earth for the 100,000th time.

Almost immediately after HST's launch, scientists discovered a problem with the mirror on board meant to aid in taking pictures. Technicians had ground the mirror incorrectly based on earlier miscalculations with one of their tools. The mistake with the mirror was miniscule—estimated to be about $1/50$th the width of a human hair. But tiny though it was, it was major in terms of how well the telescope worked. The good news was that this was not a fatal error. But it would be three years before a service mission could take place.

Astronaut John Grunsfeld: Hubble Repairman

John Grunsfeld knew as a young child he wanted to be an astronaut. The Chicago-native told *The New York Times* he finds peace in space he can't find down on planet Earth. All told, he has spent more than 800 hours in space.

Grunsfeld's biggest job during his 18-year career with NASA was fixing the Hubble Space Telescope—three different times. He has likened this to brain surgery, only in space.

In 1999, during his first repair mission, Grunsfeld spent more than eight hours in space working on the gyroscopes, which help point the telescope in the right direction. Three years later he worked on the connecters.

When NASA suspended work on Hubble after the *Columbia* tragedy, Grunsfeld nearly quit because of the decision. He remained, though, because he didn't want the other projects he was working on to fall by the wayside. The agency changed its mind on the Hubble telescope and he completed a final third repair in 2009, installing new batteries, sensors, and even a camera.

In early 2010 Grunsfeld became deputy director of the Space Telescope Science Institute in Baltimore. There he will continue to oversee Hubble, as well as work on the James Webb Space Telescope, which is set to replace Hubble in 2014.

The Hubble Space Telescope was able to take pictures in the meantime, but they were nothing like those taken after the repair. The 11-day repair mission by *Endeavour* in 1993 (which you read about in Chapter 7, Lesson 1) did not fix the mirror itself. Instead, it modified the defect. Now the Hubble Space Telescope is wearing "glasses" to help it see better. The results have been spectacular.

The Hubble Space Telescope's findings are the basis for more than 6,000 journal articles. It has taken thousands of images and pictures that give scientists insight into the universe's history. It is important to note that the images you see in this book or elsewhere from Hubble were originally in black and white. NASA scientists added colors so the details would be visible to the naked eye, or to highlight a specific element.

Star POINTS

The James Webb Space Telescope is named after NASA's second administrator. He served from 1961 through 1968, overseeing all the Mercury and Gemini launches.

Hubble has seen four additional service missions since the 1993 repair. The most recent repair was in 2009. Current plans call for NASA to pull HST from service in 2014, when the agency will replace it with the James Webb Space Telescope.

The HST's Significant Findings

If you sit at a computer and search for "Hubble discoveries," you will find millions of Web sites, journal articles, and most notably pictures. In fact, HST has returned more than 700,000 images to scientists and has changed the way scientists understand star birth, star death, and the evolution of galaxies.

With the aid of Hubble's six main tools, scientists have captured the earliest images of the universe, just 600 million years after the Big Bang. The Hubble Space Telescope sends back to Earth images showing galaxies that are hundreds of millions of years old, considered to be in their infant stages. These newer galaxies don't have the colors or spirals of older galaxies. So, the Hubble images are in a sense giving scientists a look at baby pictures unlike any they have ever seen before.

Mission specialist Kathryn Thornton replaces Hubble's solar arrays during *Endeavour* mission STS-61, which NASA launched solely to repair the space telescope in 1993. NASA built HST to be a serviceable spacecraft, meaning astronauts could repair it in space.

Courtesy of NASA

With the aid of the HST and its images of faraway galaxies, scientists have determined that the universe is roughly 13.7 billion years old. Because of the telescope's work scientists now say that there are billions of galaxies, a radical shift from the early twentieth century, when many believed the Milky Way Galaxy was the only one.

The HST has taught scientists about the Milky Way, too. They know more than they ever imagined about the rings around Uranus, the storms on Jupiter, and the objects floating in the Kuiper Belt-region of the Solar System's farthest reaches. Hubble has also taken images of far-away planets and black holes.

Edwin Hubble rocked the scientific community in the 1920s with his theories on galaxies besides the Milky Way. Eighty years later the telescope named for him is telling scientists about these other galaxies.

Hubble: The Six Instruments That Make It Tick

Before you can understand the Hubble Space Telescope's findings, it's important to understand how it does its job. The telescope has six different instruments on it. Each plays a role in helping scientists learn more about the universe.

At different times, the **Wide Field Camera** sees three different kinds of light: near-ultraviolet, visible, and near-infrared. Scientists also use it to study dark energy and dark matter, individual star formation, and to discover remote galaxies.

The **Cosmic Origins Spectrograph** is a tool that sees only ultraviolet light. A spectrograph is *an instrument that separates light from the cosmos into its component colors.* This provides a wavelength "fingerprint" of the object being observed, which tells scientists about its temperature, chemical composition, density, and motion.

Hubble houses another spectrograph as well. The **Space Telescope Imaging Spectrograph** also sees ultraviolet light, but detects visible and near-infrared light, too. It can find black holes. While the Cosmic Origins Spectrograph works best with small sources of light, such as stars or quasars, this one can map out larger objects like galaxies.

The **Advanced Camera for Surveys** sees visible light. It studies some of the universe's earliest activity. This camera helps map the distribution of dark matter, detects the most distant objects in the universe, searches for massive planets, and studies the evolution of galaxy clusters.

The **Near Infrared Camera and Multi-Object Spectrometer** is Hubble's heat sensor. Its sensitivity to infrared light—perceived by humans as heat—lets scientists see objects hidden by interstellar dust, such as stellar birth sites, and gaze into deepest space.

Lastly, the **Fine Guidance Sensors** are instruments that lock onto "guide stars" and keep Hubble pointed in the right direction. They precisely measure the distance between stars, and their relative motions.

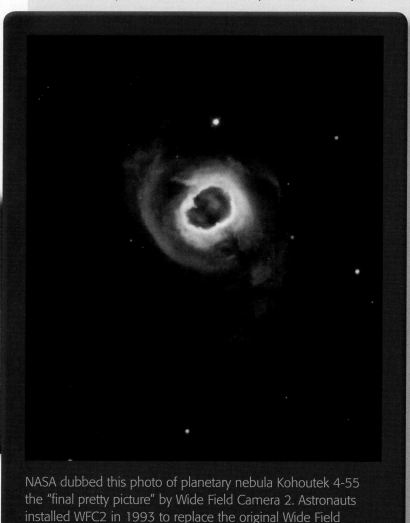

NASA dubbed this photo of planetary nebula Kohoutek 4-55 the "final pretty picture" by Wide Field Camera 2. Astronauts installed WFC2 in 1993 to replace the original Wide Field Camera. Now Hubble works with Wide Field Camera 3, which astronauts swapped out for WFC2 in May 2009.

Courtesy of NASA/ESA/Hubble Heritage Team/STScI/AURA/R. Sahai and J. Trauger/Jet Propulsion Laboratory

Unit 3: Manned and Unmanned Spaceflight

Examples of HST Findings

The Hubble Space Telescope has made countless contributions to the study of science, the Solar System, and numerous areas of astronomy. Here are examples of just a few of them.

The telescope's observations in 2002 of a pair of very distant exploding stars, called Type Ia supernovae, provided new clues about the accelerating universe and its mysterious "dark energy." Astronomers used the telescope's Advanced Camera for Surveys to help pinpoint the supernovae, which are approximately 5 billion and 8 billion light-years from Earth. The farther one exploded so long ago the universe may still have been slowing down under its own gravity.

In 1999 NASA announced that the telescope had measured precise distances to far-flung galaxies, an essential ingredient needed to determine the age, size, and fate of the universe.

NASA used the HST to observe 19 galaxies out to 108 million light-years. Scientists discovered almost 800 Cepheid variable stars, *a special class of pulsating star used for accurate distance measurements.*

In 1995 the telescope provided mankind's deepest, most detailed visible view of the universe. Representing a narrow "keyhole" view stretching to the universe's visible horizon, the Hubble Deep Field image covered a speck of the sky only about the width of a dime 75 feet away. Though the field is a very small sample of the heavens, scientists consider it representative of the typical distribution of galaxies in space because the universe, statistically, looks largely the same in all directions. Gazing into this small field, Hubble uncovered a bewildering assortment of at least 1,500 galaxies in various stages of evolution.

Astronomers using the HST first detected the atmosphere of a planet orbiting a star outside the Solar System. Their historic observations demonstrate it is possible to measure the chemical makeup of a planet's atmosphere, and search for the chemical markers of life beyond Earth. The planet they found orbits a yellow, Sun-like star called HD 209458, located 150 light-years away in the constellation Pegasus.

Scientific Discoveries Among the Outer Planets

While Hubble's telescopic eyes capture images from all over the galaxy and beyond, many unmanned missions have launched with a more specific purpose: to study the outer planets. The *Voyager* twins, along with the *Galileo* and *Cassini* missions, taught (and in some cases continue to teach) scientists about the Solar System in ways earlier astronomers could only dream of.

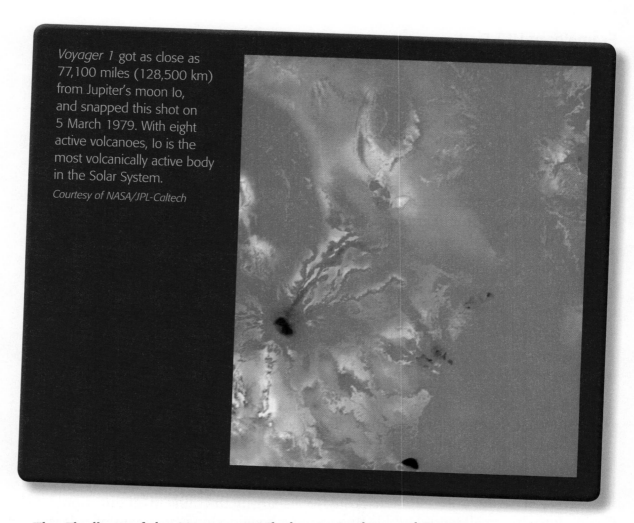

Voyager 1 got as close as 77,100 miles (128,500 km) from Jupiter's moon Io, and snapped this shot on 5 March 1979. With eight active volcanoes, Io is the most volcanically active body in the Solar System.
Courtesy of NASA/JPL-Caltech

The Findings of the *Voyager 1* Mission to Jupiter and Saturn

The *Voyager 1* space probe has traveled farther than any manmade object ever in the Solar System. It launched in 1977 to explore Jupiter and Saturn. Since then, it has outlived its life expectancy and continues to astound scientists.

Voyager 1 reached Jupiter in 1979. Right away it gave scientists new information about the largest planet. They found Jupiter has a very thin ring around it, less than 19 miles (30 km) wide. It's not as majestic as Saturn's rings, but its existence was unknown to astronomers before.

Star POINTS

Although *Voyager 1* left Earth 16 days after *Voyager 2*, its faster flight path allowed it to pass the slower craft and arrive at Jupiter more than four months ahead of *Voyager 2*. *Voyager 2*'s slower and more curved trajectory made it possible to go on to Uranus and Neptune using gravity assists.

The probe traveled within 168,000 miles (280,000 km) of Jupiter. Its most revealing details came from its exploration of the planet's many moons, including the discovery of two previously unknown moons, Thebe and Metis.

The most interesting finds came from the moon Io. The images revealed a yellow, orange, and brown surface area with eight active volcanoes. This makes Io the most volcanically active body in the Solar System.

CHAPTER 9 The Unmanned Missions of Space Probes

In 1980, after finishing its mission to Jupiter, *Voyager 1* did a similar survey of Saturn. Again astronomers were amazed by the findings. The probe came within 75,000 miles (124,000 km) of the ringed planet. Like its exploration of Jupiter's moons, *Voyager 1* found new moons around Saturn.

Voyager 1 also learned more about the rings circling Saturn than scientists had known before. These rings had many unusual qualities to them, including some "spokes" that may be particles levitated above the ring plane by electric charges. Other rings defied logic, being elliptical, discontinuous, or multi-stranded. The probe found several small satellites guiding ring material between them, offering clues to age-old questions about the formation and lifetime of planetary rings.

Voyager 1 tried to get more information on Saturn's moon, Titan. A Dutch astronomer first discovered the moon in the seventeenth century. Scientists are trying to learn more about it because it is the only moon they know of in the Solar System with thick clouds and a planet-like atmosphere. The clouds were too thick for *Voyager 1* to penetrate to get pictures of the surface, however.

Star POINTS

Voyager 1 runs on less than 300 watts of energy, less than the several light bulbs that light up your home.

NASA crafted this montage of planets and moons from the *Voyager 1* and *Voyager 2* mission photos.

Earth's Moon is in the foreground, with Jupiter, Saturn, Uranus, and Neptune and their moons in the background along with a nebula named Rosette. The *Voyager* spacecraft are still operating as of this writing in 2010.

Courtesy of NASA

Star POINTS

It's unlikely, but if anyone or anything ever recovers *Voyager 1* or *Voyager 2*, there will be a greeting waiting for them. The so-called Golden Record, included with both probes, includes greetings in more than 50 languages, sounds of life on Earth, a message from then-President Jimmy Carter, and music from various cultures. It is a space time capsule that scientists estimate *could* be found in 40,000 years if another culture exists and they are spacefaring creatures.

After leaving Saturn, *Voyager 1* began traveling past the other outer planets. In 2004 it reached the heliosheath, *the final frontier in the Solar System where the solar wind slows and meets the approaching wind in interstellar space.* This is a region where the Sun is barely a dot. Because of its remote location, it takes a message sent from Earth 14 hours to get to *Voyager 1*. *Voyager 1* is now traveling at a rate of 1 million miles (1.6 million km) a day.

The Findings of the *Voyager 2* Mission to Jupiter, Saturn, Uranus, and Neptune

An historic alignment took shape among the four outer planets in the 1970s and 1980s. This alignment happens once every 175 years. It allowed *Voyager 2* to take a journey that scientists do not expect to be repeated in your lifetime.

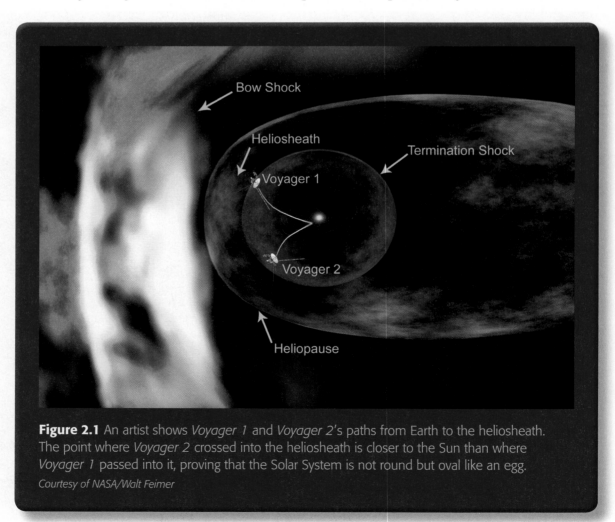

Figure 2.1 An artist shows *Voyager 1* and *Voyager 2*'s paths from Earth to the heliosheath. The point where *Voyager 2* crossed into the heliosheath is closer to the Sun than where *Voyager 1* passed into it, proving that the Solar System is not round but oval like an egg.
Courtesy of NASA/Walt Feimer

Voyager 2 launched shortly before *Voyager 1*. In 1979 it traveled to Jupiter and came within 448,000 miles (722,000 km) of the large planet. Like its twin, the probe found rings around the planet. It also observed in great detail a volcano erupting 125 miles (200 km) from the surface of the moon Io. The moon Europa also provided great detail to the probe, including a possible liquid ocean.

After traveling to Jupiter, *Voyager 2* went onward to Saturn. Here it detected temperatures ranging from −312 degrees F (−191degrees C) to 18 degrees F (−7 degrees C), and wind speeds up to 1,100 miles (1,750 km) per hour. It flew within 62,000 miles (100,000 km) of the planet and like its twin spent considerable time studying the rings and moons.

Six years later *Voyager 2* traveled to Uranus. It discovered that this seventh planet from the Sun also has a small set of rings, nothing like Saturn's but bigger than Jupiter's. It also explored Uranus's many moons, notably Miranda. This moon, which is a fraction the size of the Earth's Moon, has a strange surface area that includes large canyons—some 12 miles (19 km) deep.

Voyager 2's final planetary "stop" was Neptune in 1989. It came within 3,000 miles (4,800 km) of the farthest planet. The most interesting fact for scientists from this visit was new information about the moon Triton. *Voyager 2* found several geysers of gaseous substances on this moon.

As with *Voyager 1*, *Voyager 2* did not stop relaying information once it passed the last planet on its path. It continues to educate scientists about the Solar System's makeup more than 30 years after it was launched. In late 2007 NASA announced a critical finding from *Voyager 2*. The bubble of solar wind surrounding the Solar System is not a round ball, but an oval like an egg. It learned this because the point where *Voyager 2* crossed into the heliosheath is closer to the Sun than where *Voyager 1* passed into this "final frontier." The two probes are billions of miles from one another at this stage of their travels (Figure 2.1).

The Findings of the *Galileo* Orbiter and Probe

In 1989, NASA launched *Galileo* on a 14-year mission to further study Jupiter. Like many other space probes, *Galileo* taught scientists much, much more. Rather than launch the spacecraft directly to Jupiter from Earth—as originally planned— NASA instead used the gravity of Venus and Earth to make the long journey. The agency opted for this longer flight path, referred to as a "gravity assist," because it had to settle for a smaller launch vehicle because of the *Challenger* disaster.

The mission began in 1989. *Galileo* first traveled to Venus. Here it tested the equipment it would use on Jupiter in the years to come. As a result, scientists have a better understanding of the cloud cover over Venus, the closest planet to Earth.

Galileo took these four photos within seconds of one another of comet Shoemaker-Levy 9's collision with Jupiter.

The point of impact can just be made out by the dot of light on the bottom left of the last three images. In the first of the four pictures, there's no impact so all is black. In the next three the point of light is dim, then grows, and then fades. The probe took these pictures from 148 million miles (238 million km) away.

Courtesy of NASA/JPL

Galileo used Venus's gravity to travel back around Earth—twice—and then strike outward to the asteroid belt. In these maneuvers, scientists used Venus's and Earth's gravity to shape the probe's trajectory. The trajectory then allowed the planets to transfer orbital momentum to the probe to accelerate it—like cracking a whip. During its close encounter with Earth, *Galileo* took pictures of the Earth and the Moon together, a view rarely seen. During its time in the asteroid belt, the probe encountered some technical difficulties but scientists were able to repair most of the damage remotely.

As the probe approached Jupiter with the help of the gravity assist from Earth it took spectacular pictures of the Shoemaker-Levy 9 comet as its fragments crashed into the planet. These photos gave scientists a close-up view of the impact, something they otherwise never would have seen.

Up until this point *Galileo*'s orbiter and probe had traveled as one. Once they reached Jupiter, however, they separated and went on very different missions. In 1995 the probe began a solo five-month journey that would take it deep into Jupiter's atmosphere.

The *Galileo* probe entered the atmosphere on 7 December 1995 traveling 106,000 miles (170,000 km) per hour. The probe launched a parachute and a heat shield to protect it as it traveled through the gaseous atmosphere. *Galileo* survived for 56 minutes before the planet's atmosphere vaporized and destroyed it.

Even so, in the short time the probe functioned, scientists learned a lot about Jupiter. The probe measured wind speeds up to 450 miles (724 km) per hour. It observed fewer lightning strikes than scientists had expected and found the atmosphere to be drier than anticipated. Later scientists determined the probe had landed on a hot spot, *a cloud-free area where warmer thermal heat from elsewhere on the planet emerges.*

The *Galileo* orbiter was equally successful with its mission. It spent the next two years orbiting the four big moons that circle Jupiter, and went up to 1,000 times closer to the large planet and its moons than the Voyager missions had. In all, during this initial two-year period, *Galileo* completed 11 orbits around Jupiter, during which it also studied the moons Io, Europa, Ganymede, and Callisto.

Star POINTS

Scientists had to take a hurry-up-and-wait approach to *Galileo*. It took a week's worth of data from the orbiter up to two months to transmit back to Earth. So while the orbiter downloaded information it remained in orbit around Jupiter without collecting new data.

But *Galileo* still had life and power after these orbits, so NASA extended the mission. During the next four years the orbiter got so close to Europa it could detect images as small as a school bus. It also learned that Europa has an ocean below the surface. It saw fiery volcanic eruptions on Io, too.

Scientists were very excited when *Galileo* was able to coordinate with the *Cassini* space probe in 2000 for experiments relating to Jupiter's magnetosphere. A magnetosphere is *the region in which a celestial body's magnetic field interacts with charged particles from the Sun.* NASA also refers to magnetospheres as comparable to a giant bubble around a planet. Neither space probe could have conducted these experiments on its own.

By 2003 *Galileo* was running low on fuel. NASA's big concern was that the probe might crash into Europa, which the agency intends to explore further someday because astronomers suspect life could exist there. Scientists feared that if the spacecraft were to strike Europa, it could contaminate the moon with microorganisms from Earth and harm any native organisms. So NASA deliberately crashed *Galileo* into Jupiter's atmosphere. During its 14-year journey it traveled 2.6 billion miles (4.6 billion km) and far exceeded NASA's expectations.

The *Cassini* Mission's Purpose

Cassini is a joint venture between NASA and the European Space Agency (ESA). It launched in 1997 and is on a mission extension through September 2010.

Cassini, like *Galileo*, needed a planetary push to make it to Saturn. So it first went by Venus and later circled back past Earth, and then onward to Jupiter until reaching Saturn in 2004.

Cassini carried with it the *Huygens* probe. The two separated in late 2004. On its descent to Saturn's moon Titan—the second largest moon in the Solar System—in January 2005, *Huygens* encountered winds of up to 267 miles (430 km) per hour. On 14 January 2005, for more than three hours, it transmitted information from the atmosphere and later from Titan's surface. The images sent back from the surface are the first from a moon surface other than Earth's Moon.

The probe showed that Titan's surface is rocky and muddy. Scientists also observed lakes composed of methane and ethane. This is the same composition as the rain on Titan. The moon also has sand dunes made up of a mixture of ice grains and a dark organic compound that falls from the moon's upper atmosphere. Many similarities exist between Titan and Earth. Scientists believe greater study of the moon might help them learn more about what Earth was like before life evolved.

The *Cassini* probe has spent considerable time around Saturn's many other moons as well. The biggest surprise came from Enceladus, where geysers thought to be frozen exploded into the air. These geysers create one of the rings around Saturn.

NASA and ESA extended *Cassini*'s mission in 2004. Scientists continue to explore Saturn's moons and rings. NASA uses the Internet regularly to update scientists and the general public about *Cassini*'s ongoing advances.

Scientific Investigations of Comets and Asteroids

Besides these missions to planets, scientists have also made great efforts to study other objects in the Solar System—including dwarf planets, comets, and asteroids. These smaller objects also have important lessons to teach.

The *New Horizons* Mission to Pluto and the Kuiper Belt

The NASA probe *New Horizons* launched from Earth in 2006 and will reach Pluto in 2015. Since there are more icy, dwarf-like planets similar to Pluto, it is important to get a better understanding of Pluto's atmosphere and environment. No such mission has ever been undertaken before.

The *New Horizons* team rehearses launch at Cape Canaveral in Florida in late 2005. They are (from left) David Kusnierkiewicz, New Horizons mission system engineer; Glen Fountain, Applied Physics Lab project manager; and Alan Stern, principal investigator from Southwest Research Institute. The probe should reach Pluto in 2015.

Courtesy of NASA

Pluto holds many mysteries for astronomers. Learning more about the dwarf planet and its three moons may help explain some unanswered questions about the Solar System. *New Horizons* will spend 150 days exploring Pluto during its flyby, and will travel within 6,000 miles (9,650 km) of Pluto's mass.

One question scientists aim to answer is why Pluto's atmosphere is escaping into space like the tail of a comet. They hope the answers will explain more about the Earth's atmosphere.

New Horizons visited Jupiter in 2007 and confirmed that the moon Europa does have an ocean beneath its surface of ice. Now as the probe travels outward toward Pluto it will "sleep" for most of the next five years. Scientists will turn the probe on for less than two months out of the year, every year from now through 2015, to do regular maintenance and upload information.

Once the probe finishes exploring Pluto, scientists will send *New Horizons* to the Kuiper Belt. This ring of comets, which begins after Neptune and extends beyond Pluto, is believed to be the source of short-period comets, those having a life span of about 200 years. Scientists would like *New Horizons* to examine a few of the objects in the Kuiper Belt closely.

Missions to Explore Comets

Astronomers from many countries have put together 15 missions to explore comets over the years. The most common target of their exploration was Halley's Comet, which came close to Earth in the 1980s.

The Soviet missions *Vega 1* and *Vega 2*, launched in 1984, are two of the most ambitious ever undertaken. Not only did scientists plan to use the probes to explore Venus—they also arranged for the nearly identical probes to have close encounters with Halley's comet in 1986. Several European nations, including Bulgaria and what were then East and West Germany, took part in the mission.

The landers from both probes made it to the surface of Venus. The *Vegas* then used Venus's gravity to propel them toward the comet. According to one NASA estimate, *Vega I* traveled to within 6,214 miles (10,000 km) of Halley's Comet, while *Vega II* got even closer—1,864 miles (3,000 km). Together the two probes took more than 1,200 pictures of the comet at angles scientists on the ground never could have seen. The two probes gave scientists data about the comet's nucleus, dust production, its chemical composition, and its rotational rate.

The Japanese probe *Sakigake*, launched in 1985, was the first deep space probe from a country other than the United States or the Soviet Union. Its primary mission was to serve as a test for a second probe called *Suisei*, which also studied Halley's Comet. While its tenure covering Halley's comet was brief, *Sakigake* continued to send signals back to Earth until 1999. *Suisei* snapped a series of ultraviolet images of the huge 12.4 million-mile (20 million km) coma of dust and gas that surrounded Halley's nucleus.

Meanwhile, the ESA's contribution to the international fleet that greeted Halley's Comet was *Giotto*. European engineers designed it to fly as close as possible to the comet's nucleus. They did not expect it to survive the encounter. During the mission, *Giotto* flew within 372 miles (600 km) of the comet's core. It took hits 100 times a second from comet particles and debris. By the end of the mission to study the comet, more than 57 pounds (27 kilograms) of dust coated the probe.

Because of its success with Halley's comet, the ESA extended the probe's mission to monitor Comet Grigg-Skjellerup in 1992. This mission gave scientists a wealth of information on the dynamics of comets.

Currently three missions are under way to study comets, including ESA's *Rosetta*, which launched in 2004. It should encounter Comet 67P/Churyumov-Gerasimenko in 2014. The spacecraft will release a small probe onto the icy nucleus, and then spend the next two years orbiting the comet as it heads toward the Sun. On the way to the comet, *Rosetta* will receive gravity assists from Earth and Mars, and will fly past many asteroids.

Manned and Unmanned Spaceflight

Finally, NASA's *Deep Impact* probe flew to Comet Tempel 1 in 2005 and launched a probe to crash into the comet. *Deep Impact* then studied the material the impact ejected from the comet. It found evidence that the comet was covered in a small amount of water ice.

Missions to Explore Asteroids

By learning more about asteroids, scientists can gain a better understanding of the Solar System's chemistry. They believe this will help them develop more precise theories on Earth's formation and history.

NASA designed its *Deep Space 1* probe to test new technologies for future deep space and interplanetary missions. It launched in 1998 and was the first in a new series of technology demonstration missions under NASA's New Millennium program.

The main objectives were to test equipment such as ion propulsion, autonomous optical navigation, a solar power concentration array, a miniature camera, and an imaging spectrometer during a flyby of the asteroid 9969 Braille.

A month after its launch, scientists started the ion propulsion system (fueled by xenon gas) while the spacecraft was 2.9 million miles (4.8 million km) from Earth. The engine ran continuously for two weeks. The xenon ions provided a push to the spacecraft as much as ten times greater than possible with chemical propellants.

The probe passed within 16 miles (26 km) of 9969 Braille in 1999. This was the closest asteroid flyby up to that time. Photographs taken and other data collected during the encounter were transmitted back to Earth in the following few days. Some of the findings included the size of the asteroid, which was anywhere from 1.3 miles (2.2 km) long to less than a mile (about 1 km).

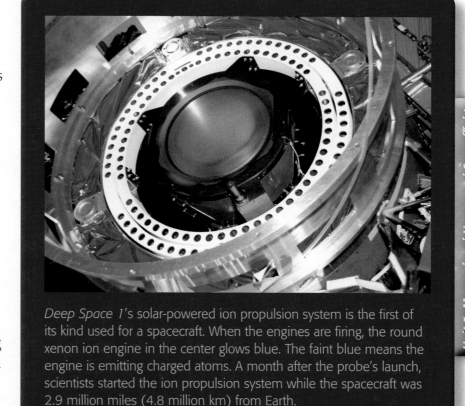

Deep Space 1's solar-powered ion propulsion system is the first of its kind used for a spacecraft. When the engines are firing, the round xenon ion engine in the center glows blue. The faint blue means the engine is emitting charged atoms. A month after the probe's launch, scientists started the ion propulsion system while the spacecraft was 2.9 million miles (4.8 million km) from Earth.

Courtesy of NASA/JPL

The mission was so successful that scientists extended it. In 2001 after scientists corrected some technical difficulties, *Deep Space 1* encountered the coma of Comet Borrelly at 10 miles (16.5 km) per second, capturing pictures of the nucleus along with other data.

Unmanned missions, such as these flybys of asteroids, comets, Jupiter, and Saturn, spark the human imagination in many important ways. They gather chemical, organic, and other data on the Solar System that reveals much about Earth.

Space missions also, quite simply, excite people and stimulate new ideas. As *Apollo 11* flight director Glynn Lunney told *The Washington Times* in 2010, "The space program has a lot to do with American pride, technology, and inspiration. … What creates true wealth for a nation in the long run? People can lose track of that. It doesn't come from the lawyers or the journalists—but [from] engineers. They have consistently contributed to the American portfolio of ideas, which must be replenished or it will shrivel up. We'll run out of innovations."

CHECK POINTS

Lesson 2 Review

Using complete sentences, answer the following questions on a sheet of paper.

1. What was wrong with the Hubble Space Telescope's original design that required an immediate correction?

2. With the aid of HST, how old do scientists now say the universe is?

3. Name one important finding from the Hubble Space Telescope and what it has done for the field of science.

4. What makes Io the most volcanically active body in the Solar System?

5. What did scientists discover in 2007 about the bubble of solar wind surrounding the Solar System?

6. What did the *Galileo* probe learn about Jupiter during its brief mission through its atmosphere?

7. Describe the landscape of the moon Titan.

8. What's the source of short-period comets? What is their average life span?

9. What was the first deep space probe to come from a country other than the United States or the Soviet Union? Who sent the mission into space?

10. What were the main objectives of the *Deep Space 1* mission?

 APPLYING YOUR LEARNING

11. List the advantages of unmanned missions over manned missions, and vice versa.

Unit 3: Manned and Unmanned Spaceflight

UNIT 4

An artist's conception of NASA's next Mars explorer, *Curiosity*—a mobile robot also known as the Mars Science Laboratory. Its task will be to discover whether Mars ever had or currently can support microbial life.
Courtesy of NASA/JPL-Caltech

Space Technology

Unit Chapters

CHAPTER 10

STS-127 crew members release the Atmospheric Neutral Density Experiment 2 (*at left*). This experiment consists of two spherical micro-satellites, which will measure the density and composition of the low-Earth orbit atmosphere.
Courtesy of NASA

Orbits and Trajectories

Chapter Outline

> We are gliding across the world in total silence, with absolute smoothness; a motion of stately grace which makes me feel godlike as I stand erect in my sideways chariot, cruising the night sky.

Michael Collins, Carrying the Fire: An Astronaut's Journey

LESSON 1 Orbits and How They Work

Quick Write

_____

_____

Can science fiction contribute to scientific progress? What other science-fiction writers can you think of who foresaw important scientific discoveries and inventions?

Learn About

- how orbits work
- different types of orbits used for different purposes

Sir Arthur C. Clarke, who lived from 1917 to 2008, was a famous British science fiction writer and futurist. Perhaps his most famous book was *2001: A Space Odyssey*. Director Stanley Kubrick made it into a popular movie, which was released in 1968. One of the film's most memorable characters is the renegade computer, HAL, which tries to take over the spacecraft and kills several astronauts before the surviving crew member can subdue it. Clarke later wrote three additional novels involving several of the same characters.

Like many of the best science-fiction writers, Clarke was a practical thinker. In 1945 he published an article in *Wireless World* calling for a system of relay satellites in geostationary orbit around the Earth. He thought these satellites would be useful for telecommunications around the world. Remember that at that time, no country had even launched a satellite. Clarke was not the first or only person to come up with the idea, but his article helped popularize it. Today some people refer to geosynchronous orbit as "Clarke orbit."

How Orbits Work

You've been reading about different orbits throughout this book. But think a minute about what an orbit is. It's a regular, repeating path that one object in space takes around another object. The Moon and man-made satellites orbit the Earth. The Earth and the other planets orbit the Sun, along with asteroids, dwarf planets, and comets. Most of these objects orbit along or close to an imaginary flat surface astronomers call the ecliptic plane, which you read about in Chapter 9.

In Chapter 1 you read about Johannes Kepler's discovery that the planets' orbits are elliptical, or oval-shaped. This is true of many satellites' orbits, too. Satellites are not always the same distance from Earth during their orbits. In Chapter 2 you read that *perigee* is when a satellite is closest to Earth. *Apogee* is when an object is at its farthest distance from Earth.

Vocabulary

- orbital velocity
- inclination
- geosynchronous Earth orbit
- Sun-synchronous orbit
- low-Earth orbit
- high-Earth orbit
- medium-Earth orbit

Momentum and Gravitational Force

To place a satellite into orbit, scientists must deal with Sir Isaac Newton's law of gravity, which you studied in Chapter 1. If it weren't for gravity, a satellite launched into space would keep right on going away from Earth in accordance with Newton's first law of motion: An object in motion will stay in motion unless something pushes or pulls it. But gravity pulls the object back to Earth. An orbit is the constant struggle between the momentum of the object in a straight line, and gravity pulling the orbiting object back toward the body it circles.

In previous chapters you learned about various missions to space, including the Pioneer missions of the mid-twentieth century. *Pioneer 1* and *Pioneer 2* both failed, in large part, because their momentum wasn't enough to escape Earth's gravity and propel them to the Moon. (*Pioneer 1* did return some useful data about Earth, however.) In 2009 a NASA satellite launched to study global warming. But it was too heavy and failed to escape Earth's gravity. It crashed in the Antarctic Ocean less than 48 hours after takeoff from California.

A space shuttle must rocket out of Earth's atmosphere to reach orbit. Yet the shuttle can't travel so far beyond Earth's atmosphere that gravity can't hold it in orbit around the planet. When momentum and gravity are balanced, a satellite is always falling into the planet. But because it's moving "sideways" fast enough, it never hits the planet.

Escape Velocity

Escape velocity—which you read about in Chapter 4—is the speed an object must reach to overcome a body's gravity and leave its orbit. Because a body's size and mass determine its gravitational pull, escape velocity differs from planet to planet, moon to moon, and asteroid to asteroid.

The location and direction from which scientists launch a satellite or spacecraft is also important. For instance, the closer an object is to the Earth's equator, the more speed it gains from the Earth's own rotation—assuming it's launched eastward. This is because the planet's surface rotates fastest at the equator. The surface rotation speed slows down the farther you get from the equator, until it's almost at a standstill at the poles. A satellite launched in a different direction does not gain this advantage, and one launched toward the west needs more velocity to overcome the rotation.

A <u>Saturn 1B</u> takes off for a <u>Skylab</u> mission in 1973.

Courtesy of NASA

The additional miles per hour that engineers gain by launching from near the equator helps them propel rockets and their payloads into space. That's one reason the European Space Agency maintains a spaceport in Latin America in French Guiana, about 300 miles (500 km) from the equator.

To reach escape velocity, engineers must decide how much thrust to apply to a launch vehicle. They can't have too little, but, depending on their destination, they can't have too much, either.

Figure 1.1 In this artist's conception, a GOES satellite orbits some 22,000 miles (35,000 km) above Earth. Once a spacecraft achieves orbit, it must reach a certain orbital velocity—the speed an object must maintain to stay in orbit.
Courtesy of NOAA

As NASA International Space Station science officer Ed Lu wrote during his *Expedition 7* mission on how to reach orbit: "[A] lot of speed and initially a little bit of aiming to make sure you don't hit the ground…." And if you get these factors right along with getting high enough to be out of Earth's atmosphere, "[Y]ou will just keep going round and round the planet."

Star **POINTS**

The escape velocity for an object from the Earth is seven miles (11.3 km) a second.

Orbital Velocity

Once a spacecraft achieves orbit, it must reach a certain orbital velocity—*the speed an object must maintain to stay in orbit*. This velocity differs, depending on how high up in orbit a spacecraft or satellite is. For example, NASA says that at 150 miles (242 km) above Earth, a spacecraft must travel about 17,000 miles (27,000 km) per hour—far less than full escape velocity.

The closer an object is to Earth, the faster it needs to travel to remain in orbit. This is because the closer an object is to Earth, the stronger Earth's gravity is pulling downward on it. The faster an object travels, the more it "rejects" the gravitational pull. As Newton reasoned more than 300 years ago, Earth's gravity pulls much less on the Moon than it does on objects in closer orbit. So the higher a spacecraft climbs from Earth, the slower it can travel and still resist gravity (Figure 1.1).

Unit 4 Space Technology

Orbital Period

The amount of time a satellite takes to orbit Earth—its orbital period—changes dramatically depending on how far above Earth it is. The closer to Earth, the stronger gravity's pull, with the result being a faster orbital period. Here are three examples for comparison:

1. NASA's Aqua satellite orbits Earth in 99 minutes from 437 miles (705 km)
2. A weather satellite orbits in 23 hours, 56 minutes, from 26,141 miles (42,164 km)
3. The Moon orbits Earth in 28 days from 238,329 miles (384,403 km).

How Height, Eccentricity, and Inclination Affect an Orbit

Many factors affect orbiting satellites. Height, as you've read, determines the speed required for a satellite to remain in orbit. But an orbit's shape and angle will affect what path a satellite will take and, therefore, determine what it sees and reports back to Earth.

The elliptical orbits of satellites come in different degrees. As you read in Chapter 1, *eccentricity* is the term scientists use to refer to an orbit's shape. A low-eccentricity orbiting object is one flying in a more-round, less-oval path. A circular orbit has an eccentricity of zero. A highly eccentric orbit is close to 1 (but never reaches it).

Inclination is *the angle an orbit takes as it circles the Earth*, especially as it relates to the equator. If an orbit is directly over the equator, it has a zero inclination. An object that revolves or orbits north to south (and then south to north) has a 90-degree inclination.

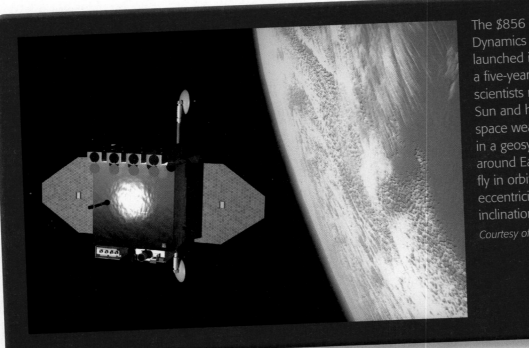

The $856 million Solar Dynamics Observatory, launched in 2010 for a five-year mission, tells scientists more about the Sun and how it impacts space weather. It flies in a geosynchronous orbit around Earth. Satellites fly in orbits with varying eccentricities and inclinations.
Courtesy of NASA

Different Types of Orbits Used for Different Purposes

Scientists not only consider speed and gravity when planning launches, they also must choose from many types of orbits. The kind of orbit they pick depends on the mission.

Some satellites maintain station over the same spot on Earth's equator. Others circle from North Pole to South Pole and back again. Spacecraft also enter a low-, medium-, or high-Earth orbit, again determined by the mission's purpose.

Geosynchronous Earth Orbit (GEO)

Satellites that seem to be attached to some location on Earth are in geosynchronous Earth orbit (GEO). Geosynchronous orbit is *an orbit around a planet or moon that places the satellite in the same place in the sky over a particular point on the surface each day.* An object in geosynchronous orbit appears to remain still all day long, but in reality it orbits at a speed equal to the Earth's rotation. So it remains fixed over the same spot at all times. When a geosynchronous orbit is over Earth's equator, it's called a *geostationary orbit.*

Geostationary Operational Environmental Satellites (GOES)

Satellites in geostationary orbit—such as Geostationary Operational Environmental Satellites (GOES)—fly very high above the Earth in the equatorial plane. GOES orbit at approximately 21,700 miles (35,000 km) high. Scientists place GOES this high so the satellites can travel slowly enough to make only one orbit a day, just as the Earth makes one revolution per day (Figure 1.2).

Since GOES and other geostationary satellites stay above a fixed spot on Earth's surface, they watch continuously for the atmospheric triggers of severe weather such as tornadoes, flash floods, hailstorms, and hurricanes. When these conditions develop, the GOES monitor the storms and track their movements. They also provide general weather monitoring to provide data for daily weather forecasts.

Figure 1.2 GOES fly very high above the Earth in the equatorial plane.
Courtesy of NOAA

These satellites provide scientists with consistent, long-term observations, 24 hours a day, seven days a week. They monitor fast-breaking storms across "Tornado Alley" (a region in the heart of the United States often hit hard by tornadoes) as well as tropical storms in the Atlantic and Pacific Oceans. Scientists also use the satellite information to monitor coral reefs, fires, and volcanic ash. Monitoring the Earth from space helps scientists better understand how the Earth works.

Data from GOES satellites also help meteorologists estimate rainfall during thunderstorms and hurricanes for flash flood warnings. Such data also help them make snowfall projections, and issue winter storm warnings and spring snowmelt advisories. Satellite sensors also detect ice fields and map the movements of sea and lake ice. They help track icebergs that threaten shipping—like the one that sank the *Titanic* in 1912.

Polar-Orbiting Operational Environmental Satellites (POES)

While GOES are high-flying orbiters, Polar-orbiting Operational Environmental Satellites (POES) fly lower and closer to the Earth. This means they have a much faster orbital period. In fact, they monitor the poles and circle the Earth once every 100 minutes. Combining the data from three POES over a six-hour period allows scientists to compile an image covering nearly every inch on Earth.

POES also pass over the same latitudes at the same times each day. This way scientists can more easily note any changing conditions on the ground.

These polar orbits are Sun-synchronous orbits. A Sun-synchronous orbit is *an orbit coordinated with Earth's rotation so that the satellite always crosses the equator at the*

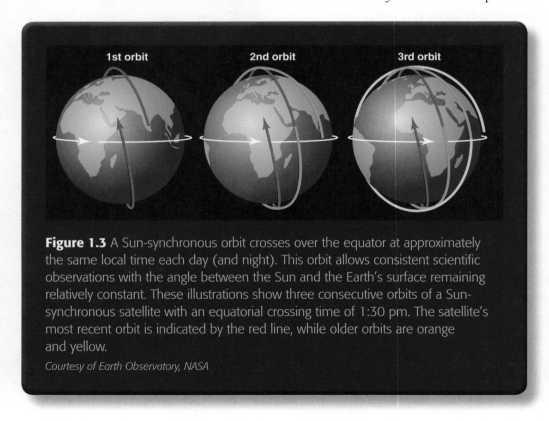

Figure 1.3 A Sun-synchronous orbit crosses over the equator at approximately the same local time each day (and night). This orbit allows consistent scientific observations with the angle between the Sun and the Earth's surface remaining relatively constant. These illustrations show three consecutive orbits of a Sun-synchronous satellite with an equatorial crossing time of 1:30 pm. The satellite's most recent orbit is indicated by the red line, while older orbits are orange and yellow.

Courtesy of Earth Observatory, NASA

POES and You

You and your neighbors benefit from POES satellites, particularly if you're planning on flying, boating, fishing, or farming. More than 50 percent of the US public uses the three-to-five day forecasts issued by the Weather Service to plan business or pleasure activities. These are based on POES data. And your local governments use the information from these satellites to plan expansion and to monitor growth. Finally, search-and-rescue equipment on board the satellites has helped save more than 24,000 lives.

The military benefits from the satellites as well, tactically and strategically. As a result, the military takes advantage of good weather conditions to maneuver, and exercises caution when the weather is bad.

same local time on Earth. The orbital planes of these satellites rotate one degree a day because the Earth bulges at its equator. This bulge exerts extra gravitational pull on the satellites as they pass overhead and gradually rotates the satellites' orbital planes. As a rule, a Sun-synchronous satellite's orbital plane will complete one full rotation about Earth's axis in one year's time due to this force of gravity (Figure 1.3).

POES see everything from environmental conditions to real-time weather. They provide global coverage of the Earth's weather, atmosphere, oceans, land, and near-space environment. NASA and the National Oceanic and Atmospheric Administration (NOAA) share responsibility for these satellites, as they do for GOES. But NOAA operates them.

The POES system monitors the entire planet and provides information for long-range weather and climate forecasts. The data gathered by these satellites saves lives by allowing more-efficient disaster planning and faster response times to severe weather conditions, such as tornadoes and floods.

These satellites also collect a large amount of specific information about the Earth's surface, as well as atmospheric and space environmental measurements. This lets scientists and forecasters monitor and predict weather patterns with greater speed and accuracy.

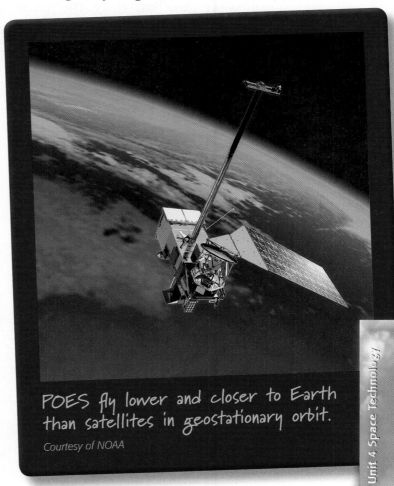

POES fly lower and closer to Earth than satellites in geostationary orbit.

Courtesy of NOAA

Low-Earth Orbit (LEO)

POES usually travel in a low-Earth orbit—*an orbit up to about 1,240 miles (2,000 km) above the Earth*. This area is the easiest to get to when orbiting the Earth.

The International Space Station and space shuttles travel in this zone as well. A complete revolution in low-Earth orbit can take as little as 90 minutes.

If it were possible to drive to the International Space Station in low-Earth orbit, it would take you only a few hours to get there. It's only about 250 miles (400 km) high. By comparison, a geosynchronous high-Earth orbit—*an orbit at an altitude of about 22,300 miles (35,900 km)*—begins about one-tenth of the way from the Earth to the Moon.

A medium-Earth orbit is *one with an altitude of about 12,400 miles (20,000 km)*. Satellites in this orbit take about 12 hours to orbit the Earth. Such an orbit is outside Earth's atmosphere and very stable. Radios across the globe can receive signals from satellites at this altitude. Along with the orbit's stability, this makes it ideal for navigation satellites, although such satellites have used both higher and lower orbits.

Of course, once engineers and scientists have decided which orbit to use for a mission, they have to get the spacecraft there. For manned and some research flights, they have to get it back as well. Maneuvering into space and in space is a lot more complicated than driving a car or even flying a plane. The next lesson will examine some of the challenges.

CHECK POINTS

Lesson 1 Review

Using complete sentences, answer the following questions on a sheet of paper.

1. Give an example of what can happen when the speed of an object racing toward space is not balanced properly with Earth's gravity.

2. To determine escape velocity, what must engineers decide?

3. What is the orbital velocity needed to stay 150 miles above the Earth?

4. When an orbit is directly over the equator, what inclination does it have?

5. What is a geostationary orbit?

6. Why do scientists place GOES high above Earth?

7. What advantage do scientists gain by having POES pass over the same latitudes at the same times each day?

8. What is low-Earth orbit?

APPLYING YOUR LEARNING

9. Why do engineers like to launch satellites from near the equator?

Quick Write

_____

_____

A lot of scientists with higher degrees and a lot more experience were working on the problem Michael Minovitch solved. What does his story tell you about scientific discovery and technological progress?

Learn About

• trajectories in space travel
• maneuvering in space
• navigation data

UCLA graduate student Michael Minovitch gets credit for developing gravity assist trajectories in the 1960s. He was working at the Jet Propulsion Laboratory at the California Institute of Technology.

Before Minovitch came along, scientists figured that the only way to send satellites to the outer planets was to use more-powerful launch vehicles. They even concluded that possibly only nuclear reactors could produce enough thrust. This would have come with a big price tag.

But Minovitch took principles astronomers already knew about comet trajectories and applied those same laws to spacecraft travel. Many missions since then have benefited from this grad student's discovery.

Trajectories in Space Travel

Getting a spacecraft where you want it to be is a lot more complicated than simply flipping a switch and having a rocket carry it into space. Many factors come into play to place a spacecraft on the correct trajectory; that is, the right path to its target. Depending on where they want a spacecraft to travel—from one planet to another, or from one orbit around Earth to another, or from low-Earth orbit (where the International Space station is) to the Moon—scientists must calculate and plan the proper trajectory to get it there.

In determining a mission's trajectory, scientists must take into account two immediate concerns—fuel and cost. Less propellant can mean a slower, therefore longer and more expensive trip. That means finding other energy sources to boost the spacecraft and keep it on course. They must find ways for these other energy sources—Earth's revolution or a planet's momentum, for example—to work in conjunction with the propellant.

Vocabulary

- periapsis
- apoapsis
- Type 1 trajectory
- Type 2 trajectory
- Hohmann transfer orbit
- nanosecond
- right ascension
- declination

Star POINTS

Strictly speaking, a trajectory is the same thing as an orbit—but scientists usually speak of trajectories in terms of getting from one location to another, while orbit usually refers to a spacecraft's path around another object.

How much energy a mission uses depends in part on its target and where it is located in relation to Earth at any given time. Getting to the Moon, Mars, and a distant asteroid will each require a different trajectory. In space, you can't just fly in a straight line from Point A to Point B. Remember that everything is in orbit around the Sun—and that includes all the spacecraft launched from Earth, regardless of where they are going.

The trick in getting from one planet to another is to develop a trajectory in which a spacecraft leaves Earth at the right time so that when it arrives in the other planet's orbit, the planet is actually there. For example, spacecraft can go to Mars' orbit any time, but they have only very specific windows of opportunity when they can leave Earth and actually get to Mars itself. The distance between Earth and the target object at the arrival time obviously determines the length of the trip, and therefore the amount of energy the mission requires.

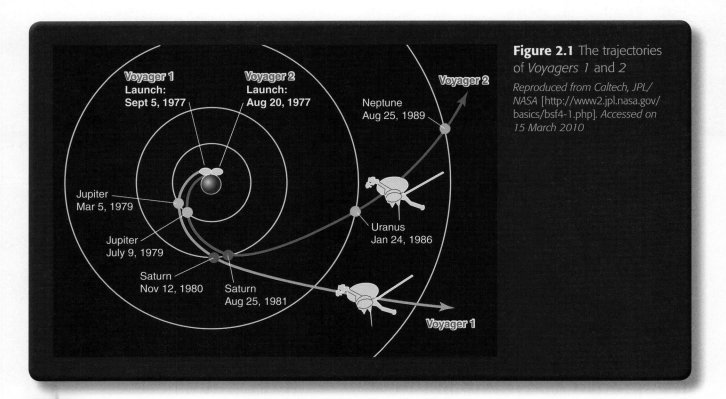

Figure 2.1 The trajectories of *Voyagers 1* and *2*

Reproduced from Caltech, JPL/ NASA [http://www2.jpl.nasa.gov/ basics/bsf4-1.php]. *Accessed on 15 March 2010*

Gravity-Assist Trajectories

One type of trajectory space scientists use is the gravity assist trajectory. As you have read, every object in space exerts a gravitational force. NASA often uses the gravity of a planet and its orbital momentum to save energy and money, and add to a spacecraft's speed.

Gravity-assist trajectories draw on a planet's angular momentum—the motion that comes from a body's rotation around its axis. You read about the conservation of angular momentum in Chapter 3. This angular momentum can be transferred from an orbiting planet to a spacecraft approaching from behind the planet in its path around the Sun.

Spacecraft use this angular momentum either to speed up when behind a planet, or slow down when in front of a planet. Stealing or giving back angular momentum on a gravity assist trajectory also allows a spacecraft to change its direction. So a spacecraft such as *Voyager 2* swings around Jupiter to reach Saturn next. It then whips around Saturn to head to Uranus and make use of its gravity. *Voyager 1* skipped getting close to Neptune because NASA scientists didn't need it to gain speed at that point in its travels (Figure 2.1).

When a spacecraft's goal is to fly by another planet, scientists must time missions to coincide with that planet's orbit. The planet's gravity and orbital momentum then accelerates the probe free of charge, so to speak. When this happens, the planet actually slows down—it has given up a tiny bit of momentum to the spacecraft. (This loss of speed is too small to measure.)

As the spacecraft approaches the planet and reaches periapsis, *the point where an orbiting body is closest to the object it is orbiting*, gravity accelerates it. As the object moves farther away from the planet toward apoapsis, *when an orbiting body is farthest from the object it is orbiting*, it begins to escape gravity's pull.

Scientists have used the gravity assist method numerous times. In 1973 *Mariner 10* was the first spacecraft to use it when it flew by Venus en route to Mercury.

Type 1 and Type 2 Trajectories

Scientists divide trajectories into three types. A Type 1 trajectory is *a route that is less than 180 degrees around the central body*. A Type 2 trajectory is *a path that is more than 180 degrees*. A Hohmann transfer orbit is *a trajectory that travels exactly 180 degrees around the central body* (Figure 2.2).

A trajectory's shape and the relative alignment of the start and end points drive how much of a change in velocity scientists need. That determines the amount of propellant needed to make the change. This amount ultimately determines a spacecraft's size. The launch vehicle and the spacecraft must hold enough propellant to give a satellite the necessary velocity.

Scientists often use the Hohmann transfer orbit because it requires the minimum amount of energy transfer between any two points in space—particularly from one orbit to another. In other words, it's the most fuel-efficient trajectory. You only need enough fuel to move the spacecraft into a solar orbit in which its closest approach to the Sun is at the distance of Earth's orbit—and its farthest distance from the Sun aligns with the target object's orbit.

However, a Hohmann transfer orbit is not always the speediest, nor does it always meet a mission's particular needs. Therefore, scientists usually employ a Type 1 or Type 2 trajectory. The choice has mostly to do with how much time it takes to get from one place to another and when the window of opportunity presents itself. For example, going to Mars requires scientists to use a Type 1 trajectory, which typically uses more propellant and energy but travels there quickly.

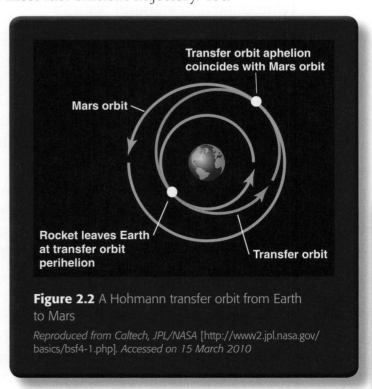

Figure 2.2 A Hohmann transfer orbit from Earth to Mars

Reproduced from Caltech, JPL/NASA [http://www2.jpl.nasa.gov/basics/bsf4-1.php]. Accessed on 15 March 2010

Unit 4 Space Technology

Cassini took this photo of Saturn's moon Mimas against the backdrop of the planet's northern latitudes.

Courtesy of Cassini Imaging Team/SSI/JPL/ESA/NASA

If the planets line up perfectly, scientists can use a near-Hohmann transfer to spend as little energy as possible. This saves on propellant but puts spacecraft in transit for longer periods. Scientists typically plan robotic mission launches to take advantage of those alignments as much as possible. With Mars, this alignment happens once every 20 months or so.

For a manned mission to Mars, on the other hand, scientists might want a quicker trip. So they might prefer to use a Type 1 trajectory—that way, the astronauts would spend less time in deep space. Scientists also used this type of trajectory to speed the Mars rovers to the red planet.

Type 2 trajectories are less desirable because they take much longer to reach the target. Sometimes they are necessary, however, because scientists are trying to fly past several planets to get somewhere else using gravity assist.

Take the *Cassini* spacecraft, for example, which you learned about in previous chapters. When it went to Saturn, it flew by Venus twice, and Earth and Jupiter once, following several Type 2 trajectories along the way.

Outbound and Inbound Velocity

Recall the earlier discussion about *Voyager 2*. As the spacecraft approached Jupiter, it sped up because of the planet's gravitational pull. Having picked up some of Jupiter's orbital momentum, it passed by Jupiter, with a velocity greater than Jupiter's escape velocity. (Had its velocity been less than Jupiter's escape velocity, it would have entered into orbit around the giant planet or even crashed into it.) As *Voyager 2* traveled away from Jupiter, it began to slow down again as the planet's gravity pulled on it.

But *Voyager 2* left Jupiter carrying an increase in angular momentum stolen from the planet. Jupiter's gravity served to connect the spacecraft with the planet's ample reserve of angular momentum. This allowed the spacecraft to change the direction of its trajectory to get to Saturn. *Voyager 2* repeated this technique at Saturn and Uranus.

These varying speeds and angles are known as the *inbound* and *outbound velocity*. Think of them like on and off ramps on a freeway. When driving onto a freeway, you speed up as you travel along the on ramp so that you are traveling at the corresponding speed to other traffic on the road. Likewise, when you are exiting the highway, your speed slows. (In this analogy, of course, you are causing the change in speed. The spacecraft, on the other hand, is using natural forces.)

Hyperbolic, Parabolic, and Elliptical Paths

Once out of Earth's atmosphere, an orbiting object can take one of three types of paths: hyperbolic, elliptical, or parabolic. An elliptical path is one that travels a trajectory similar to a racetrack, closed in on itself. It is the path orbiters take if they are destined to return to Earth. If left on its own with no other forces involved, a spacecraft could travel this path indefinitely.

A hyperbolic path is one that takes the object much farther out into space. Spacecraft that will not return to Earth use this kind of trajectory. Such a satellite's goal is to reach the farthest corners of the universe and monitor space in an area where limited gravity pulls exist. This path stretches to infinity.

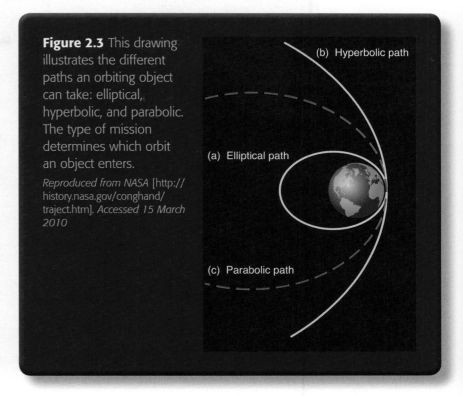

Figure 2.3 This drawing illustrates the different paths an orbiting object can take: elliptical, hyperbolic, and parabolic. The type of mission determines which orbit an object enters. *Reproduced from NASA* [http://history.nasa.gov/conghand/traject.htm]. *Accessed 15 March 2010*

(b) Hyperbolic path

(a) Elliptical path

(c) Parabolic path

Finally, a parabolic orbit is the middle ground between the two. It travels between the two other paths. Like the hyperbolic path, however, objects in a parabolic path do not return to Earth (Figure 2.3).

Maneuvering in Space

Once scientists set a spacecraft on its proper trajectory, they must occasionally nudge it one way or another to keep it on the right path. They may need to either slow it down, speed it up, or turn it. The satellite *Aqua* provides a good case study.

Mission Control Tune-Up for the Satellite Aqua

NASA launched *Aqua* in May 2002. Its name is Latin for *water*, and the list of things the satellite monitors relating to water on Earth is nearly endless: atmosphere, clouds, soil moisture, sea ice, snow cover, vegetation cover, water temperatures, and much, much more.

Aqua is on a Sun-synchronous orbit, which means it passes over the exact same spot at nearly the same time every day. For example, *Aqua* passes over the equator between both 1:30–1:45 a.m. and 1:30–1:45 p.m. It travels north to south and back north, covering nearly all of Earth daily. By traveling this route the satellite gives consistent and precise measurements, something that is crucial as NASA and others investigate theories such as climate change.

The Aqua satellite flies around Earth in a Sun-synchronous orbit.
Courtesy of Earth Observatory, NASA

As with all Sun-synchronous orbits, gravity gradually pulls the satellite off course. So scientists in Greenbelt, Maryland, must work within a narrow window to adjust its path. They can only send signals to the satellite at certain times of the day when the Earth is shielding *Aqua* from the Sun's harmful rays.

In 2009 scientists acted to fix the satellite's orbit. By firing its thrusters, they corrected *Aqua*'s direction (it was pointing the wrong way), as well as moved its orbit $^1/_{100}$th of a degree closer to the North and South Poles.

The Use of Ion Propulsion to Maneuver in Space

While scientists use thrusters to correct *Aqua*'s orbit, some spacecraft such as *Deep Space 1* employ ion engines. *Aqua*'s more traditional solid-fuel rocket motors work by pushing propellant away from the spacecraft. The action of the propellant leaving the engine causes a reaction that pushes the spacecraft in the opposite direction.

An ion engine uses this same principle, but this engine's great innovation is in how efficiently this happens. The gas xenon, which is like helium but heavier, flows into the ion engine. Here it receives an electrical charge. Once xenon atoms change to xenon ions (ions are charged atoms), an electrical voltage pushes them around. A pair of grids in the ion engine, electrified to almost 1,300 volts, accelerates the ions to a very high speed and shoots them out of the engine. As the ions race away from the engine, they push back on the spacecraft, propelling it in the opposite direction.

CHAPTER 10 Orbits and Trajectories

The xenon ions travel at about 77,000 miles per hour (124,000 km per hour). This is about 10 times faster than the exhaust from conventional rocket engines. So xenon gives about 10 times as much of a push to the spacecraft as chemical propellants do for the same amount of mass expelled. That means that it takes only one-tenth as much propellant for an ion engine to work as it does a chemical propulsion system.

For some missions, NASA can't afford to build and launch boosters large enough to carry the chemical propellants that a mission would require. Ion propulsion, therefore, is one of the ways to get around this problem. Ion engines can gradually build up high rates of speed that would be necessary in deep space missions.

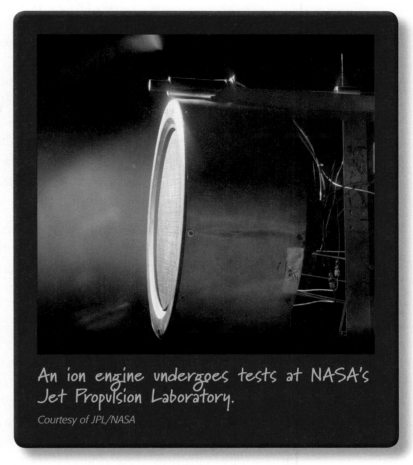

An ion engine undergoes tests at NASA's Jet Propulsion Laboratory.
Courtesy of JPL/NASA

Another advantage of the xenon ion is its longevity. Chemical engines can be operated for minutes, or in extreme cases, for an hour. Ion engines, however, can be operated for years. NASA's *Deep Space 1* spacecraft, which operated from October 1998 to December 2001, tested an ion propulsion system. The system carried about 179 pounds (81.5 kg) of xenon propellant, and could increase the spacecraft's speed by about 10,000 miles (16,100 km) per hour. The engine operated for more than 16,000 hours. Although retired, the spacecraft remains in orbit around the Sun with its radio receiver turned on in case future generations wish to contact it.

Orbit Determination and Flight Path Control

Whether scientists use solid rocket motors or ion engines, they must monitor their spacecraft with the same two concerns in mind: *orbit determination* and *flight path control*. Orbit determination is knowing and predicting a spacecraft's position and velocity. Flight path control is firing the rocket motor to change the spacecraft's velocity.

Remember that a spacecraft en route to a distant planet is actually in orbit around the Sun. And the portion of its solar orbit between the launch and the destination is its trajectory. Orbit determination involves finding the spacecraft's orbital details and taking into account any difficulties along the way.

Unit 4 Space Technology

Flight path control involves telling the spacecraft to change its velocity. Scientists decide how much to change a spacecraft's velocity, or speed, by comparing its trajectory with the destination object's orbit. They call these types of changes *trajectory correction maneuvers*. These maneuvers usually range from a couple meters to tens of meters per second. Engineers back on Earth command the spacecraft to fire its thrusters to make these adjustments. Changing the spacecraft's speed also changes the trajectory. When it goes faster, the trajectory straightens out a bit. When the spacecraft slows down, the trajectory curves more. These small changes will send the spacecraft to different points in space.

Changes made once a spacecraft has reached its target destination—around a distant planet, for instance—scientists refer to as *orbit trim maneuvers*. These are very minor adjustments. Engineers may order these maneuvers because they need to point an instrument camera on board a satellite toward a planet's surface, for instance.

Space Shuttle Rendezvous Maneuvers

Two extremely important maneuvers, which were absolutely essential to the manned mission to the Moon, are rendezvous and docking maneuvers. The space shuttle and Russian *Soyuz* capsules rendezvous and dock with the International Space Station, which is in permanent orbit around Earth. That is, they meet up (rendezvous) with the station and then the two lock together (dock) so crews and supplies can flow from one vessel to the next.

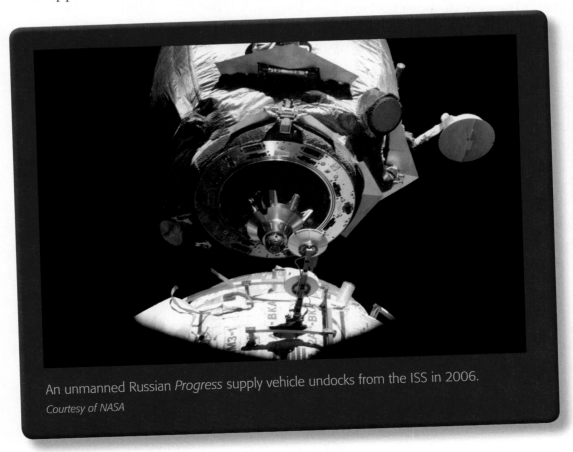

An unmanned Russian *Progress* supply vehicle undocks from the ISS in 2006.
Courtesy of NASA

CHAPTER 10 Orbits and Trajectories

To rendezvous and dock two spacecraft together is one of the more difficult tasks astronauts perform. NASA has instructors who train astronauts for six months ahead of a mission before allowing them to pilot these operations.

Rendezvous maneuvers are tricky and dangerous because both vehicles are traveling at about 17,500 miles (28,000 km) per hour. The pilot must learn precision. Instructors begin lessons in a classroom and then move the astronauts to simulators to practice docking.

To meet up in space, says shuttle rendezvous instructor Steve Gauvain, is not like driving a car on Earth. On the planet's surface, a car might accelerate to meet up with another car ahead of it. But in space, the higher a spacecraft climbs the more slowly it orbits. Gauvain says, "Even though you speed up [in space], you slow down relative to the other vehicle, because your altitude increases."

Therefore, astronauts must make sure that they don't fire their thrusters too long or they just might shoot past their target (Figure 2.4). As Gauvain says, the process is "backward" from what you might intuitively think. (As you read in the previous lesson, the closer an object is to Earth, the faster it travels to remain in orbit. You slow down to speed up because when you slow, you drop into a lower orbit where your velocity is higher and your orbital path shorter.) That's why the astronauts need so much training.

Navigation Data

When astronauts aren't at the helm, the engineers back at mission control must handle maneuvers in space. They do this regularly for spacecraft orbiting Earth and heading out to deep space. They draw on a variety of data to make their calculations. These navigation data include a spacecraft's velocity, distance, and angular measurements.

Spacecraft Velocity and Distance Measurement

Scientists interpret much of this information from their perspective on Earth. For instance, they measure velocity based on a spacecraft's speed toward or away from Earth. They measure distance based on a spacecraft's distance from the planet.

For missions that are closer to home, scientists will also consider a spacecraft's position in Earth's sky. And for longer missions, sometimes they use what they call optical navigation. This method captures images of a target planet or body against background stars and steers a spacecraft toward that destination.

Scientists test spacecraft velocity using the Doppler shift, which you read about in Chapter 4. Remember that as a sound or electromagnetic source approaches the receiver, its frequency increases. As the source moves away from the receiver, the frequency gets lowered. So scientists send a radio signal to the spacecraft. They compare the frequency of the original signal with the signal the spacecraft returns. The difference is the Doppler shift, which tells scientists how fast the craft is moving.

Figure 2.4 This painting shows a French orbiter approaching Mars with its thrusters firing, solar panels deployed to provide power, and a heat shield (white) to protect the spacecraft as it enters orbit around the planet. Scientists use thrusters to correct satellite orbits and trajectories.

© David Ducros/Science Photo Library

To measure a spacecraft's distance from Earth, scientists track what's called a *ranging pulse*. A station in NASA's Deep Space Network (DSN), a series of antennas placed around Earth to communicate with spacecraft, sends this pulse to a probe. When the spacecraft receives the ranging pulse, it returns it to the DSN station. Scientists measure the time it takes the spacecraft to turn the pulse around to determine the vehicle's distance from Earth. They include into this measurement a certain amount of time it will take a spacecraft to process the pulse before sending it back.

For example, *Cassini* takes approximately 420 nanoseconds—a nanosecond is *one billionth of a second*. The system has built-in delays, including several microseconds needed to go from the computers to the antenna within the Deep Space Network. Scientists measure these delays before each use. They determine the true elapsed time at light speed when the Deep Space Network receives the pulse. Finally they apply corrections for known atmospheric effects and compute the spacecraft's distance from Earth.

In Her Own Words: Linda Spilker, PhD

I have always enjoyed science, most especially studies of the stars and planets. My parents bought me my first telescope when I was nine years old. The first thing I did was to use it to look at Jupiter and its moons. As I was growing up I read many books about astronomy, our Solar System, and the missions that NASA flew to the planets. I always hoped that some day I would be able to work on NASA missions. For a while I even wanted to be an astronaut and go to the Moon!

In junior high and high school I took as many advanced math and science classes as they offered. I felt some pressure not to go into science because, at that time, it was not a field that women traditionally pursued. I was told that science would be too hard for me in college (even though I had done very well in all my classes!). I really liked science, however, and decided to pursue my interests in spite of what other people said. My parents also encouraged me to pursue my interest in science.

Linda Spilker, *Cassini* deputy project scientist
Courtesy of JPL/NASA

I was an undergraduate at Cal State Fullerton where I got my degree in physics. Over two summers I had funding from a National Science Foundation grant to do meteorite research. I worked at the California Institute of Technology with Professor Dorothy Woolum. She encouraged my development as a scientist.

After I graduated from college I applied for job at the Jet Propulsion Laboratory. In 1977 I was hired to work on the *Voyager* mission. A few months later I was thrilled to watch *Voyager* launch from Florida.

I worked for 13 years for the *Voyager* Infrared Team as *Voyager* flew by Jupiter, Saturn, Uranus, and Neptune. *Voyager*'s discoveries were astounding! I used some of the *Voyager* data on Saturn's rings to write my master's thesis. I really enjoyed the research aspect of my job. I realized, however, that in order to conduct my own research I needed to get my PhD I went to graduate school at UCLA and got a PhD in geophysics and space physics in 1992. Now I am able to pursue my own research on rings and join science teams on planetary missions like the *Cassini* mission to Saturn.

I am currently the *Cassini* Deputy Project Scientist and a member of the *Cassini* Composite Infrared Team. I have worked on *Cassini* since 1988. My responsibilities include helping the project put together the best science possible for the [time] *Cassini* will spend in orbit around the planet Saturn. Studying planets like Saturn helps us understand more about the Earth. By studying the atmosphere and winds of a giant planet like Saturn we may be able to better predict the Earth's weather. Saturn's moon Titan has an atmosphere that contains hydrocarbons and other compounds that may represent the building blocks for life. By studying Titan we may get a better understanding of how life evolved on the early Earth. *Cassini* will be making observations of Saturn, Saturn's rings, icy moons, the large moon Titan, and the magnetosphere.

In Her Own Words, *continued*

As a research scientist, my primary interest is in understanding ring systems and how they work. Saturn's rings are made up of millions of particles ranging in size from dust to large boulders. Many of these ring particles are affected by the moons that orbit outside them. The gravity from the moons causes the ring particles to bump into each other and create interesting patterns in the rings such as waves and wakes. Science and math are involved in modeling ring systems.

During a typical day I spend some time in meetings discussing how best to use the *Cassini* spacecraft. I also spend time talking to other scientists about what they would like to see *Cassini* do at Saturn. For my research, I run computer models and try to match the models to the data from the rings that the *Voyager* spacecraft sent back to Earth in the 1980s. One of the most enjoyable aspects of my job is the opportunity to study the rings and learn new things about how they work.

I would encourage young women to listen to themselves and follow their dreams. If I had listened to those around me who tried to discourage my interests in science, I would have missed the opportunity to be part of the scientific search for new knowledge.

Spacecraft Angular Measurement

In guiding and maneuvering a spacecraft, scientists have to know where it is. Just as maps have coordinates, scientists have terms and markings for space travel. They note the location of a spacecraft in space—its angular measurement— by giving its coordinates on the celestial sphere, that limitless imaginary sphere in the sky with the Earth as its center.

Right ascension is *the celestial sphere's equivalent of longitude on Earth.* Declination is *the celestial sphere's equivalent to latitude on Earth.* On Earth longitude is measured in degrees east and west, and latitude is measured in degrees north and south. In space, right ascension is measured in hours, and declination in degrees plus or minus. Earth's equator is zero, the North Pole is plus 90 degrees, and the South Pole is minus 90 degrees. Astronomers and NASA use these terms and locations to describe where an object is in space.

NASA monitors the Deep Space Network's antennas with an accuracy of thousandths of a degree. But this is not an accurate enough measure for navigational purposes. DSN tracking antenna angles are useful only for pointing the antenna to the specifications given for receiving a spacecraft's signal.

Fortunately, scientists have other ways of getting angular measurement. They can independently process the data by having two separate DSN stations track the same spacecraft at the same time.

Before you can maneuver in space, of course, you must first get there. That means you need a launch vehicle, or rocket. The problem of how to escape Earth's gravity and the machines and vehicles needed to do it are the subjects of the next chapter.

CHECK POINTS

Lesson 2 Review

Using complete sentences, answer the following questions on a sheet of paper.

1. What did *Voyager 2* "steal" from Jupiter?

2. Describe gravity assist trajectories and name the first spacecraft to use a gravity assist in space.

3. What are the advantages of the Hohmann transfer orbit?

4. Which orbital path is best for an orbiter returning to Earth?

5. Name three things *Aqua* monitors.

6. How many times faster is the exhaust from a xenon ion engine than from a chemical propellant engine?

7. Scientists must compare which two factors when deciding how much to change a spacecraft's velocity?

8. What are two maneuvers shuttle pilots perform when meeting up with the ISS?

9. How do scientists measure spacecraft velocity?

10. What are the coordinates scientists use for the celestial sphere that are similar to longitude and latitude on Earth?

APPLYING YOUR LEARNING

11. Based on what you have learned, why do you think scientists would prefer to use the faster, Type 1 trajectory for a manned mission to Mars? What difference does it make how fast the spacecraft gets there?

Unit 4 Space Technology

A *Delta II* rocket sends the *Phoenix* spacecraft on its journey to Mars in 2007. Engineers designed the *Delta II* to lift medium-sized satelllites and robotic explorers into space.
Courtesy of NASA

Rockets and Launch Vehicles

Chapter Outline

LESSON 1 It *Is* Rocket Science: How Rockets Work

LESSON 2 Propulsion and Launch Vehicles

> " [The rocket] will free man from his remaining chains, the chains of gravity which still tie him to this planet. It will open to him the gates of heaven. "
>
> *Wernher von Braun*

LESSON 1 | It *Is* Rocket Science: How Rockets Work

Quick Write

Can you identify any modern "toy" which may be as underappreciated as Hero's steam engine was in the first century? What do you imagine its use could be in the future?

Learn About

- the history and principles of rocket science
- different types of rockets
- the propulsion and flight of rockets

An ancient Greek mathematician named Hero discovered the main principle behind rocket and jet propulsion in the first century AD. It was a fine example of Newton's third law of motion—action and reaction—put into practice.

Historians can't know for sure what Hero's steam engine looked like. But they say that he made it out of a copper bowl set over a fire. The fire heated water in the bowl, and the steam from the bowl rose up two pipes into a hollow sphere. Inserted into the sphere's sides were two L-shaped tubes that allowed the steam to escape. The sphere then spun around in the opposite direction of the steam escaping from the L-shaped pipes.

According to historians, people called Hero's invention a toy rather than recognizing what a terrific revelation it really was.

The History and Principles of Rocket Science

Vocabulary

- magnitude
- nozzle
- oxidizer

A Greek philosopher named Archytas built and flew the first rocket back in the fourth or fifth century BC. That's nearly 2,500 years ago. It was a primitive device, though. Historians assume it must have flown round and round on a wire and was propelled by steam or compressed air.

But the Chinese are credited with developing the first practical chemical-fueled rockets. In this case, the fuel was gunpowder, invented by Chinese alchemists in the ninth century. By the eleventh century, the Chinese had incorporated black powder into fireworks. And by the thirteenth century, the Chinese military was using rockets in battle, including two-stage versions used by the Chinese Navy (Figure 1.1).

The greatest advances in rocketry, however, have occurred in the last 100 years. Some people argue that the first rocket launched into space was the German V-2, in 1942. The Nazis lobbed thousands of these at the Allies during World War II. But its ceiling generally topped out at about 55 miles (88 km) from Earth's surface, whereas outer space begins around 60 miles (96 km). This leaves some engineers and scientists arguing that the V-2 wasn't really mankind's first spaceflight. As *Universe Today* observed, "The significance of the V-2 launch is that it proved that rockets *could* be used to enter space."

In the late 1940s, the Army tested V-2 rockets with a WAC Corporal rocket as a second stage. These test flights eventually reached an altitude of 250 miles (400 meters). This was the first two-stage space rocket.

The first rocket that actually did launch into space *and* achieve orbit (a decisive marker that scientists point to as true spaceflight) was the rocket booster that carried *Sputnik 1*, the Russian satellite, in 1957. The Soviets took the next step in reaching space as well with *Vostok 1*, which sent the first human into orbit in 1961—as you read in Chapter 6.

This lesson looks at how rockets operate. When all the necessary forces align properly, like a perfect storm, rockets achieve flight.

Figure 1.1 An early Chinese rocket
Courtesy of Civil Air Patrol National Headquarters

How Force, Mass, and Acceleration Apply to Rockets

Rocket science may seem complicated. But for spaceflight purposes, a rocket designer actually is asking one basic question: How much force must be applied to a mass to accelerate that mass to speeds that will allow it to reach space and fall back to Earth, put it into orbit, or have it leave Earth's orbit completely?

If those terms—force, mass, and acceleration—sound familiar, they should. Newton's second law of motion makes good use of them. As you may recall from Chapter 1, the law reads: "Force equals mass times acceleration," or $f = ma$.

As a quick review: The force in question here is *thrust*, the power produced by a rocket engine. Mass is the amount of matter that is shot out of the engine— usually gases from combustion. The thrust is equal to the mass of the gases shot out of the engine times those gases' acceleration. Acceleration is a change in motion, which can be a speeding up, a slowing down, or a change in direction.

NASA's <u>Ares I-X</u> test rocket soars into blue skies at the Kennedy Space Center in 2009.

Courtesy of NASA/Sandra Joseph and Kevin O'Connel

How Action and Reaction Apply to Rockets

Newton's third law gives you an additional insight into how rockets work. This law, which you first read about in Chapter 1 reads: "For every action there is an equal and opposite reaction."

With rockets, the action is the thrust produced as the hot exhaust gas accelerates out the rocket's nozzle. The rocket's ascent is the reaction to the action of thrust. In short, the exhaust goes one way, and the rocket goes the other.

The Importance of Thrust for Rocket Flight

Now for a closer look at forces, which brings this discussion to Newton's first law: "Objects at rest remain at rest and objects in motion remain in motion in a straight line unless acted upon by an unbalanced force."

What might these "unbalanced" forces be? They are weight, thrust, lift, and drag. Each of these forces has both a magnitude— *an amount, size, speed, or degree that can be measured*—and a direction.

A rocket, whether on the ground or in flight, always weighs something. The space shuttle, for instance, weighs about 120,000 tons just before launch, including the external tank and its fuel, as well as the orbiter's two solid-rocket motors. Once a rocket begins its ascent, it also is subject to *mechanical* forces: thrust, lift, and drag. (You'll read more about mechanical forces later in this lesson.)

A rocket's weight depends on its mass and gravity's pull. Remember that although mass and weight sound about the same, they are not. Mass is a measure of how much matter an object contains. An object's mass remains the same unless you add or remove matter. Weight, however, changes in response to gravity's pull. Your mass is the same, whether you stand on Earth or the Moon. But your weight is far different. If you weigh 120 pounds (54 kg) on Earth, you would weigh only 20 pounds (9 kg) on the Moon, which has one-sixth of Earth's gravity.

Thrusters launch Explorer 1 toward space from Cape Canaveral in 1958.
Courtesy of NASA

A rocket's thrust depends on three factors:

1. The rate at which the mass flows through the engine
2. The velocity with which the exhaust gas flows
3. The pressure the exhaust gases encounter as they leave the nozzle.

The nozzle—*a rocket's end that releases gas, smoke, and flame to produce thrust*—determines how all these factors combine to produce thrust. Rocket nozzles have a narrow throat that opens into a bell-shaped bottom. That narrow throat boosts the pressure of the hot gas flowing through it.

Different Types of Rockets

The principles of rocket science are the same whether you're speaking about a model rocket or a space shuttle. Nonetheless, it's useful to look at different types of rockets to better understand their parts and how they operate.

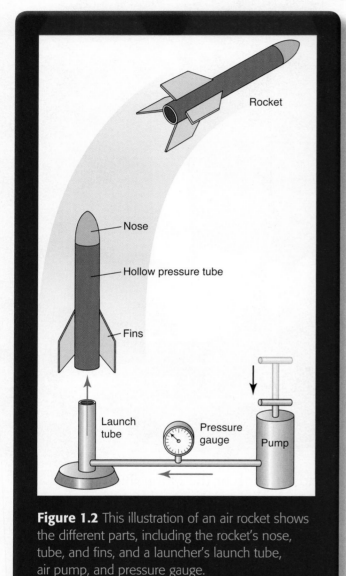

Figure 1.2 This illustration of an air rocket shows the different parts, including the rocket's nose, tube, and fins, and a launcher's launch tube, air pump, and pressure gauge.

Adapted from: *NASA* [http://exploration.grc.nasa.gov/education/rocket/rktstomp.html]. *Accessed 15 March 2010.*

Air Rockets

An air rocket is the simplest type of rocket. It uses compressed—or squeezed—air produced by a pump to thrust the rocket into the air. Air is its "working fluid."

The air rocket consists of two parts: the rocket and the launcher. The rocket is a hollow tube with a nose cone on top and an opening at the bottom. A launch tube and a launch base make up the launcher. Like the rocket, the launch tube is also hollow. This launch tube inserts into the hollow rocket. In this way the rocket becomes a closed pressure tube. The launch base has a hose, or feeder line, that connects to an air pump that gives the rocket its power.

By pumping or "stomping" the air pump with a hand or foot, you force air through the hose and up into the rocket. This increases the air pressure in the rocket, and the only place for that air to escape is at the bottom. Air forced out from the rocket bottom pushes it upward. Since one blast of air at the beginning of flight spends all of an air rocket's thrust, its flight is similar to that of a bullet. So during flight, only weight, lift, and drag act on it. Fins on the rocket's body give it stability during flight (Figure 1.2).

Bottle Rockets

Bottle rockets, or water rockets, are relatively safe and inexpensive. Like air rockets, bottle rockets teach lessons about force. Their two main features are also a rocket and a launcher. The two types of rockets share the same type of launcher. But the rockets themselves are different.

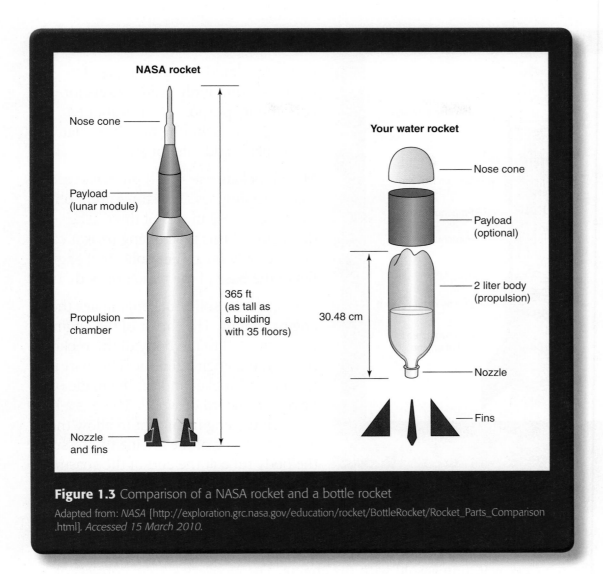

Figure 1.3 Comparison of a NASA rocket and a bottle rocket

Adapted from: *NASA* [http://exploration.grc.nasa.gov/education/rocket/BottleRocket/Rocket_Parts_Comparison .html]. *Accessed 15 March 2010.*

Usually people use a two-liter soda pop bottle for the rocket. Before launch, water fills the bottle to a specified point. This water propels the launch. Since water is more than 800 times heavier than air, the expelled water produces more thrust than compressed air alone. As with an air rocket, the bottle's base is only slightly larger than the launch tube. The body tube, or rocket, becomes a closed pressure vessel when the rocket goes on this launch tube. The pressure inside the body tube equals the pressure produced by the air pump. Fins attached to the bottom of the body tube stabilize the flight (Figure 1.3).

A water rocket's flight is similar to the flight of a compressed air rocket, with one important exception. The bottle rocket's weight varies during the flight because of the exhausting water plume. A bottle rocket uses water as its working fluid and pressurized air to accelerate the working fluid. Because water is much heavier than air, bottle rockets generate more thrust than air rockets.

Model Rockets

A standard model rocket is one of the most commonly used rockets for educational purposes. It has about a dozen parts, including fins. It is relatively inexpensive and safe for students.

Model rockets use small, pre-packaged, solid-fuel engines. Toy stores and hobby stores sell these in sets of three. The downside of this type of engine is it can be used only once. The solid-fuel engine fits in the base of the rocket or body tube.

On top of the solid-fuel engine lies the engine mount. The thrust of the engine is transmitted to the body of the rocket through the engine mount. This part is fixed to the rocket and can be made of heavy cardboard or wood. There is a hole through the engine mount to allow the ejection charge of the engine to pressurize the body tube at the end of the coasting phase and eject the nose cone and the recovery system.

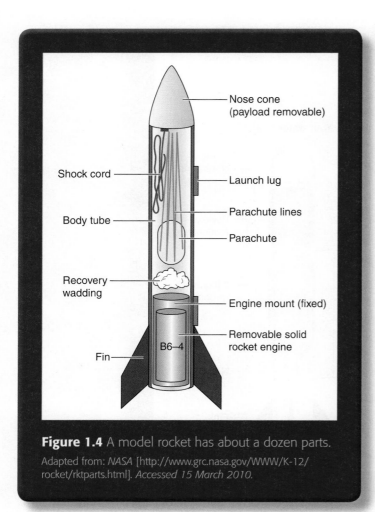

Figure 1.4 A model rocket has about a dozen parts.

Adapted from: *NASA* [http://www.grc.nasa.gov/WWW/K-12/rocket/rktparts.html]. *Accessed 15 March 2010.*

Recovery wadding is inserted between the engine mount and the recovery system to prevent the hot gas of the ejection charge from damaging the recovery system. The recovery wadding is sold with the engine. The recovery system consists of a parachute (or a streamer) and some lines to connect the parachute to the nose cone. Parachutes and streamers are made of thin sheets of plastic. The nose cone can be made of balsa wood or plastic, and may be either solid or hollow. You insert the nose cone into the body tube before flight. An elastic shock cord is connected to both the body tube and the nose cone and is used to keep all the parts of the rocket together during recovery. Launch lugs, or small straws, are attached to the body tube. The launch rail is inserted through these tubes to provide stability to the rocket during launch (Figure 1.4).

Full-Scale Boosters

Full-scale boosters are in the same league as the German V-2, the space shuttle, and *Delta* boosters. These machines are so complex that engineers group their parts into four major systems: structural, payload, guidance, and propulsion (Figure 1.5).

The *structural system* simply refers to the frame. Lightweight materials like titanium or aluminum "stringers" run the full length of the rocket. Hoops up and down the body tube hold the structure together. They strengthen the outside walls of the rocket. This allows engineers to build rockets of lightweight materials, reducing the mass of the launch vehicle and allowing it to lift bigger payloads. The outside walls of the rocket that carried John Glenn aloft, for example, were about as thin as a dime. Pressurizing the rocket gave it the strength required.

Missions determine the *payload system*. Engineers may build a rocket to carry astronauts up to the International Space Station or to launch a satellite or nuclear warhead. The payload generally sits up in the nose cone.

The *guidance system* rests just below the payload system and connects to parts elsewhere on the rocket— perhaps including at the bottom or along the sides. Engineers use this to control a rocket's flight. These days they rely more often than not on onboard computers, radars, and sensors. But back when the Germans launched their V-2s, they used small vanes in the nozzle's exhaust area to deflect and, therefore, control the engine's thrust. V-2s also had rudders on their fin tips. These were ballistic missiles—missiles driven by thrust but which head into a free fall back toward Earth once they've reached their peak (Figure 1.6).

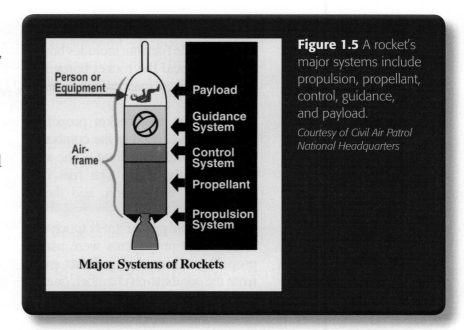

Figure 1.5 A rocket's major systems include propulsion, propellant, control, guidance, and payload.

Courtesy of Civil Air Patrol National Headquarters

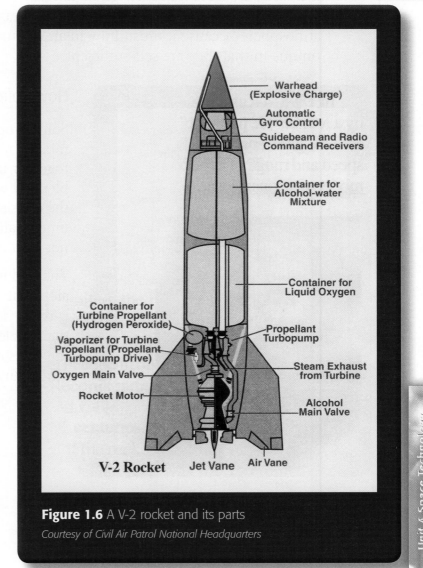

Figure 1.6 A V-2 rocket and its parts

Courtesy of Civil Air Patrol National Headquarters

A full-scale booster's *propulsion system* easily takes up at least two-thirds of the rocket. Propulsion systems come in two classes: liquid rocket engines (as in the German V-2) and solid rocket motors (as in the space shuttle's two solid rocket boosters). A liquid-fueled rocket's propulsion system includes fuel and oxidizer tanks, fuel pumps, a combustion chamber, and nozzle. Solid-fuel motors are long cylinders packed with propellant. In some rockets, this has the consistency of hard rubber, but in others it may be a more rigid material. For guidance, they can sport gimbaled nozzles—nozzles that can be swiveled from side to side. But otherwise, they have few moving parts.

The Propulsion and Flight of Rockets

Both solid and liquid propellants are a mix of fuel and oxidizer. An oxidizer is *a substance that includes oxygen to aid combustion.* Because a rocket carries its own oxidizer, it can travel in space. By contrast, a jet engine cannot leave Earth's atmosphere because it must draw in air to burn fuel. Therefore, space vehicles are unique in that they are self-sufficient.

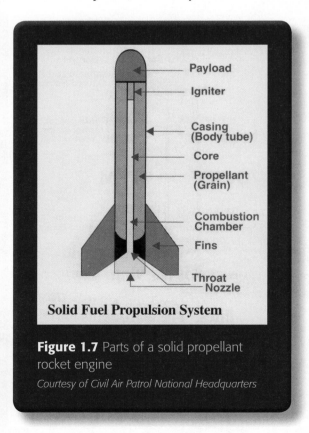

Figure 1.7 Parts of a solid propellant rocket engine

Courtesy of Civil Air Patrol National Headquarters

The labels in the figure read:
Payload
Igniter
Casing (Body tube)
Core
Propellant (Grain)
Combustion Chamber
Fins
Throat
Nozzle

Solid Fuel Propulsion System

How Solid Propellant Rocket Engines Work

Solid propellant rocket engines are made by mixing and then packing fuel and oxidizer together into the rocket's body. The mixture is blended with a binder, giving the hardened fuel the consistency of rubber. The rocket casing's inner walls are insulated so that the casing does not melt as the fuel inside burns.

The fuel is ignited from the top and burns along the entire length of a central shaft. This runs the length of the fuel assembly toward the nozzle. The fuel burns outward toward the sides of the casing from that central shaft. Once burning starts, it continues (Figure 1.7).

How Liquid Propellant Rocket Engines Work

Liquid propellant rocket engines also rely on a mix of fuel and oxidizer. But these rocket engines, a twentieth-century invention, are much more complex than the solid variety. The oxidizer in liquid propellant rocket engines is liquid oxygen, a form of pure oxygen cooled to −297.3 degrees F (−183 degrees C). The chilled temperatures condense the pure oxygen into a liquid.

Liquid propellant rocket engines store the two ingredients in separate tanks until it's time for launch. Then pumps shoot the fuel and oxidizer into a combustion chamber where they mix and ignite. From there they shoot exhaust through a throat and nozzle. Some liquid fuel rockets require an igniter to begin the engine's burn. Others use two liquids that burn spontaneously when mixed. These are often used in upper-stage motors (Figure 1.8).

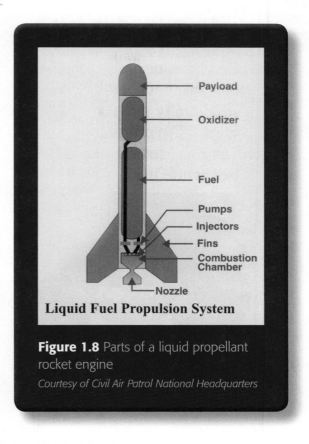

Figure 1.8 Parts of a liquid propellant rocket engine

Courtesy of Civil Air Patrol National Headquarters

One of the advantages a liquid propellant rocket engine has over the solid type is it can be started, stopped, and started again later. A solid propellant rocket engine really just has one shot at thrust. Once combustion begins, it's difficult to stop. Furthermore, the flow of liquid propellants into a combustion chamber means that engineers can much more easily control the amount of thrust delivered by a liquid propellant rocket engine.

Of course there's much more to how rockets fly and maneuver. Rocket science excites some people so much that they spend years in school studying it. Then they go on to work in the many space industries around the country, both government and private. If you find it interesting, maybe rocket science is in your future. You'll learn more about the different types of launch vehicles in the next lesson.

Leland D. Melvin, Astronaut

Leland D. Melvin grew up in Lynchburg, Virginia, and never thought about being an astronaut. He graduated from the University of Richmond and afterward the Detroit Lions drafted him to play football in the NFL. But he pulled a hamstring, and the team dropped him. He later went to play for a Canadian team and finally the Dallas Cowboys. But when the hamstring pulled again, his football career was over.

Not many people "fall back" on chemistry, but Melvin did. He had been taking classes during his down time with the NFL. So when his football career ended, he was able to hit the ground running. He began working for NASA in 1989. He never dreamed of being an astronaut until a friend suggested he apply. In 1998 the astronaut program accepted him.

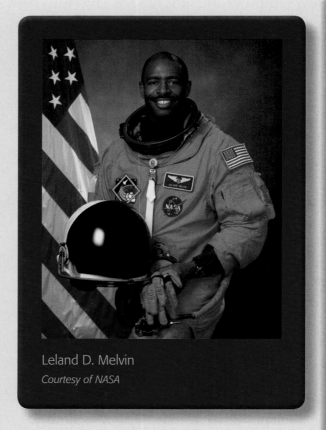

Leland D. Melvin
Courtesy of NASA

Melvin's career has taken him from studying the damage that chemicals inflict to aerospace materials during space travel to twice visiting the International Space Station and going on spacewalks. During his November 2009 mission, he took along the coin that was later used for the New Orleans Saints vs. Indianapolis Colts Super Bowl XLIV coin toss.

CHECK POINTS

Lesson 1 Review

Using complete sentences, answer the following questions on a sheet of paper.

1. What produces a rocket's action force? What is the rocket's reaction?

2. Which four "unbalanced" forces act on a rocket in flight?

3. The amount of thrust a rocket engine produces depends on what three things?

4. What is an air rocket's "working fluid"?

5. How much heavier is water than air?

6. Where does the recovery wadding go in a model rocket and what does it do?

7. Because full-scale boosters are so complex, engineers group their parts into which four systems?

8. In a solid propellant rocket engine, what are the fuel and oxidizer mixed with?

9. What is one advantage a liquid propellant rocket engine has over the solid type?

APPLYING YOUR LEARNING

10. When fuel and oxidizers combust in a solid or liquid propellant rocket engine, and then exhaust through the throat and nozzle, how does this illustrate Newton's second and third laws of motion?

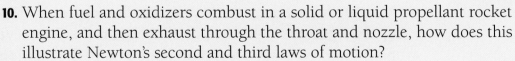

Unit 4 Space Technology

Propulsion and Launch Vehicles

Quick Write

_____

_____

What kind of internship would interest you, either in the science field or out of it? How would you go about applying for that internship?

Learn About

- the evolution of rocket technology
- the types of launch vehicles
- the factors and features of a rocket launch

One college student got to live out a dream over the summer in 2009. Brandon Lojewski, a student at the University of Central Florida, earned a spot as an intern at NASA's Kennedy Space Center in Cape Canaveral, Florida. While he'd launched his own paper rockets for school competitions beginning with a high school physics class, watching massive *Delta IV* and *Atlas V* rockets lift off gave him a whole new perspective.

"I had a headset and a computer console to click around in. The headsets have about 30 different voice channels you can tune in to and hear all the steps of the mission and any problems that are encountered," Lojewski says. "Being 'behind the scenes' is really an honor because it has made me realize and appreciate the depth, size, and complexity of our nation's space program."

Lojewski did more than just watch launches. He helped with projects to improve the launch process. One of his projects was to design, develop, and publish what are called Iris pages for display on launch consoles. Lojewski explains that Iris pages compile telemetry measurements (measurements such as temperature and pressure) from a spacecraft and launch vehicle. Engineers then look over this data during launch countdown, liftoff, and ascent.

"Planning for a launch starts years in advance," says Lojewski. "Every single component of the rocket is analyzed and monitored every second until the end of the mission. Some missions can even last a few decades. The stuff I am exposed to here is absolutely incredible."

The Evolution of Rocket Technology

Over the millennia, not only the technology of rockets but also their purpose has evolved. The first rocket-like devices were more like toys, invented by Archytas and Hero, two Greek men you read about in the previous lesson. But as the sophistication of rockets grew and human beings began to understand their power, rockets changed from toys to weapons.

Today countries around the world use rockets for both military and peaceful purposes. The United States, the European Union, Russia, and others use rockets to launch humans and satellites into space. The results include an improved understanding of Earth and better communication here on the planet. But as the United States proved in 1985 and China proved in 2007—when they used missiles to destroy satellites in space—rockets may also help introduce war to space.

The Early Use of Rockets

The Chinese were, in fact, instrumental in developing rockets. They reportedly had a very rudimentary version of gunpowder in the first century. In the thirteenth century they created fireworks. They filled bamboo and leather tubes with saltpeter, sulfur, and charcoal. In 1232 the Chinese used these rockets to ward off Mongol invaders (Figure 2.1). Some historians suggest that these were the first true rockets.

Vocabulary

- intercontinental ballistic missile
- launch window
- ALTO
- clean room
- payload shroud
- launch vehicle adapter

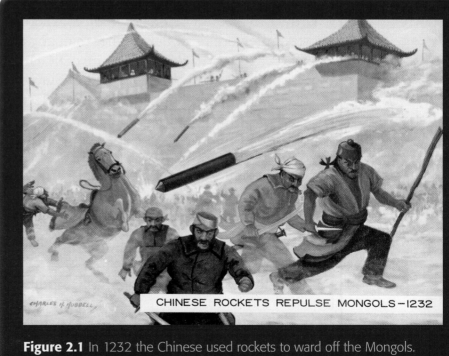

CHARLES H. HUBBELL

CHINESE ROCKETS REPULSE MONGOLS-1232

Figure 2.1 In 1232 the Chinese used rockets to ward off the Mongols.
Courtesy of NASA

Figure 2.2 An Italian invented a surface-running torpedo in the 1400s.

Courtesy of NASA

The Chinese rockets so impressed the Mongols that the Mongols began work on their own rockets. They in their turn likely introduced Europeans to rockets. Europeans soon began introducing their own modifications. Studies by a thirteenth-century English monk named Roger Bacon improved gunpowder. His efforts increased the range of rockets. He showed he understood the uses of gunpowder and the rocket when he wrote that it "is possible with it to destroy a town or army."

In the fifteenth century an Italian named Joanes de Fontana developed the first surface-running torpedo, which coasted on top of water and set enemy ships on fire (Figure 2.2). And in the same century Frenchman Jean Froissart discovered that launching rockets through a tube increased their accuracy. His finding led to the modern bazooka.

For a time starting around the sixteenth century, Western civilization set aside rockets in warfare and mostly shot them off for fireworks. During this lull, a German named Johann Schmidlap made a critical improvement. He invented the step rocket. As you read in the last lesson, rockets often have a first stage that launches a spacecraft to low-Earth orbit and a second stage that breaks it out of Earth's orbit. Schmidlap gets credit for these multistage rockets because of his two-stage fireworks that flew higher than any ever had before. And Kazimierz Siemienowicz (KAH-zee-meer Sye-mye-NOH-vich), a seventeenth-century commander in the Polish Royal Artillery, wrote about multistage rockets in a way that also laid the foundation for their development in the centuries ahead.

Gravesande, Congreve, and Hale: The Early Rocket Scientists

Two qualities eluded scientists in their earliest rockets: accuracy and power. Without these features commanders on the battlefield relied instead on lobbing great numbers of rockets to overwhelm their enemies. Nations resumed using rockets in a serious way in war beginning in the late eighteenth century and early nineteenth century.

Willem Gravesande was a Dutch professor in the early eighteenth century. He propelled model cars with jets of steam. Experimenters in Russia and Germany began working with rockets that had exhaust flames powerful enough to bore holes in the ground before liftoff.

A Tale of 47 Rockets

A sixteenth-century Chinese legend tells the tale of an official named Wan Hu, who wanted to fly into space. He placed 47 gunpowder rockets at the base of a chair, sat in it, and asked his assistants to light the fuses. When they did, a huge explosion took place. When the air cleared, Wan Hu was gone. Some suggested he made it to space and can be seen as the Man in the Moon.

Figure 2.3 The legend of Wan Hu

Courtesy of NASA

Then, after the British succumbed to Indian rocket attacks in battles in 1792 and 1799, an English colonel named William Congreve began designing rockets. Congreve's rockets could travel three miles (4.8 km). He developed different models: Some showered the enemy with shot (small metal pellets); others were incendiary rockets for burning ships and buildings. He also pioneered launching rockets from ships. You hear of his work every time you sing the National Anthem.

The British battered Fort McHenry with Congreve rockets during the War of 1812. Here, Defenders Day fireworks light up the garrison flag in September 2009.

Courtesy of Timothy Ervin/Fort McHenry National Monument and Shrine/NPS

Francis Scott Key's lyrics that include "the rockets' red glare" refer to Congreve rockets pounding Fort McHenry in Baltimore during the War of 1812. Still, Congreve rockets lacked accuracy.

William Hale, an Englishman, invented a stickless rocket in the middle-nineteenth century that revolutionized battle. Hale's spin stabilizer, which eliminated the need for the long guide sticks that Congreve had used to provide stability in flight (Figure 2.4), improved that much sought-after quality: accuracy. Exhaust gases struck vanes at the rocket's base and made it spin like a bullet. The United States used these rockets in the Mexican War of 1846–48. They also appeared in the Civil War.

More-effective breech-loading cannon (cannon loaded from the rear) with rifled barrels (cut with spiraled grooves that improve accuracy) and exploding warheads eventually caused armies to once again set rockets aside for a time. But variations of Hale's principle are still in use today.

The Contributions of Tsiolkovsky, Goddard, and Oberth to Modern Rocket Science

Modern rocket scientists focused not only on accuracy and power but also on distance and altitude. Russian Konstantin Tsiolkovsky earned the title of "father of astronautics" for his contributions to rocketry. After nearly going deaf at age 10, he educated himself. In 1898 Tsiolkovsky suggested using rockets to travel to space. In 1903 he wrote in his most famous work, *Research Into Interplanetary Space by Means of Rocket Power*, that liquid propellant could increase how far rockets could travel. He also said that the velocity of exhaust gases determined how far and how fast a rocket would travel.

American Robert Goddard built the first liquid propellant rocket in 1926. It traveled only 41 feet (12.5 meters) but was as important to the future of spacecraft as the Wright Brothers and Kitty Hawk were to the development of aircraft in 1903. Goddard based his rocket on the theory that a stable flight could be attained by mounting the engine ahead of the fuel tanks, with the tank shielded from the flame by a metal cone. This work was instrumental in the development of the *Saturn V* Moon rocket in the 1960s.

Among other ideas, Goddard developed a gyroscope system to control his rockets as well as a parachute recovery system. This "father of modern rocketry" also demonstrated that rockets will fly in a vacuum (no air and no gas)—that they don't need air to push them.

Figure 2.4 A British ship launches Congreve rockets at Copenhagen in 1807.
© *Science Photo Library*

Finally, Hermann Oberth of Germany wrote a book in 1923 titled *By Rocket to Space* that delved into the math of spaceflight as well as rocket designs and space stations. His book inspired many to study rockets, including a German group called the Society for Space Travel. A young German engineer named Wernher von Braun joined this society along with Oberth. The German military eventually recruited Von Braun along with many society members in the 1930s to help develop a rocket for its arsenal. The V-2 rocket was the result. It weighed 12 tons, had a range of 200 miles (300 km), flew more than 3,500 miles (5,600 km) per hour, and could carry a one-ton payload.

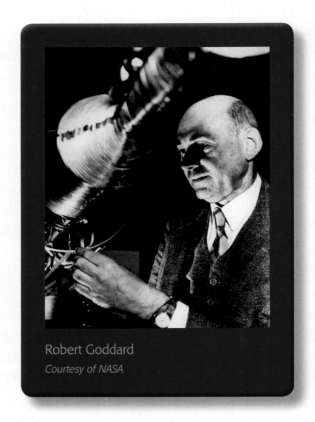

Robert Goddard
Courtesy of NASA

The Types of Launch Vehicles

After World War II the United States and the Soviet Union seized abandoned V-2 rockets for study back home. Hundreds of Germany's top scientists accompanied them. The United States became home to the brilliant Von Braun, who surrendered to American troops along with 500 of his fellow scientists.

Thus both the United States and the Soviet Union based their space programs on the work done in Germany with V-2 rockets. Until 1960 Von Braun and his scientists worked for the US Army. Then the US government appointed Von Braun as first director of NASA's Marshall Space Flight Center, a position he held until 1970. Under his watch, NASA developed the *Saturn* rocket that would take man to the Moon. But the contest of rocket development between the United States and the Soviet Union first really heated up back in 1957 with the USSR's launch of *Sputnik*.

The R-7 Intercontinental Ballistic Missile (ICBM) That Launched the *Sputnik* Satellite

Sputnik launched atop the R-7, a Soviet intercontinental ballistic missile. An intercontinental ballistic missile (or ICBM) is *a missile designed to deliver a payload to another spot on Earth several thousand miles away*.

The R-7 is a *base rocket*. Engineers add to it upper stage rockets as needed to "build" different launch vehicles. Since the R-7 launched *Sputnik*, the Russians have launched more than 1,600 R-7-derived rockets, more than any other launch vehicle in the world.

CHAPTER 11 Rockets and Launch Vehicles

The R-7 consisted of a core rocket (which the Russians refer to as the second stage) surrounded by four boosters (the first stage), each shaped like a tapered cylinder. Both first and second stages ignite at launch. Each of the strap-on boosters had one engine producing tons of thrust at sea level. The four boosters separate from the core about two minutes after liftoff, leaving the core to continue firing. After the core finishes firing, additional upper stages fire to insert the payload into orbit.

Voskhod, *Soyuz*, *Tsiklon*, and *Proton*: Russian Launch Vehicles

The Russian *Voskhod* and *Soyuz* launch vehicles have relied on "R-7 plus upper stage" combinations for launches. So atop an R-7 (or a rocket developed from the R-7 design) sits an upper stage referred to as a "block." Starting in 1963 the *Voskhod*, using an "R-7 plus Block 1" combination sent reconnaissance satellites into space.

A couple years later *Soyuz* rockets lifted off with satellite payloads on a somewhat revised version of the R-7 plus Block 1 mix. The *Soyuz* launch vehicle interestingly enough also gets credit for launching not only manned *Soyuz* spacecraft into orbit starting in 1967 but also the manned *Voskhod 2* in 1965 as well. The Russians still use *Soyuz* today to get cosmonauts and astronauts to the International Space Station. And they continue to develop new *Soyuz* rockets.

An R-7 rocket carrying a <u>Soyuz</u> capsule rolls to the launch pad at the Baikonur Cosmodrome in 2003.

Courtesy of Scott Andrews/NASA

The Soviets started using their more powerful *Tsiklon-3* launch vehicle in 1977 to send weather and military satellites into orbit. They based it on their two-stage R-36 ICBM, a liquid-propellant military missile that was the first missile the United States viewed as a threat to its own ICBM arsenal. The *Tsiklon* could deliver about 3.5 tons (3,175 kg) into a polar orbit. To deliver that payload, it relied on a restartable third stage. The Soviets used an earlier two-stage version of *Tsiklon*, the *Tsiklon-2*, as far back as 1967 to launch high-security military payloads into space. These payloads included anti-satellite weapons and ocean reconnaissance satellites.

Russia's largest launch vehicle in use today is *Proton*. Unlike *Voskhod*, *Soyuz*, and *Tsiklon*, the Soviets did *not* base *Proton* on one of their ballistic missiles. From the beginning stages of development, they intended to use *Proton* for space missions.

Proton's most basic package, the *Proton-K*, comprises three or four stages depending on the mission. Six engines make up the first stage, four engines make up the second, and one engine makes up the third stage. If a mission heads to deep space or geostationary orbit, it calls for a fourth stage called the Block DM, which is restartable and can carry as much as 4.9 tons (4,445 kg). In the 1970s and 1980s the Russians launched a three-stage version to send all of the parts for the *Mir* and *Salyut* stations into space. In more recent years they've used *Proton* to send such things as the module *Zarya* (Dawn) to the International Space Station (ISS), as they did in 1998.

Transporting a <u>Tsiklon</u> rocket to the Plesetsk space center in Arkhangelskaya Oblast, far north of Moscow
© RIA Novosti/Alamy Images

Delta, Titan, and Atlas: US Expendable Launch Vehicles (ELVs)

The United States produces two types of launch vehicles: expendable and reusable. The country's only partially reusable launch vehicle is the space shuttle, which you read about in Chapter 7. Expendable US booster rockets—models that can be used only once—include *Delta*, *Titan*, and *Atlas* (Figure 2.5).

Delta

The *Delta* launch vehicle was America's answer to the R-7 and *Sputnik*. NASA engineers purchased a dozen of the US Air Force's *Thor* medium-range ballistic missiles, added their own modified second stage, and called this new package *Delta*. The space agency launched its first *Delta* in 1960. With it, NASA has launched many weather, scientific, and communications satellites over the years.

Initially *Deltas* could support a payload of only 100 pounds (45 kilograms). Over time, improvements allowed the *Delta* to become NASA's go-to spacecraft. Engineers increased thrust by adding solid rocket boosters and making room for more propellant to add burn time. A *Delta II* is about 127 feet (39 meters) tall, while the *Delta IV* can range from between about 206 feet (63 meters) and 235 feet (72) meters tall. Today the *Delta IV Heavy* can launch a payload weighing more than 50,000 lbs (23,000 kg) into low-Earth orbit. It features three common booster cores joined together and topped with a second-stage engine.

The first space shuttle flight in 1981 seemed to spell doom for *Delta*. NASA stopped placing orders, expecting that the shuttle would launch payloads the rocket had carried previously. By 1986 NASA had only three *Deltas* left in its inventory. After the *Challenger* explosion that same year, however, President Reagan decided that space shuttles would no longer carry commercial or military payloads. *Delta* was back in demand. Today these launch vehicles carry about 34 percent of all the commercial satellites in the world. The United States, Russia, Europe, and China compete for this commercial business, so *Delta* plays a crucial role in winning this business for the United States.

Star POINTS

The *Delta II* first flew in February 1989 and is still in use today. *Deltas* have launched the entire Global Positioning System satellite fleet, which provides accurate guidance for the smart bombs used in Afghanistan. It also provides location information for the GPS in your car, boat, and even your cell phone.

Titan

Titan was another critical US expendable launch vehicle (ELV). It was the country's first two-stage ICBM, and first successfully test-launched in 1959. This liquid-propellant rocket gave the United States strategic comfort during the Cold War. And it was the first missile stored in a hardened underground silo.

Later versions achieved other firsts. *Titan II* was the first US launch vehicle to use fuels that could sit for long periods in the missile's fuel tanks. This meant that if war broke out between the United States and the Soviet Union, the US military could quickly launch its nuclear-armed *Titans* without wasting valuable time fueling them. NASA adopted *Titan II* to launch its two-man *Gemini* spacecraft for the 10 *Gemini* missions between 1965 and 1966 that prepared for the Apollo program.

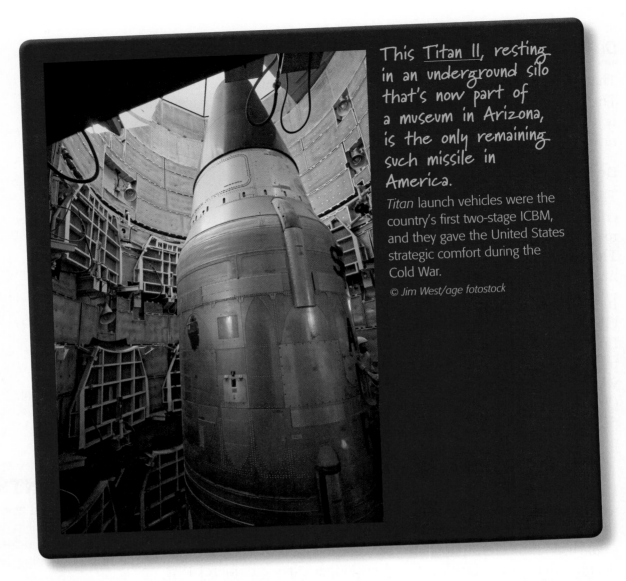

This Titan II, resting in an underground silo that's now part of a museum in Arizona, is the only remaining such missile in America.

Titan launch vehicles were the country's first two-stage ICBM, and they gave the United States strategic comfort during the Cold War.

© Jim West/age fotostock

The US Department of Defense (DoD) next requested the development of *Titan III* for space missions, not ICBMs. DoD needed launch vehicles to fly their intelligence-gathering satellites into space. This version launched satellites from 1966 until 1987. Meanwhile NASA turned to *Titan III* for numerous missions, including *Viking* and *Voyager 1* and 2.

And the most costly rocket of all time, *Titan IV*—at $400 million each—propelled NASA's *Cassini* and *Huygens* probes on their way. *Titan IIIs* and *IVs* continued to use a liquid-propellant core but also had strapped-on solid rocket motors. Because they cost so much to operate, the United States retired its *Titan* rockets in 2005 in favor of the *Atlas V*.

Atlas

The *Atlas* rocket, like *Titan*, launched exploratory probes but also had a national defense mission. The United States Army Air Forces, the predecessor to the US Air Force, first ordered what would become the *Atlas* in 1945 after World War II's end. The liquid-propellant rocket's design evolved over the years, but in 1959 it became the first American ICBM.

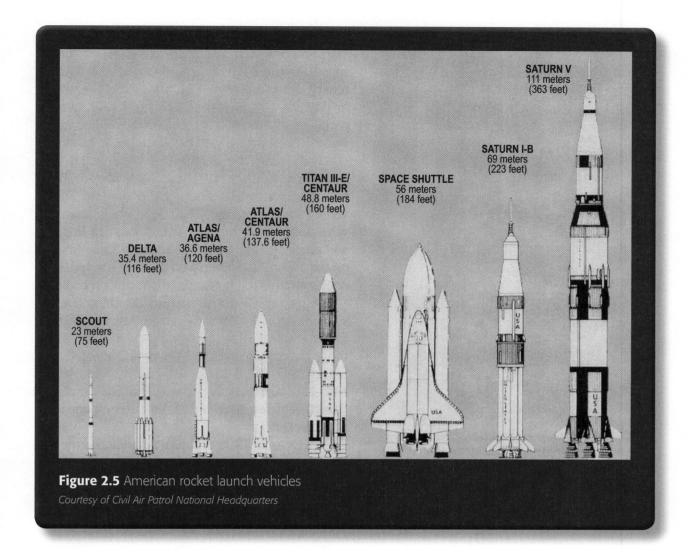

Figure 2.5 American rocket launch vehicles

Courtesy of Civil Air Patrol National Headquarters

While no longer used as an ICBM—the military phased it out in 1966 to replace it with the solid-propellant Minuteman ICBM—*Atlas* is still used today for a variety of other military, space, and commercial missions. And its design has not changed much in its 50-plus years. It is lightweight and can carry large payloads.

Of historic significance, NASA sent astronaut John Glenn on man's first Earth orbit in the *Friendship 7* Mercury capsule in 1962 with an *Atlas* launch vehicle. Other Mercury missions also launched atop the *Atlas*.

Recent *Atlas* upgrades include the *Atlas IIA* and *IIAS*. They can carry large payloads, including communications satellites, weighing almost 8,500 lbs (3,856 kg), to a geostationary orbit 22,300 miles (35,888 km) above the equator.

Interestingly, the *Atlas V*, introduced in 2006, contains an engine built in Russia. This is an interesting development for a rocket whose initial purpose was to launch nuclear warheads against the former Soviet Union.

On 16 July 1969 a *Saturn V* launch vehicle lifts off, sending astronauts Neil Armstrong, "Buzz" Aldrin, and Michael Collins to the Moon.

The *Apollo 11* mission launched from Cape Canaveral, Florida.

Courtesy of NASA/Stennis Space Center

Saturn V: Launch Vehicle for the Apollo Program

NASA needed a more powerful launch vehicle to send astronauts to the Moon. None of its previous ELVs would do. With Von Braun as chief engineer, the Marshall Space Flight Center developed the *Saturn V*. NASA built 15 of these rockets.

In December 1968 the *Apollo 8* spacecraft atop a *Saturn V* launched a crew into orbit around the Moon. And in July 1969, propelled by another *Saturn V*, the *Apollo 11* crew lifted off for mankind's first landing on the Moon. The *Saturn V* also launched the *Skylab* space station. NASA launched a smaller version of the *Saturn V*, called the *Saturn 1B*, to carry crews to *Skylab* in 1973 as well as for the 1975 *Apollo-Soyuz* docking mission. This largest and most powerful of all rockets ever launched had a first stage that alone weighed more than a space shuttle. It had three stages in all. Each of its first two stages contained five engines. And no *Saturn* ever failed, despite the fact that it was made up of some 3 million parts.

Star POINTS

The *Saturn V* rocket that sent astronauts to the Moon was as powerful as the energy created by 85 Hoover Dams.

The Space Transportation System (STS): US Reusable Launch System

As you read in Chapter 7, the space shuttle's official name is Space Transportation System (STS). This reusable launching system relies on a combination of liquid and solid propellants. Fed by the external tank, the shuttle main engines burn liquid hydrogen and oxygen. The strap-on boosters burn solid propellant. Only the external tank isn't reusable. After each flight, scientists retool and repair the main components of the STS for use on future flights.

The STS has many duties. It launched *Galileo*, *Magellan*, and *Ulysses*. It has taken spacecraft to orbit, performed satellite rescues, and assembled and serviced the ISS. It also carries out a wide variety of scientific missions ranging from the use of orbiting laboratories to small self-contained experiments. And it was designed to transport loads of up to 66,000 lbs (30,000 kg) into low-Earth orbit.

Astronaut Stephanie Wilson

Stephanie Wilson was born in Massachusetts in 1966. She graduated from Harvard University with an engineering degree and went to work for Martin Marietta, where she worked on *Titan IV*. After two years, she went back to graduate school at the University of Texas for her master's in aerospace engineering. When she graduated, she got a job with the Jet Propulsion Laboratory in California and worked on software development, as well as various aspects of *Galileo*.

Wilson says that she chose engineering because it offers so many career choices. "As a mechanical engineer, I could work on automobiles if the bottom fell out of aerospace. I could work building designs. I could work on city planning. I always felt like engineering was a good career move."

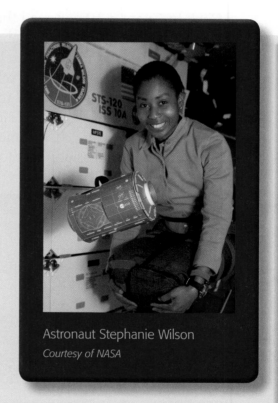

Astronaut Stephanie Wilson
Courtesy of NASA

NASA accepted her into the astronaut program in 1996. But before she showed up for training, she had to learn how to swim, since so much astronaut training takes place underwater. She spent her whole summer training with a coach. And it paid off. Wilson has been to space twice and took part in the Spring 2010 launch of *Discovery*. On previous missions she has traveled to the International Space Station where she performed maintenance. Her missions have also transported various astronauts to and from the ISS.

Wilson was the second African-American woman in space. Two other women— mission specialist Dottie Metcalf-Lindenburger and Japanese astronaut Naoko Yamazaki— joined Wilson and ISS flight engineer Tracy Caldwell Dyson in the first mission to feature four women aboard the same spacecraft for the first time.

Once NASA retires the space shuttle program—intended for September 2010—it plans to replace the shuttles with the *Ares I* launch vehicle for human spaceflight and the *Ares V* for cargo launches. But as of this writing in 2010 the Obama administration has dropped its support for the idea, putting the *Ares*'s future in doubt.

The Factors and Features of a Rocket Launch

Once NASA has picked its launch vehicle based on the mission, it still has many choices to make, from where to launch it to when and how. NASA has names for each of these steps: launch sites (the "where"), launch windows (the "when"), and preparation and integration (the "how").

Launch Sites

Back in Chapter 10 you read that launches from the equator can take advantage of Earth's rotational speed of about 1,040 miles (1,675 km) per hour. This means that even though the launch vehicle is only sitting on the launch pad, it is already moving at those great speeds relative to Earth's center. NASA applies these miles per hour to the speed needed to orbit Earth (about 17,350 mph or 28,000 km per hour). A spacecraft launched close to the equator calls for less propellant and can launch a larger vehicle than one launched farther away. But launches at the equator help only those missions that follow an orbit close to Earth's equator.

An *Atlas V/Centaur* rocket sits at a launch pad at Cape Canaveral Air Force Station. It carries NASA's Lunar Reconnaissance Orbiter and Lunar Crater Observation and Sensing Satellite. Circling the launch pad are protective lightning towers.
Courtesy of NASA/Jack Pfaller

Missions headed for a high-inclination Earth orbit (such as a North Pole to South Pole orbit) don't benefit from an equator launch. In such a case, it is up to the launch vehicle to provide all the energy needed to reach orbit.

For interplanetary launches, a launch vehicle takes advantage not only of Earth's rotation but also of its orbital motion about the Sun. Using these readily available sources of free power allows NASA to make up for the limited energy available from today's launch vehicles. The launch vehicle accelerates in the direction of Earth's orbital motion, which averages about 62,000 miles (100,000 km) per hour. This is in addition to the launch vehicle's using Earth's rotational speed.

Launch sites must also have a clear path downrange so the launch vehicle will not fly over populated areas, in case of accidents. Space shuttles require a landing strip with acceptable wind, weather, and lighting conditions. They also need landing sites overseas, in case of an emergency landing.

The Kennedy Space Center at Cape Canaveral, Florida, stages many East Coast launches. Others are conducted at nearby Patrick AFB. However, these sites are suitable only for low-inclination orbits (those nearer the equator). This is because major population centers underlie the trajectory required for high-inclination launches. On the West Coast, NASA launches high-inclination missions from Vandenberg Air Force Base in California. This location is suitable because the trajectory for high-inclination orbits avoids population centers.

Finally, heavy launch vehicles call for complex ground facilities. Smaller vehicles use mobile facilities. And some even launch from airplanes.

The Launch Window

The launch window is *the specific timeframe during which a launch can take place.* Many factors determine when the launch window occurs, including safety and mission objectives.

An interplanetary launch has a limited number of weeks in which it must take place. The timing of the launch window depends on Earth's location in its orbit about the Sun and the target planet's position in its own orbit about the Sun. The timing must permit the launch vehicle to use Earth's orbital motion for its trajectory, while timing it to arrive at its destination when the target planet is in position. The launch window may also be constrained to a number of hours each day to take best advantage of Earth's rotational motion.

Actual launch times must also consider how long the spacecraft needs to remain in low Earth orbit before its upper stage places it on the desired trajectory toward a target planet.

In addition, a launch that will rendezvous with another vehicle in Earth's orbit must time its liftoff with that object's orbital motion. This was the case with the Hubble Space Telescope repair missions.

NASA crews lift an <u>Atlas V</u> booster into a Vertical Integration Facility at Cape Canaveral.

Courtesy of NASA

Preparations for a Launch

NASA has an acronym that stands for launch preparations: ALTO stands for *assembly, test, and launch operations*. This process is very precise.

The plan requires delivery of a spacecraft's parts to a large clean room. A clean room is *a workspace with a constant temperature and humidity and low levels of contaminants such as dust*. Engineers put the parts together and test them in this clean room using computer programs that are nearly identical to those used in flight.

NASA then transfers the spacecraft to an environmental test lab. Engineers place it on a shaker table and subject it to launch-like vibrations. Additionally they install it in a thermal-vacuum chamber and test its thermal properties—how it responds to extreme temperatures. The engineers make adjustments as needed in thermal blanketing to protect spacecraft from the harsh space environment.

Star POINTS

Pyrotechnic devices have a variety of uses, such as separating one stage of a spacecraft from another once in flight. An example of this is when a space shuttle and an emptied external tank separate several minutes after launch.

Once complete, NASA moves the spacecraft to the launch site. Engineers seal it inside an environmentally controlled carrier, either a truck or an airplane, and carefully monitor the spacecraft during the trip.

More testing takes place at the launch site. Technicians load propellants on board and arm any pyrotechnic devices. They then mate the spacecraft to its upper stage, and finally hoist and mate it to the launch vehicle.

The Process of Launch Vehicle Integration

NASA refers to the phase of mating the spacecraft with the launch vehicle as *launch vehicle integration*. It is a long and detailed process. Engineers maintain clean-room conditions on top of the launch vehicle while they put the payload shroud in place. A payload shroud, also called a payload fairing, is *the thin metal cover, or nose cone, that protects a spacecraft and upper stages during a launch when aerodynamic forces can batter the rocket*. The shroud gives the rocket nose an aerodynamic shape (Figure 2.6).

Finally they double-check that the launch will place the spacecraft on the proper trajectory so it gets where it needs to go.

About three months before launch, the engineers transfer the spacecraft and launch vehicle to the launch site. They then attach the spacecraft to a launch vehicle adapter, which is *a physical structure used to connect a spacecraft to a launch vehicle*. They place this whole package into the payload shroud, move it to the launch pad, and hoist it by crane onto the top of the launch vehicle. This final stage takes place about 10 days before liftoff. A countdown then begins.

The countdown helps everyone orchestrate the many operations needed to get everything ready. People have to perform their various tasks at very specific times during the countdown so they don't interfere with one other. During this period, many final operations on the spacecraft take place, including removing instrument covers and other "remove before flight" items, installing arming plugs, and generally getting everything buttoned up for the big day.

Pauses in the countdown, or "holds," are built in. These allow the launch team to target a precise launch window, and to provide a cushion of time for certain tasks and procedures without affecting the schedule. For the space shuttle countdown, built-in holds vary in length and always occur at the following times: T minus 27 hours (that is, 27 hours before liftoff), T minus 19 hours, T minus 11 hours, T minus 6 hours, T minus 3 hours, T minus 20 minutes, and T minus 9 minutes.

The final hold is always at T minus 9 minutes. It often lasts 20 minutes, although this can vary, depending on the mission. During this time, NASA officials:

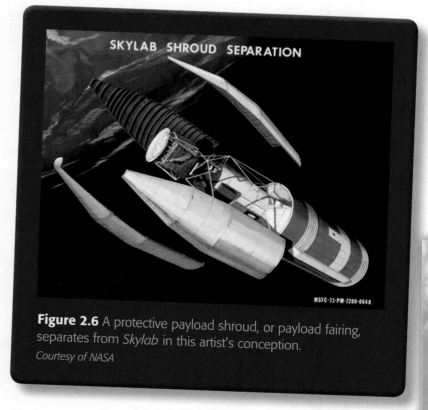

SKYLAB SHROUD SEPARATION

MSFC-71-PM-7200-064A

Figure 2.6 A protective payload shroud, or payload fairing, separates from *Skylab* in this artist's conception.
Courtesy of NASA

- determine the final launch window
- activate the flight recorders
- conduct the final "go/no-go" launch polls and decide whether to launch.

Specialists monitor the spacecraft's health at all times. If a problem arises, they can stop the launch. The Deep Space Network begins tracking immediately after the launch.

Despite these finely tuned procedures, launches remain too complex and dangerous for the everyday person to take part in. NASA's goal is to find ways to make space more accessible. The space agency says that launch vehicles must be less expensive and more reliable than they are today. They also must be reusable—the agency can't afford to throw away expensive hardware after each launch. That's been the beauty of the reusable space shuttle program.

Engineers see a future that allows more people to travel into space and encourages commercial development as well. Some people believe these commercial companies could replace NASA in delivering astronauts to places like the International Space Station. NASA itself may eventually develop aerospace planes that will "take off from runways, fly into orbit, and land on those same runways, with operations similar to airplanes," the agency says. It may sound like the stuff of science fiction—but it could become a reality in your lifetime.

Kennedy Space Center engineers work on a launch vehicle adapter (*center*) and a Mars rover (*left*).
Courtesy of NASA

CHECK POINTS

Lesson 2 Review

Using complete sentences, answer the following questions on a sheet of paper.

1. Which people were instrumental in developing rockets?

2. Which type of rocket does the National Anthem lyric "the rockets' red glare" refer to?

3. Who built the first liquid propellant rocket?

4. Which launch vehicle has been used more than any other since its invention in the 1950s?

5. Which is the largest launch vehicle the Russians use today?

6. Which launch vehicle was America's answer to *Sputnik*?

7. Which ELV launched human beings to the Moon?

8. What is the Space Transportation System more commonly known as?

9. Which type of orbit doesn't benefit from an equator launch?

10. If a launch is to rendezvous with another spacecraft, it must time which two elements?

11. What does ALTO stand for?

12. When does a countdown begin?

APPLYING YOUR LEARNING

13. Explain why NASA has not needed a rocket as powerful as *Saturn V* since the end of the Apollo program.

Unit 4 Space Technology

The Mars rovers *Spirit* and *Opportunity* launched in 2003 to search for water on Mars. By 2004 scientists had their answer: the rovers found rocks that showed signs liquid water had once flowed over them. And water may still flow on occasion there.
Courtesy of NASA

Robotics in Space

> To efficiently explore the Moon and Mars, flight crews will have to be much more self-reliant than before. Development of such self-reliance requires machine intelligence, coupled tightly with human direction.
>
> *David Korsmeyer, NASA*

Learn About

- the purpose of using robots in space
- the history of robots in space
- the current robotic missions in space

I f robots could have a "best friend," they've got one in software engineer T. Adrian Hill. Hill is the fault protection lead for not just one, but two robotic missions: *MESSENGER* and *New Horizons*. *MESSENGER*, launched in 2004, is currently studying Mercury. It's already completed three flybys of the planet, and NASA plans to insert it into Mercury's orbit in 2011. The space agency launched *New Horizons* in 2006 to examine Pluto and its moons and other worlds in the Kuiper Belt. As of 2010, it was halfway to Pluto with an expected arrival sometime in 2015.

Hill is one of the critical players for NASA whose mission is to make sure these robots make it safely through their long, lonely, and difficult journeys. A fault protection engineer like Hill programs spacecraft so that even in the deep reaches of space the robots can detect and isolate problems with their onboard systems and then make corrections. Spacecraft such as *MESSENGER* and *New Horizons* are so far into deep space that no manned mission can reach them for repairs. So Hill has to program these vessels so that they can self-repair, and be somewhat self-sufficient.

Hill says of his two big responsibilities: "I feel very fortunate that I am able to work on two spacecraft that are going to opposite ends of the Solar System. How many people can make a claim like that?" He's worked for more than a dozen years on NASA-sponsored projects, and adds that one of the most memorable experiences he's had was as flight software lead for the Hubble Space Telescope. "I had the opportunity not only to lead the software development, but to actually watch on the video as the astronauts removed the old computer and installed our computer with our new software," he said. "Watching from the ground as the software came up and running and started operating—it was a satisfying moment." Since 2000, Hill has been working for The Johns Hopkins University Applied Physics Laboratory's Space Department.

The software engineer doesn't just troubleshoot in space. He also troubleshoots here on the ground as a college football referee. He travels around the country moonlighting as a ref on his Saturdays during the season. "It's my big passion. I started ref-ing football back in 1990— Little League youth ball, then high school ball, and now I've been fortunate enough to work college football."

Hill sounds like a fortunate person all around: a thrilling daytime job that puts him in touch with the deep reaches of space and fulfilling weekends on the gridiron.

Vocabulary

- robots
- robotics
- end effector
- Robonaut
- bearing strength
- conjunction

Software engineer T. Adrian Hill
Courtesy of the Johns Hopkins University Applied Physics Laboratory

The Purpose of Using Robots in Space

Advances in space would not be possible without robots—*machines that operate automatically or by remote control to perform tasks*. Scientists have built the International Space Station (ISS) with the help of robots. They also use robots to repair satellites and explore the far reaches of the Solar System and beyond. Astronauts aboard the space shuttle and the ISS maneuver robotic arms and hands to build, repair, and more. Quite simply put, NASA would not know as much as it does about outer space or be able to accomplish as much without robots.

Star POINTS

Robot comes from the Czech word *robota,* which means *servitude or forced labor*. The Czech playwright Karel Capek coined the term in 1923.

On Earth, typical industrial robots do jobs that are difficult, dangerous, or dull. They lift heavy objects, paint, handle chemicals, and perform assembly work. They perform the same job hour after hour, day after day with precision. They don't get tired and they don't make errors associated with fatigue. So they are ideally suited to performing repetitive tasks.

Space-based robots at NASA fall within three specific mission areas: exploration, maintaining science payloads, and on-orbit servicing. Today, NASA has two important proven space robots. One is the Remotely Operated Vehicle (ROV) and the other is the Remote Manipulator System (RMS). An ROV can be:

- an unmanned spacecraft that remains in flight
- a lander that lands on an object in space and operates from a stationary position
- a rover that can move over terrain once it has landed. It is difficult to say exactly when early spacecraft evolved from simple automatons to robot explorers or (ROVs).

How Spacecraft That Explore Other Planets Are Robots

You have likely heard the expression "on autopilot." This can refer to a computer doing the work that humans otherwise would have to do. This is also a premise behind robots. Once programmed, some robots can work by themselves. Others must be controlled by a specialist using a device such as a joystick at all times. Robotics is *the study of robots*.

NASA uses robots to explore the Solar System in ways humans often cannot. Even the earliest and simplest spacecraft operated with some preprogrammed functions monitored closely from Earth.

The space probes that travel to the Moon or Mars are robots. The *Lunar Reconnaissance Orbiter* launched in 2009 is one example. The Mars rovers *Spirit* and *Opportunity* are robots that study the ground and environment on the fourth planet from the Sun.

Figure 1.1 In this artist's depiction, *Cassini* enters Saturn's orbit by firing its thrusters to reduce its velocity.

Courtesy of NASA/JPL

You have learned about other spacecraft that orbit or fly by planets and send detailed information to scientists on Earth. These, too, are robots—such as the *Cassini* spacecraft studying Saturn and its rings (Figure 1.1). Even the *Voyager* and *Pioneer* spacecraft traveling millions and millions of miles away outside the Earth's Solar System are robots.

The space probes and rovers that explore other planets, moons, and celestial bodies are autonomous robots. This means they can work by themselves and don't need astronauts or scientists to manually move them. They follow the commands people send from Earth. NASA and other space agencies use computers and powerful antennas to transmit messages to the spacecraft. The robots have antennas that receive the messages and transfer the commands into their computers.

The Use of Robotic Arms in Space

The most common type of robotic device in space is the robot arm often used in industry and manufacturing. The mechanical arm recreates many of the movements of the human arm. It can move not only side-to-side and up-and-down, but also 360 degrees at the wrist, which humans cannot do.

In December 1993 the space shuttle's Canadarm moves astronaut Jeffrey A. Hoffman into position to work on the Hubble Space Telescope. Astronauts also use the Canadarm to capture satellites. For instance, Canadarm has grabbed the Hubble Space Telescope during five different repair missions.

Courtesy of NASA

Robot arms are of two types. One is computer-operated and programmed for a specific function. The other requires a human to actually control the strength and movement of the arm to perform the task. To date, the NASA Remote Manipulator System (RMS) robot arm has performed a number of tasks on many space missions—serving as a grappler, a remote assembly device, and as a positioning and anchoring device for astronauts working in space.

Robotic arms relieve astronauts from having to take so many spacewalks, which are difficult and risky activities. They are able to move very large objects in space. The space shuttle has a robotic arm called Canadarm. Its name comes from the fact that it was built in Canada. It first flew on the shuttle's second mission in 1981.

Canadarm can bend and rotate in seven different directions to perform its tasks. It has a shoulder joint, an elbow joint, a wrist joint, and a gripping device called an end effector. This is *a device located at the end of a robotic arm that can grasp and snare objects much as human hands do.* Controlled by an astronaut inside the shuttle, the Canadarm becomes an extension of the controller's own arm.

Astronaut Joan Higginbotham

Joan Higginbotham applied to become an astronaut because her boss kept encouraging her to. She never expected to go into space.

Higginbotham graduated from Southern Illinois University, and only two weeks afterward started working for NASA in Florida. She later received two master's degrees from Florida Institute of Technology, all the while keeping her full-time job at the Kennedy Space Center. Higginbotham stayed with NASA for nine years. She participated in more than 50 launches in supporting roles on the ground.

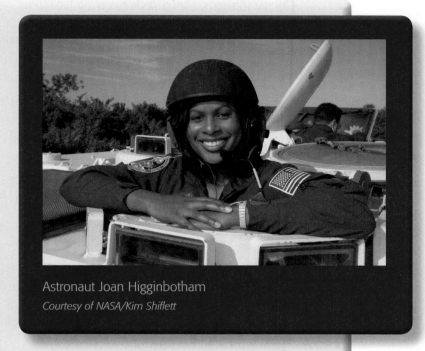

Astronaut Joan Higginbotham
Courtesy of NASA/Kim Shiflett

The Chicago native finally got her chance to go on her own shuttle mission in 2006 with mission STS-116. When *Discovery* went to the International Space Station, mission specialist Higginbotham was the person responsible for controlling the robotic arm. She was part of a team improving the space station's solar array, which provides power to the ISS.

A year after her mission on *Discovery*, Higginbotham retired from NASA for work outside the government.

NASA has used this robotic arm to release and recover satellites. For instance, Canadarm has grabbed the Hubble Space Telescope during five different repair missions.

The International Space Station also has a robotic arm, a larger one called Canadarm2. Scientists have used both Canadarm2 and the shuttle's smaller arm to build the ISS. The arms work like construction workers and cranes, moving the proper pieces into place. The robotic arms also help move astronauts around the station on spacewalks.

The arm on the ISS can move around to different parts of the station. It creeps along the station's outside like an inchworm, attached at one end at a time. The arm also includes a robotic hand named Dextre, which the astronauts use to move smaller objects. Astronauts maneuver joysticks, similar to the ones you play videogames with, to move the arm and hand around. Personnel back at Mission Control on Earth can also control the arm and hand this way.

How Robots Help Astronauts

In addition to aiding astronauts at the International Space Station, robots help scientists investigate new worlds. Scout robots, such as the *Lunar Reconnaissance Orbiter*, can act like members of an advance team, taking pictures and measuring the terrain of a potential landing site. This allows engineers and scientists to prepare future missions accordingly. These robots can also look for various dangers and note the best places for astronauts to walk, drive, and explore, as well as to avoid. This work saves astronauts valuable time once on a mission, and allows them to work in as safe an environment as possible. Humans working together with robots is essential to studying other worlds.

NASA is constantly working to improve its field of robots, keeping up with advances in technology. A new idea technicians from NASA and General Motors Corporation are exploring is the Robonaut. A Robonaut is *a robot that has an upper body shaped like a human, with a head, chest, arms, and hands.* It could work outside a spacecraft, doing the work an astronaut would do on a spacewalk. It might have wheels or, as engineers continue to tinker with it, some other way to move about. This mobility would allow it to work on the Moon or some other planetary object.

Robonauts wouldn't replace human astronauts unless a mission were too dangerous for a human. Rather, they would work alongside them, complementing the work they do. NASA and General Motors are currently working on a second-generation Robonaut called Robonaut2 or R2 that is stronger and has an easier time handling such things as tools.

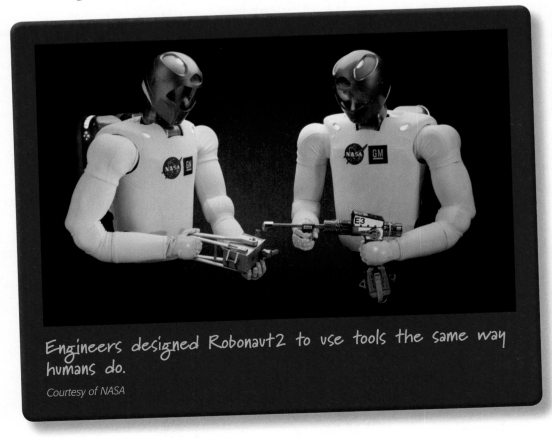

Engineers designed Robonaut2 to use tools the same way humans do.

Courtesy of NASA

SPHERES are bowling-ball-size robots still in their testing phase. They have been on several missions aboard the ISS. Engineers designed them to float in microgravity and move about using compressed CO_2 thrusters.

Courtesy of NASA

R2's head contains two camera "eyes," which give it a form of depth perception, like human eyes. Its hands have five fingers, including a thumb, that it can manipulate just as you move your hands and fingers to pick up and carry different objects. But R2's arm has a greater range of motion than a human arm.

NASA is also working on another type of robot, known as SPHERES. Instead of looking like people, these small robots look a little like soccer balls. NASA uses SPHERES on the ISS to test how well they move in microgravity. Engineers hope that in the future similar robots could fly around inside the station helping astronauts.

Scientists hope to develop a smaller robotic arm for inside the ISS. They could use it in the event of a medical emergency. If an astronaut became sick or was injured, doctors on Earth could maneuver the robotic arm to perform surgery. This technology would also be helpful on Earth. Doctors could place these robotic arms in remote locations where there are no doctors, and in the event of an emergency, these robotic arms could also perform surgery.

The History of Robots in Space

Three years before President Kennedy made his famous speech to Congress about sending a man to the Moon, the United States entered space for the first time. *Explorer 1* launched on 31 January 1958 with instruments on board that detected radiation in Earth's atmosphere. Scientists eventually called this trapped energy the Van Allen Radiation Belt, which you read about in Chapter 2.

> ### *Star* POINTS
>
> NASA decided early on to use a humanlike frame for its Robonauts because so much space equipment is already designed for use by human beings, particularly for the human hand.

The Successes and Failures of the *Explorer* and *Ranger* Missions Into Space

Explorer 1 launched atop a four-stage rocket. This 30.8-pound satellite was the fourth stage. It was also the first American robot launched into space because it housed scientific instruments programmed to gather data. *Explorer 1* orbited Earth more than 58,000 times until it burned up in the planet's atmosphere on 31 March 1970.

The first five *Explorers* launched before NASA's formation (which took place on 1 October 1958). More than 70 Explorer missions—making such discoveries as solar wind, micrometeoroid properties, solar plasma, and more—have launched over the years. Only two have failed: *Explorers 2* and *5*, both before Congress created NASA. The Army Ballistic Missile Agency sent the pre-NASA *Explorers* into space.

The Explorer program continues to this day. NASA continues to use these missions for scientific study. The space agency keeps them on a small- to mid-size scale to hold down costs and to send missions to space at a fairly rapid pace.

Juno I carried Explorer 1 to space.

Courtesy of NASA

Once the United States and the Soviet Union began their race to the Moon, both relied on robots to lead the way to manned landings. NASA called its exploratory robotic program Ranger.

Rangers 1 and *2*, which were to scout the Moon's surface for a safe landing spot, never made it out of Earth's orbit. *Ranger 3* did make it out of the Earth's orbit, but was traveling too fast to enter the Moon's orbit.

Ranger 4 gave off the first hint of success when it crashed into the Moon's far side on 26 April 1962. Although the spacecraft's power failed, it was already well enough on its way that it managed to complete its intended collision course. It was the first American spacecraft to reach a celestial body.

With *Ranger 7*, NASA finally found total success. Three days after a 28 July 1964 launch, *Ranger 7* sent back more than 4,000 pictures of the Moon's surface. Furthermore, it crash-landed on target. The images were nearly 1,000 times better than previous ones taken from Earth. This mission also helped lay the groundwork to determine a safe landing zone for future manned missions to the Moon.

Rangers 8 and *9* were also successes. *Ranger 8* captured more than 7,000 images. And *Ranger 9* took pictures that transmitted live to TV. In an age of wireless Internet and YouTube, this may not seem monumental, but in the 1960s this was a tremendous accomplishment. *Ranger 9* also crashed only four miles (6.5 km) from its target.

The Accomplishments of the NASA *Surveyor* Spacecraft

Still relying wholly on robotic technologies, NASA next attempted soft landings on the Moon. Such landings had to be soft enough for astronauts to survive them. That's where the *Surveyor* spacecraft series came in during the late 1960s.

The seven robotic missions cost NASA $469 million. But when all was said and done, scientists had more than 80,000 pictures of the Moon's surface, as well as data from soil samples, information on the terrain, and the knowledge that a soft landing was possible.

Unlike the Ranger program, which had successful failures, the Surveyor program was almost entirely successful. Five of the seven missions accomplished what they set out to do. In June 1966 *Surveyor 1* made America's first soft landing ever. The combination of retrorockets and thrusters slowed down the spacecraft from a speed of 6,000 miles (9,600 km) per hour to three miles (5 km) per hour until *Surveyor 1* went into a freefall at 14 feet (4.3 meters) from the surface. Furthermore, it landed within 46 feet (14 meters) of its target. This first mission sent more than 11,000 pictures back to Earth.

A couple of months later *Surveyor 2* crashed into the Moon's surface after one of its thrusters failed and sent the spacecraft into a spin. But the program found success again in 1967 with *Surveyor 3*. It scooped up lunar soil to test its bearing strength, *the ability to support a certain amount of weight*—in this case the weight of an *Apollo* lunar lander. Scientists found it similar to wet sand, solid enough for the planned manned missions. Before this, many had feared that the Moon might have a very thick layer of soft dust that could swallow a lander—or prevent it from standing up straight.

NASA lost contact with *Surveyor 4* only 2.5 minutes before the expected landing on the Moon, but the space agency did well with *Surveyor 5* in late 1967. While a thruster leak made the mission's outcome dicey for a time, engineers figured out a

Surveyor 6 lifts off on its mission to the Moon.

Courtesy of NASA/JPL

different way to brake and land the spacecraft. As a result, *Surveyor 5* became the first US spacecraft to conduct soil analysis on another planet. *Surveyor 5* also fired its main engine for less than a second to see what, if any, affect it had on the soil. NASA was starting to test takeoffs from the lunar surface to ensure that a manned mission could return to Earth.

Surveyor 6 took this one step further by actually lifting off. It was the first lander to lift off a surface other than from Earth. It rose about nine feet (three meters) and headed westward about eight feet (2.5 meters). Now the stage truly was set for the manned *Apollo* missions. *Surveyor 7*, the final in the series, was notable because it landed in an area completely different from the previous missions. It was an entirely scientific mission as NASA had accomplished all of its Moon-mission tests with the other *Surveyor* lunar landers.

The Significance of the *Mariner 9* Mission to Mars

By the 1970s robotic technologies had improved enough that scientists could begin to contemplate missions even more distant and difficult than a Moon landing. NASA set its eyes on Mars and Venus with the *Mariner* series. Here again its first—and thus far only—attempts depended on robots, not manned spacecraft.

Mariner 9 lifted off for Mars on 30 May 1971. On 14 November it became the first spacecraft to orbit another planet, an event you read about in Chapter 9. While the Soviets had been the first to fly by Mars as far back as 1962 with their probe *Mars 1*, their newer spacecraft *Mars 2* didn't enter Martian orbit until two weeks after *Mariner 9*. However, the Soviet *Mars 2* released a probe for the planet's surface on 27 November. Although it crash-landed, the Soviet lander still became mankind's first spacecraft to reach the Martian surface.

Mariner 9's goal was to map 70 percent of the Martian surface. A massive dust storm engulfed the planet, however. Once conditions improved around mid-January 1972, *Mariner 9* mapped more than 85 percent of the surface.

Mariner 9 also was significant because it helped lay the groundwork for the historic Viking missions a few years later.

More-Recent Achievements of Robots in Space

In the 1980s countries around the world launched 13 robotic missions. Eleven of these were successful. They returned so much information that they kept scientists very busy.

This decade was historic because the space race was no longer limited to the United States and the Soviet Union. For the first time ever, other countries began sending up spacecraft. Japan was the first. In January 1985 it launched *Sakigake*, which you read about in Chapter 9. It was part of an international fleet of five spacecraft en route to Comet Halley. Between 6 and 13 March 1986 each flew by its target. The Soviet Union arrived first with *Vega 1* followed by Japan's *Suisei* (a second Japanese mission), the Soviets' *Vega 2*, then *Sakigake*, and finally the European Space Agency's *Giotto*. In addition to examining Comet Halley, *Sakigake* also taught scientists about the Earth's magnetic tail.

While the United States didn't take part in the international flyby of Comet Halley, it still managed to be the first country in the world to study a comet. In 1978 NASA launched the *International Sun-Earth Explorer 3*, or ISEE-3. In the early 1980s NASA reprogrammed the spacecraft to intercept and study Comet Giacobini-Zinner, which it did on 11 September 1985. In honor of its new mission, the space agency renamed ISEE-3 the *International Cometary Explorer*, or ICE (Figure 1.2).

Figure 1.2 An artist shows the International Cometary Explorer (ICE), conducting the first satellite study of a comet. In the early 1980s NASA reprogrammed the spacecraft to intercept and examine Comet Giacobini-Zinner.

Courtesy of NASA

ICE confirmed comets are "dirty snowballs" of ice, gas, and dust. It also found that Giacobini-Zinner's tail had less dust than expected. NASA next sent ICE to study Comet Halley. It arrived in late March after the international group of five spacecraft. ICE has been out of service since 1997, but scientists are making plans to retrieve it if it passes near Earth's orbit as expected in 2014.

During most of the 1980s America directed its energies toward manned missions aboard the space shuttle. But in May 1989, after an 11-year break from robotic spacecraft launches, NASA sent a new deep space probe on a mission. *Magellan* launched from the space shuttle *Atlantis*. It marked the first time that a probe headed for another planet launched from a space shuttle while in orbit. *Magellan* gave scientists the most informative pictures of Venus's surface assembled to date. With its help, scientists learned that lava flows cover 85 percent of the planet's surface. The probe used powerful radar to penetrate Venus's thick clouds and map surface features. In October 1994 NASA deliberately destroyed *Magellan* in Venus's atmosphere.

Achievements in the 1990s

In the 1990s improved robotic technologies meant sharper, crisper images of the Solar System returning from deep space missions. Twenty-one robotic spacecraft launched around the world studied comets, planets, asteroids, and moons. But plenty of probes launched in previous decades also contributed. All of these unmanned robots acted as the eyes of the scientists on the ground.

For instance, *Voyager 1*, launched in 1977, in 1990 took a "family portrait" of the Solar System from 4 billion miles (6.4 billion km) away. And *Galileo*, launched in 1989, snapped 150 photos of the asteroid Gaspara in 1991. This spacecraft, which you read about in Chapter 9, was the world's first to fly by an asteroid. From Gaspara, *Galileo* headed toward another groundbreaking asteroid encounter. In 1993 it flew past the asteroid Ida and discovered the first moon orbiting an asteroid.

In 1994 the world watched Jupiter's gravity capture the Comet Shoemaker-Levy 9 with the aid of *Galileo* and numerous other robotic spacecraft. The comet smashed into the planet's atmosphere. One fragment of it named Fragment A hit the planet with a resulting explosion equal to 225,000 megatons of TNT. This created a plume rising about 620 miles (1,000 km) above Jupiter's clouds. This was the first time humans had observed direct evidence of such a large series of collisions between a celestial object and a planet.

In 1996 NASA launched its Mars *Pathfinder* with the robotic rover *Sojourner* aboard. A year later *Sojourner* rolled off the Mars *Pathfinder* science station onto the Martian surface. This was the first time a rover had moved around on the surface of another planet. The earlier *Viking* probes were stationary. Data gathered by *Pathfinder* and *Sojourner* made possible the next decade's *Spirit* and *Opportunity* rover missions to the red planet.

CHAPTER 12 Robotics in Space

Achievements in the Twenty-First Century

It's still early in the new millennium, but robotic spacecraft continue to score a number of firsts in space exploration. In 2000 NASA's NEAR spacecraft was the first to orbit an asteroid. A year later mission controllers tried something else new: they landed it on the asteroid Eros.

The orbits of Mercury and Mars in the first decade of the twenty-first century also provided some unique opportunities. In 2003, for instance, NASA's *Mars Global Surveyor* snapped images of the Earth, the Moon, and Jupiter from its orbit around Mars. This was the first-ever planetary conjunction—*when two or more celestial bodies appear to be close to one another*—photographed from another planet.

The ISS has been in orbit since 1998 and has been permanently occupied since 2000. It is a success in large part because of its work with robots.

The Mars *Spirit* and *Opportunity* rovers are two other great achievements of this century, but you'll read about those in depth in the next lesson.

The *Mars Global Surveyor* captured this image of (*from the top*) the Moon, Earth, and Jupiter in conjunction.
Courtesy of NASA/JPL/Malin Space Science Systems

The Current Robotic Missions in Space

The economic downturn that began in 2008 put many plans and projects on hold. The Obama Administration made budget requests in early 2010 that would eliminate the Constellation program that aimed to have Americans back on the Moon by 2020. Manned missions to Mars are almost certainly postponed under this proposal.

Manned missions aren't the only programs facing the global economic realities of recent years. Robotic missions are also coming under increased scrutiny, with many experts weighing the pros and cons.

The Advantages and Disadvantages of Using Robots Instead of Humans in Space

Humans have some obvious advantages over robots when exploring space. Astronauts can describe what they see and feel. It's not possible to do this in a similar manner through the lens of a robotic camera. On the flip side, the advantages of sending a robot into space on an unmanned mission are just as obvious: It costs less and it's much, much safer. Aside from a few engineers who constructed the robot, no one will cry if a piece of metal blows up. But the loss of human life is a pain that never goes away.

The Hubble Space Telescope offers a good example of how these competing ideas, even in an economic crunch, can play out. As you learned in Chapter 9 the Hubble Space Telescope has returned some of the most incredible pictures taken from outer space. Without its images, scientists would know much less about the Solar System and the universe. The telescope—a robot—took all these pictures. This illustrates that astronomers can obtain meaningful information for analysis without putting humans at risk, and at a reduced cost.

But as you also read, the Hubble Space Telescope has needed a series of repairs and upgraded instruments. Hubble launched in 1990 and almost immediately scientists determined there was a problem with its mirror that caused images to blur. NASA had no way to fix this problem without sending humans into space to do the work.

One astronaut is tethered to the space shuttle's robotic arm while working on the Hubble Space Telescope. The other is holding on to the telescope's handrail.

Courtesy of NASA

Astronauts have made five different trips to the Hubble Space Telescope to make repairs and install newer equipment. They went on spacewalks and worked with tools that they had to hold in their hands. But robots helped: The shuttle crew used Canadarm, the shuttle's robotic arm to capture Hubble and move it into the cargo bay.

During the spacewalks, one astronaut was attached to the arm with a foot restraint to keep from drifting away. An astronaut in the shuttle maneuvered the arm as needed to move the astronaut around the telescope. In this way, the arm became a space version of the "cherry picker" that utility workers use on Earth.

These were costly trips, and especially in light of the more recent *Columbia* tragedy, trips with great risk. Had these manned trips not taken place, though, Hubble would be far less useful than it is.

The Use of Robotic Arms in Assembling the International Space Station

While the ISS has been under construction since the launch of its module *Zarya* in 1998, the robotic Canadarm2 has played a critical role in space station assembly and maintenance since its arrival in 2001. It launched aboard *Endeavour* for mission STS-100.

Expedition 11's crew maneuvers Canadarm2 from 225 miles above Cape Horn.
Courtesy of NASA

Canadarm2 is actually part of a larger system called the Mobile Servicing System, or MSS, built in Canada. It has three parts:

1. Canadarm2, a bigger, better, smarter version of the space shuttle's robotic arm. It is 57.7 feet (17.6 meters) long. It weighs 3,968 pounds (1,800 kilograms). It also operates with seven motorized joints. Canadarm2 handles large payloads and helps the ISS dock with the space shuttle.

2. Mobile Base System, a work platform that moves along rails covering the length of the space station. It allows Canadarm2 to move from end to end. Mobile Base System arrived in June 2002 aboard *Endeavour* for STS-111.

3. Special Purpose Dexterous Manipulator, or Dextre, a two-armed robot that can carry out delicate assembly tasks.

Unit 4 Space Technology

Japan provided another robotic arm in 2009 for use with its lab module Kibo. The ISS crew first tested it out on 23 July. It helps with such things as experiments and installing hardware.

As of this writing in 2010, ISS partners plan to add one more robotic arm. The European Space Agency will contribute the European Robotic Arm (ERA) to move small payloads into and out of a new Russian airlock. Like a cherry picker, it will also be able to move astronauts and cosmonauts during spacewalks. It has two wrists, two limbs, and one elbow joint. Each end of the arm can act as a hand or can attach to the space station to serve as a base from which the robotic arm can operate. NASA has scheduled to ferry ERA aboard *Atlantis* for mission STS-132 sometime in 2010.

The Use of the Wide-field Infrared Survey Explorer (WISE)

The last time an infrared survey of the sky was taken from space back in 1983, Ronald Reagan was president. In December 2009 NASA launched a new robotic spacecraft, the Wide-field Infrared Survey Explorer, or WISE, atop a *Delta II* rocket. The aim of the expected 10-month mission was to search for the origins of galaxies. WISE was in a polar orbit 326 miles above Earth.

Star **POINTS**

WISE takes a picture every 11 seconds. In the first six months of its exploration it will have taken more than 1.5 million pictures.

WISE is creating an infrared map of the entire sky. It does this by detecting the infrared, or heat, signatures of objects. Because of this, the WISE instrument must always be kept cooler than the objects it is studying. WISE will catalog a diverse group of astronomical targets, including near-Earth asteroids, stars, planet-forming disks, and distant galaxies.

During its first few months in space it sent back astonishing pictures of comets, bursting stars, and faraway galaxies. Information gathered from WISE will help with the planning of other missions, notably Hubble's replacement, the James Webb Space Telescope.

In the next lesson you'll read about two more current robotic missions: *Spirit* and *Opportunity*. NASA launched these rovers to Mars in 2003. The lesson will also look at the space agency's plans for future robotic missions.

CHECK POINTS

Lesson 1 Review

Using complete sentences, answer the following questions on a sheet of paper.

1. What does it mean if a robot is "autonomous"?

2. Name one thing a robotic arm can do.

3. What is an end effector?

4. What is a Robonaut?

5. Which was the first "total success" *Ranger* mission? What were its two major accomplishments?

6. Which spacecraft made America's first soft landing on the Moon, and which first successfully lifted off the lunar surface?

7. Which was the first robotic spacecraft to orbit another planet?

8. Which five spacecraft traveled in an international fleet to study Comet Halley? Which countries launched them?

9. Which was the first spacecraft to fly by an asteroid? Which was the first to orbit and land on one?

10. Name an advantage to robotic over manned missions and vice versa.

11. What's the name of the work platform that allows Canadarm2 to move from end to end on the ISS?

12. Why must WISE always be kept cooler than the objects it is studying?

APPLYING YOUR LEARNING

13. You read that *Ranger 3* was traveling too fast to enter Moon orbit. Explain this in terms of *escape velocity* and *orbital velocity*.

LESSON **2** The Mars Rover and Beyond

Quick Write

_____

_____

If you could name
a spacecraft or a rover,
what would you name
it and why?

Learn About

- the history of the Mars Rover Expedition
- the results of the Mars Rover Expedition
- the goals for future rover expeditions

Sofi Collis was born in Siberia, a part of Russia. For the first two years of her life, she lived in an orphanage. She dreamed of getting out and exploring. An Arizona family adopted her and she moved to the United States.

When NASA decided on a new mission to Mars in 2003, the agency held a contest to name the new spacecraft. Legos, the manufacturer of children's toys, administered the contest to students nationwide. NASA hoped the contest would instill a sense of creativity and exploration in a new generation of students in the twenty-first century. The agency received more than 10,000 entries.

Sofi, age 9, entered the contest. She told NASA that her story, and the life she has in America, should be the basis of the names for the new missions.

"I used to live in an orphanage. It was dark and cold and lonely," Sofi wrote in her winning entry. "At night, I looked up at the sparkly sky and felt better. I dreamed I could fly there. In America, I can make all my dreams come true. Thank you for the 'Spirit' and the 'Opportunity.' "

She won the contest.

The History of the Mars Rover Expedition

NASA first began exploring Mars in the 1960s. Over the years, a series of orbiters, fly-bys, and landers have helped give scientists clues about the red planet. But in 2004 *Spirit* and *Opportunity* began their robotic exploration of the planet, giving even more insight into its history and providing clues to the ultimate question: Is there or was there ever life on Mars?

Spirit and *Opportunity*

After the success of the *Pathfinder* mission in 1997 NASA considered what should be the next step in exploring Mars. Scientists debated whether this new mission, slated to coincide with the favorable position of Mars in relation to the Earth and the Sun, would be an orbiter or a lander. If it were a lander, they asked, how should it be different from previous missions, such as *Pathfinder*?

The Mars rover Sojourner (far right) inspects a rock dubbed Yogi.

Courtesy of NASA/JPL

In 2000 NASA announced the new mission would be a pair of twin rovers. They would be able to travel in one day the distance that *Sojourner*, the rover from the *Pathfinder* mission, was able to travel during its entire mission. The mission's main goal would be to determine if water activity—such as precipitation and evaporation—existed, or had ever existed on the planet.

The two rovers would spend their first 90 days exploring as much as they could. *Spirit* launched on 10 June 2003, with *Opportunity* following a few weeks later on 7 July 2003. The timing was important: Mars came closer to Earth in August of that year than it had in thousands of years. They arrived on Mars in January 2004 after a seven-month journey. After a successful drop, bounce, and roll landing—in which the rovers were packaged in something like giant air balls attached together like a peanut cluster—the two robots began an extraordinary adventure that continues to fascinate scientists today (Figure 2.1).

The Challenges of the Flight to Mars

Spirit and *Opportunity* had several course corrections during their voyages from Earth to Mars. *Spirit* made four and *Opportunity* made three. Both arrived at Mars on schedule, but along the way they encountered numerous hiccups, including intense solar flares and high-energy particles. Because of these challenges, NASA rebooted the rovers' computers, something scientists had not planned on doing in mid-flight.

For the journey, scientists securely wrapped both *Spirit* and *Opportunity* inside intricately designed packages. This design kept the rovers safe from outside forces and prepared them for a challenging landing. Both *Spirit* and *Opportunity* were each

Figure 2.1 An artist's drawing shows *Spirit*—wrapped in airbags—as it bounces along the Martian surface.
Courtesy of NASA

CHAPTER 12 Robotics in Space

tucked inside a folded-up lander. There was a protective shell on the outside as well. NASA technicians also carefully sterilized each rover before launch to ensure it couldn't carry any Earth microbes to Mars. Such microbes might reproduce on Mars or interact with any life that might already be there.

Landing on Mars is no easy task. Many landings there since exploration began 50 years ago have failed. The rovers entered the Martian atmosphere going nearly 12,000 miles (19,300 km) per hour. The heat-shield portion of the protective shell was facing forward. During the first four minutes atmospheric friction slowed the rovers by 90 percent. Two minutes before each landed, a parachute deployed and the heat shield released. A camera began taking pictures to show the landing region up close.

When it landed, *Spirit* bounced 27 feet (eight meters) high. It went on to bounce 27 more times before stopping permanently 900 feet (274 meters) away from its initial landing spot. *Opportunity* took a similar path. Both landed very close to the initial target areas NASA had aimed for when the rovers launched. Scientists were so pleased with *Opportunity*'s nearness to its target landing site that they called it a "planetary hole in one."

The Basic Features of the Mars Rovers

The rovers are robots equipped with special tools to do their jobs. These tools include cameras, spectrometers, and equipment to drill and chip rock (Figure 2.2). They give scientists detailed information about Mars' surface. Each rover has five main instruments. The first, and perhaps most important, is the panoramic camera. This high-resolution device shows the surrounding terrain in each new location. It has two "eyes" that are located about a foot (30 centimeters) apart. It is perched about five feet (1.5 meters) above the surface. Fourteen different filters on the camera allow for color pictures and spectral analysis of minerals and atmosphere. Spectral analysis is *identifying an object based on the spectrum of light that it reflects, absorbs, or emits*. The camera is in large part responsible for determining future rock and soil targets for the rover to explore.

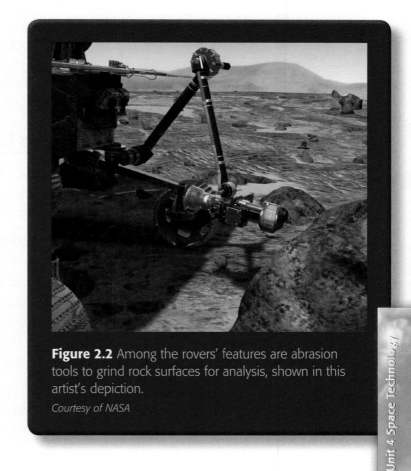

Figure 2.2 Among the rovers' features are abrasion tools to grind rock surfaces for analysis, shown in this artist's depiction.
Courtesy of NASA

Second, the Miniature Thermal Emission Spectrometer views the surrounding scene in infrared wavelengths. A spectrometer is a device that separates light signals into different frequencies. This determines the different types and amounts of many different kinds of materials. NASA uses this instrument as the rovers search for the distinctive minerals formed by the action of water. Scientists also use the spectrometer with the camera to select new science targets and areas to explore. Like the camera, the spectrometer also studies the atmosphere.

Third, the Mossbauer Spectrometer is located on the rovers' arms. It specifically looks for iron deposits. It touches rocks and soil and identifies the minerals it finds that contain iron. Scientists use this to determine what role, if any, water played in the formation of the targets and how much they have been weathered.

Fourth, the Alpha Particle X-Ray Spectrometer is similar to the one used on *Sojourner* and what scientists use in geology labs on Earth. It measures the concentrations of most of the major elements in the rocks and soil. The more scientists learn about these ingredients, the more they understand the samples' origins and how time has altered them.

Fifth, the Microscopic Imager helps scientists look at the fine-scale details of Martian rocks and soils. It searches for smaller, more delicate clues about how the rocks and soils were formed. For example, the size and angularity—*the sharpness of edges and corners*—of grains in sediments laid by water can reveal how they got there.

In addition to these five tools, other accessories on the rovers help guide and operate them, as well as provide geological information. This includes navigation and hazard-avoidance cameras. In addition, the wheels help move the rovers around and dig trenches to study the soil.

The Anticipated Duration of the Mars Rover Expedition

When the rovers launched in 2003, NASA expected their missions to last 90 days. After the primary mission concluded, however, the rovers were still going strong. So scientists extended the missions.

Star POINTS

By 2010 the rovers had worked 24 times longer than originally expected.

As this book is written, both *Spirit* and *Opportunity* were still functioning six years after they first landed on Mars. Unfortunately, a sand trap snared *Spirit* in 2009. Months of efforts to free it were unsuccessful. By early 2010 scientists determined it was best to just leave *Spirit* where it was in a stationary capacity. In the spring of 2010, scientists were waiting to see whether the rover would survive the Martian winter.

Opportunity, as of this writing, was still crawling around on the red planet. Like *Spirit*, it is solar powered and continued to operate on Mars because of the Sun's energy. It is on a different side of the planet than *Spirit*, and has not encountered the same problems. But whether the rovers lasted a few more days or several more months, they had already racked up successes far beyond scientists' expectations.

A NASA team in California tests ways to free a rover stuck in loose soil.

Courtesy of NASA/JPL-Caltech

The Results of the Mars Rover Expedition

Scientists' plan for the rovers was to build on the research from the *Pathfinder* mission. The information the rovers have gathered is helping further NASA's understanding of the Solar System, its formation, and whether other life may exist—or has existed in the past.

The Science Goals of the Mars Exploration Program

It took years of planning to get *Spirit* and *Opportunity* ready for their 2003 launch. After careful consideration, scientists decided on robot-like rovers that could explore up close the minerals and composition of Martian rocks and soil. In looking for evidence of past liquid water, the rover missions would support the four main science goals of NASA's long-term Mars Exploration Program:

- to determine whether life ever existed on Mars
- to learn about Mars' climate
- to learn about Mars' geology
- to prepare for human exploration of Mars.

NASA directed the rovers to land in places that may have held an abundant supply of liquid water millions of years ago. Scientists believe information gained at these locations gives them the best chance at determining where and whether life may have existed before.

Opportunity snapped this photo inside a crater in Mars' Meridiani Planum region.

Courtesy of NASA/JPL/Cornell

NASA gave the two rovers geologically different landing sites. Scientists sent *Opportunity* to an area known as Meridiani Planum, and it landed inside a crater scientists previously did not know existed. This region is located on the Martian equator. NASA sent the rover there because previous explorations of Mars, notably by the *Mars Global Surveyor*, discovered the presence of hematite—*an iron oxide that on Earth usually forms in an environment containing liquid water.*

On 4 January 2004 *Spirit* landed in the Gusev Crater, a 90-mile- (145 km) wide hole south of the equator that appears to have held a lake. Scientists say it formed more than 3 billion years ago. This crater was likely the result of an asteroid crashing into the planet. The region has a channel system that scientists surmise likely carried water or ice at some point. It would be unusual for a pattern like this to exist without the presence of water.

The Findings of *Spirit*

During *Sprit*'s first month on the red planet it encountered serious computer and communications problems, and some wondered if the mission were doomed. Scientists determined they could fix the problem if they erased some of the unneeded data stored on *Spirit*'s computers. Within a month, *Spirit* was back to full working mode.

Once corrected, *Spirit* began to immediately send information back to Earth that would forever change the way scientists viewed Mars. Since its landing the rover has sent more than 127,000 pictures back to scientists, giving them highly detailed information. In broad terms, it has illustrated a rock-filled, dusty planet. But it has also presented evidence that liquid water—and perhaps life—once existed on Mars.

The mission was scheduled to last 90 days. Shortly after its initial problem was corrected, *Spirit* used its rock-drilling tool to take a first-ever look inside a rock on Mars. Scientists named this rock Adirondack, after the mountains in upstate New York. During the next six years *Spirit* found numerous clues that water had existed on the planet. The Alpha Particle X-Ray Spectrometer found bromine, sulfur, and chlorine inside the Gusev Crater where *Spirit* landed. These are strong signs that liquid water was at one point present there.

NASA unofficially named a region of seven small hills inside the Gusev Crater the Columbia Hills, in honor of the seven members of the ill-fated space shuttle *Columbia* mission. Each of the hills is named for one of the members. On Husband Hill, named after *Columbia* commander Dick Husband, *Spirit* detected evidence this region of Mars was once a hot and violent place, with volcanic eruptions and asteroid impacts.

During its travels, *Spirit* also uncovered evidence of a long-ago explosion at a bright, low plateau called Home Plate in 2006. Scientists observed bulbous, or rounded, grains overlaying finer material. This fits the pattern of accumulated material falling to the ground after a volcanic or impact explosion. Scientists observed such rocks for the first time ever on Mars. The rocks revealed more evidence of the crater's violent history.

A rock that scientists named Adirondack was *Spirit's* first target after arriving on Mars. The rover drilled Adirondack to take a first-ever look inside a rock on the red planet.

Courtesy NASA/JPL

Spirit explored this low plateau called Home Plate in 2006.

NASA/JPL-Caltech/Cornell

In other parts of the crater where *Spirit* traveled, scientists discovered silica, *a white compound made of crystals that occurs naturally on Earth as quartz, sand, or flint, and is the main ingredient of window glass.* This material is often found either in a hot-springs environment like Iceland, or in an area of steam vents where acid rises into the cracks of rocks. On Earth, both settings teem with living microbes. This discovery was the mission's biggest scientific achievement.

Spirit also made several movies of dust devils in motion. These images provided the best view of how the wind affects the Martian surface in real time.

In 2006 *Spirit* uncovered bright Martian soil at a place named "Tyrone." This region contained large amounts of sulfur and a trace of water. This finding intrigued scientists. They theorize that the material could be a volcanic deposit formed around ancient gas vents. Or it could have been left behind by water that dissolved these minerals underground, then came to the surface and evaporated.

In 2009 *Spirit* got caught up to its belly in a sand trap. For months scientists tried to maneuver it out of the trap, but complications with the wheels hindered their efforts. In early 2010 NASA decided *Spirit* would cease being a mobile rover and instead become a stationary observer on Mars. It continues to conduct research, including examinations of the tilt of Mars' axis and its planetary "wobble." NASA was uncertain, however, if the solar powered vehicle would have enough power to survive once the Martian winter began in May 2010.

The Findings of *Opportunity*

Opportunity arrived on Mars a few weeks later, on the other side of the planet. NASA unofficially named the landing region the Challenger Memorial Station, in honor of that ill-fated space shuttle mission.

This rover, too, has been wowing scientists ever since, taking more than 133,000 images. Unlike *Spirit*, it had not encountered any serious problems and continued to be functional and mobile as of March 2010.

Opportunity found the strongest evidence yet that liquid water once existed on Mars' surface. Scientists believe its landing site, known as the Meridiani Planum, was once the shoreline of a salty sea.

Almost immediately *Opportunity* made key discoveries. It found layered rocks, some layers no thicker than a finger, which indicated that sediments were deposited in this region by wind, water, or volcanic ash. Shortly thereafter, in a region named Stone Mountain, *Opportunity* discovered patterns that indicated a moving current, such as a volcanic flow, wind, or water.

Opportunity was the first of the two rovers to dig a trench on Mars. The trench measured 20 inches (50 centimeters) long and four inches (10 centimeters) deep. Scientists were surprised to discover the clotty texture of the soil. *Opportunity* also discovered the first meteorite on another planet. Because it was located near the impact site of the rover's heat shield, NASA scientists named the meteorite Heat Shield Rock.

The Challenges of Operating the Rovers

Two of the biggest challenges facing the rovers are dust and sand. Dust storms occur regularly on Mars. Often these storms cover critical parts of the rovers, interfering with their movement. These storms are also problematic because when the solar-powered rovers are covered in dust and sand, they can't receive energy from the Sun. At the same time, storms also can be a blessing—they also clear some of the debris off the rovers when the wind hits at the right angle.

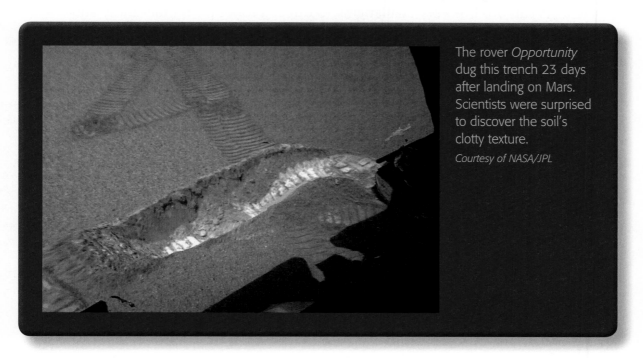

The rover *Opportunity* dug this trench 23 days after landing on Mars. Scientists were surprised to discover the soil's clotty texture.
Courtesy of NASA/JPL

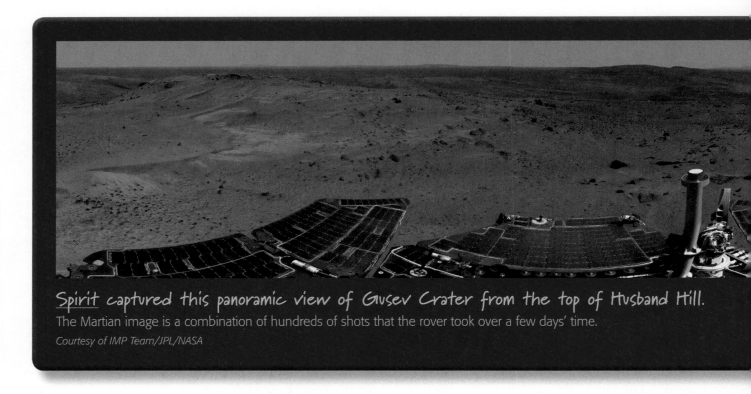

Spirit captured this panoramic view of Gusev Crater from the top of Husband Hill.
The Martian image is a combination of hundreds of shots that the rover took over a few days' time.

Courtesy of IMP Team/JPL/NASA

NASA has models of the rovers on Earth that engineers test when the real ones encounter trouble millions of miles away on Mars. They work on these rovers to find solutions to the Mars rovers without exerting precious energy to do so. But sometimes, as when *Spirit* got stuck in the sand in 2009, there is nothing scientists can do. It is very challenging to direct a robot from miles away and not be able to move it or brush it off when a problem arises. It takes 20 minutes for a message from the rover communicating a problem to reach engineers on Earth. The engineers then have to carefully write instructions (which itself takes time) and then send them back to Mars. That transmission takes another 20 minutes. The rover conducts the action, then communicates the results—using up another 20 minutes. The fact that the rovers have long outlived their expected life spans, however, demonstrates that NASA engineers and scientists have been very successful in handling the challenges Mars presents.

The Goals for Future Rover Expeditions

The more scientists learn about the Moon, Mars, and other objects in the Solar System, the more they want to know. And that means more unmanned missions.

In the lengthy tale of cosmic creation, one of the most interesting chapters is the formation and nature of the Solar System—humanity's celestial backyard. The Solar System has always inspired a sense of wonder and raises many fundamental questions: Are humans alone in a cold, impersonal cosmos? Are there habitable worlds other than Earth? How did Earth and its complex oasis of life come to be?

The Goals of the Planetary Science Program

NASA has designed a planetary science program to seek answers to these and other questions:

- How did the Sun's family of planets and minor bodies originate?
- How did the Solar System evolve to its current diverse state?
- What characteristics of the Solar System led to the origin of life?
- How did life begin and evolve on Earth and has it evolved elsewhere in the Solar System?
- What are the hazards and resources in the Solar System environment that will affect establishing a human presence in space?

NASA's program addresses the following research objectives:

- to understand the history and future of habitability in the Solar System
- to determine if there is or ever has been life elsewhere in the Solar System
- to explore the space environment to discover potential hazards to humans
- to search for resources that would enable a human presence beyond Earth.

One of NASA's primary goals is to conduct future missions using sophisticated robots such as the Mars rovers. These could conduct scientific experiments and help lay the foundation for human travel beyond the Moon. Potential targets for robotic missions in the next 30 years or so include Jupiter's moons, various asteroids, and other bodies.

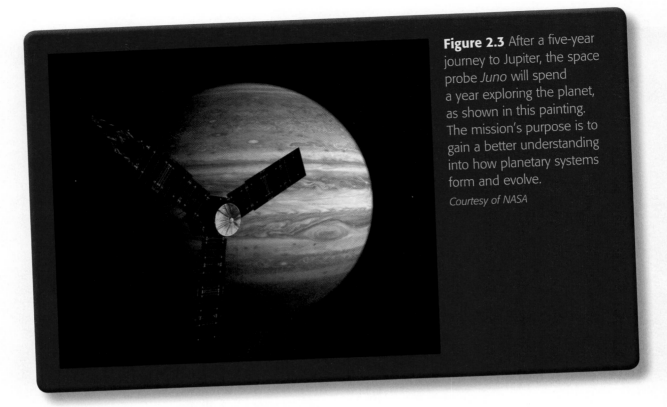

Figure 2.3 After a five-year journey to Jupiter, the space probe *Juno* will spend a year exploring the planet, as shown in this painting. The mission's purpose is to gain a better understanding into how planetary systems form and evolve.

Courtesy of NASA

The *Juno* and Europa Jupiter System Missions

To understand how planetary systems form and evolve, NASA is sending the probe *Juno* to Jupiter in 2011. The robotic spacecraft will arrive in Jupiter's orbit in 2016 and examine the largest planet in detail for a year. Scientists hope this mission will offer even more clues and insight into the Solar System's formation (Figure 2.3).

Unlike Earth, Jupiter's giant mass allowed it to hold onto its original composition, providing scientists with a way of tracing the Solar System's history. Understanding Jupiter's formation is essential to understanding the processes that led to the development of the rest of the Solar System. It's also key to understanding the conditions that led to Earth and human beings. Like the Sun, Jupiter is composed mostly of hydrogen and helium. A small percentage of the planet is composed of heavier elements. However, Jupiter has a larger percentage of these heavier elements than the Sun.

Star POINTS

In Roman mythology, the goddess Juno was Jupiter's wife. Jupiter was the king of the gods and god of the sky and thunder. He was the equivalent of the Greek god Zeus.

Juno will measure the amount of water and ammonia in Jupiter's atmosphere and determine if the planet actually has a solid core. This will aid in understanding the giant planet's origin and that of the Solar System. By mapping Jupiter's gravitational and magnetic fields, *Juno* will reveal the planet's interior structure and measure the mass of its core.

Meanwhile, NASA and the European Space Agency (ESA) are planning another mission to Jupiter and its moons. The Europa Jupiter System Mission would use two robotic orbiters to study Jupiter and its moons Io, Europa, Ganymede, and Callisto. NASA would build one orbiter, initially named Jupiter Europa. ESA would build the other orbiter, initially named Jupiter Ganymede. The probes would launch in 2020 on two separate launch vehicles from different launch sites. They would reach the Jupiter system in 2026 and spend at least three years conducting research (Figure 2.4).

Europa has a surface of ice, and scientists theorize it has an ocean of water beneath that could provide a home for living things. Ganymede, the largest moon in the solar system, is the only moon known to have its own internally generated magnetic field and may have a deep undersurface water ocean. Scientists long have sought to understand the causes of this magnetic field. Callisto's surface is heavily cratered and ancient, providing a record of events from the early history of the Solar System. Finally, Io is the most volcanically active body in the Solar System.

Figure 2.4 Solar panels will power scientific gear aboard the ESA's *Jupiter Ganymede* orbiter, shown in this artist's conception.
Courtesy of NASA

The orbiters would spend nearly a year orbiting Europa and Ganymede. NASA's probe would investigate whether Europa might harbor life. ESA's spacecraft would orbit Ganymede to conduct investigations of the surface and interior of this satellite, to better understand the formation and evolution of the Jovian system.

The Three Types of Planetary Missions

NASA's exploration plans include three types of unmanned planetary missions: Discovery Missions, New Frontiers Missions, and Flagship Missions.

Discovery Missions will let scientists address a specific question about space in an expeditious manner. These small missions will investigate comets and asteroids, among other things. They will help scientists understand impact hazards over time, the Solar System's chemistry, and the architecture of its planetary system.

New Frontiers Missions are slightly larger missions. They will explore a wide range of objects, including other planets in the Solar System. These missions will be limited in scope though, in terms of what they do once they arrive at these bodies. Hence, they too will play key roles in Solar System exploration. But they cannot achieve all of the measurement and exploration objectives necessary to answer the basic questions that can be better answered with robotic exploration of the planets.

The largest missions, known as *Flagship Missions*, will be crucial in allowing scientists to reach and explore difficult, but high-priority targets. These targets could help establish the parameters of habitability, not just for the Solar System, but for planetary systems in general. The Europa Jupiter System Mission is an example of a Flagship Mission. The targets of flagship missions may include journeys to:

- the clouds and surface of Venus
- the lower atmosphere and surface of Titan
- the surface and subsurface of Europa
- the deep atmosphere of Neptune and the surface of its moon Triton.

Exploring the Solar System is a technically challenging effort. Success is not always guaranteed. It requires tenacity and perseverance. Yet the United States and some other countries are meeting this challenge with resolve. Today NASA is planning space missions that may reveal whether other life exists or has ever existed in places beyond the Earth. It is engaged in research that probes from the very cores of planets to the atomic processes that occur high up in their atmospheres. The agency is carrying out surveys to find potentially hazardous objects in near-Earth orbits that could affect the future of all human beings.

It's an exciting time. Answers to some of the most profound questions— Are humans alone? Where did people come from? What is their destiny?— may be within humanity's grasp. As a future technician, engineer, or scientist, you could be involved in helping answer them.

CHAPTER 12 Robotics in Space

CHECK POINTS

Lesson 2 Review

Using complete sentences, answer the following questions on a sheet of paper.

1. What was the main goal of the rover missions?

2. How fast were the rovers going when they entered Mars' atmosphere?

3. Name the five main instruments onboard the rovers.

4. How long were the rovers scheduled to examine Mars' surface?

5. What are the science goals of NASA's Mars Exploration Program?

6. Why was the *Opportunity* landing site selected? What was the surprise about this site?

7. What was *Spirit*'s biggest scientific achievement?

8. What do scientists believe Meridiani Planum once was?

9. Why is dust a problem for the rovers?

10. List the four research objectives of NASA's planetary research program.

11. What is *Juno*'s mission?

12. What are NASA's three types of unmanned planetary missions?

APPLYING YOUR LEARNING

13. What reasons can you give for making Jupiter and its moons a focus of planetary exploration?

CHAPTER 13

Virgin Galactic's *WhiteKnightOne* carries *SpaceShipOne*, the spacecraft that in 2004 made the first privately funded human spaceflight in history. Space may become more accessible to more people in the coming decades.

© 2010 Jim Koepnick/Virgin Galactic

Commercial Use of Space

Chapter Outline

> " The greatest gain from space travel consists in the extension of our knowledge. In a hundred years this newly won knowledge will pay huge and unexpected dividends. "

Wernher von Braun

Quick Write

_____

_____

Should NASA allow artists to visit space, as millionaire Dennis Tito urges, or should trips be limited to scientists? Defend your position.

Learn About

• commercial satellites and launches
• the possibility of space tourism
• the potential of mining asteroids and moons

Dennis Tito studied aeronautics at New York University, and for a while after graduation worked for NASA. But his love of space was eclipsed by his desire to earn more money. So he left NASA and started his own company. That company today manages billions of dollars in investments.

Tito never lost sight of his desire to go into space, though. In 2001 he paid Russia something between $12 million to $20 million for the chance to fly to the International Space Station. Tito spent a week in space working on experiments. When he returned he said more people should go to space. He said in an interview before his launch that if artists such as musicians or writers could go up and express what they see, more people would be fascinated by space. He testified before Congress, urging for greater access to space for ordinary people.

The trip was not all a walk in the park for Tito. NASA strongly objected to it, saying the International Space Station was no place for amateurs. The agency refused to help train him alongside the Russian cosmonauts traveling with him. But NASA finally agreed only four days before launch that he could visit the space station.

Commercial Satellites and Launches

Vocabulary

- redundancy
- *Telstar*
- line of sight
- *Anik*

It's become commonplace to hear that another satellite has been launched into space. Indeed, NASA isn't even the only group of Americans sending satellites into orbit. Private companies have been building and launching them for years.

The Satellite Development of RCA, AT&T, and the Hughes Aircraft Company

When the earthquakes hit Haiti and Chile in 2010, people around the world instantly saw images of the destruction. You and your friends don't think anything of this. But 60 years ago, live television reports from the far corners of the world were a rare event.

Half a century ago, a newspaper journalist reporting from another country would send his or her article by underwater cable to an editor back in the United States. Today, that same journalist simply types up the story on a computer, hits the Send button, and a satellite almost instantly transmits it back to a US-based editor. TV news reporters transmit images by satellite routinely from places as far away as Afghanistan and China. Satellites have changed the way much in the world operates.

In the 1950s and 1960s several commercial companies began exploring ways to improve national and international communications using satellites. What began with enhanced radio and telephone communications now allows you to watch the World Cup live from South Africa.

RCA

RCA, or Radio Corporation of America, grew from the US government's recognized need for a domestic radio and telegraph company toward the end of World War I. The company owned the NBC radio and TV networks for many years. It eventually became instrumental in transmitting images from around the world, as well as from outer space, to eager audiences inside American homes.

RCA satellites broadcast the first images between the United States and Europe. They conducted the first radio communications between the United States and Latin America. An RCA camera filmed the first live pictures from space, during the *Apollo 7* mission. And an RCA radio, which astronaut Neil Armstrong carried in a backpack, was the first to transmit spoken words from the Moon back to Earth.

In December 1957 an *Atlas* missile took an RCA-manufactured satellite into space. This satellite delivered the first successful satellite radio relay to Earth. The company developed new technologies for NASA, and in 1962 sent six weather satellites into orbit.

In 1961 when NASA requested bids on a medium-orbit active communications satellite, RCA won the contract. It built three, named *Relay*. RCA designed them to have redundancy—*duplication of each critical part*. So each satellite had two sets of every major system of circuits. NASA launched *Relay 1* in 1962, and *Relay 2* in 1964 atop *Delta-Thor* rockets with the following goals in mind:

1. to test transoceanic—across ocean—communications

2. to measure radiation

3. to determine how much damage radiation might do to a satellite.

This photo of a *Relay* satellite was taken in 1964. RCA designed its *Relay* spacecraft to have redundancy—duplication of each critical part.

Courtesy of NASA

From their elliptical orbits the satellites successfully retransmitted television, telephone, and digital signals. When a part failed on *Relay 1*, mission control turned to one of its redundant parts, which fixed the problem. *Relay 1* went on to transmit live television signals of Winston Churchill's honorary US citizenship ceremony from Britain to the United States in March 1963. The life expectancy of the satellite was a year. It exceeded expectations and worked for well more than two. *Relay 2*, with improvements over the first satellite, broadcast part of the 1964 Winter Olympics taking place in Innsbruck, Austria. NASA never used the backup satellite, *Relay 3* because of the success of the first two.

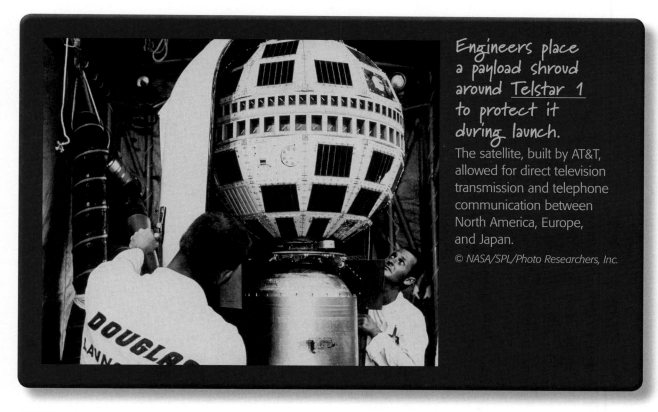

Engineers place a payload shroud around Telstar 1 to protect it during launch. The satellite, built by AT&T, allowed for direct television transmission and telephone communication between North America, Europe, and Japan.

© NASA/SPL/Photo Researchers, Inc.

AT&T

AT&T—originally the American Telephone and Telegraph Company—was another commercial company involved in communications satellite development. It achieved many milestones. Unlike RCA, the business paid out of its own pocket to develop its medium-orbit satellite named *Telstar*. And it paid NASA $3 million to launch it. *Telstar* was 34.5 inches (84 centimeters) long and weighed 170 pounds (77 kilograms).

When it launched in July 1962, Telstar was *the first privately sponsored satellite launched into space*. Many other firsts would follow, including the first live television broadcast across the ocean to France. Also, *Telstar* was responsible for the first telephone call transmitted through space, from AT&T President Frederick Kappel to then–Vice President Lyndon Johnson. Like *Relay*, *Telstar* was a huge success. It even inspired a popular song. It went out of service on 21 February 1963. AT&T launched another satellite, *Telstar II*, which also was successful. But the company turned its attention away from satellites after that second experiment.

The Hughes Aircraft Company

While *Telstar* and *Relay* were big hits, their orbits around the globe called for expensive antennas on the ground. These antennas were costly because they constantly had to rotate to follow the satellite's orbital path. Hughes Aircraft Company, yet another big player in developing commercial satellites, had an idea to bring the price down. It would place a satellite into geosynchronous orbit.

The famous American aviator Howard Hughes founded the company in 1935. In 1961 NASA awarded Hughes a contract to build its geosynchronous communications satellite, called *Syncom*. By 1964 two had operated successfully in space. Antennas could remain fixed in place on the ground as the satellites "remained in place" overhead.

Star POINTS

Early Bird provided 10 times the capacity of submarine telephone cables at about one-tenth the price. A submarine cable is one placed under the ocean, as in a cable running under the Atlantic Ocean from the United States to Britain.

Then in 1965 the company launched the communications satellite *Early Bird*. Hughes based its design on *Syncom*. A private company named Communications Satellite Corporation (COMSAT) had contracted with Hughes for *Early Bird*. The satellite, also known as *Intelsat 1*, was placed in synchronous orbit from Cape Canaveral on 6 April. In late June it sent its first transmissions back to Earth from 22,300 miles (35,700 km) up in space. It could deliver several kinds of signals, including telephone and television.

Early Bird weighed 85 pounds (39 kilograms). Engineers estimated it would work for 18 months, but it lasted four years. One of *Early Bird's* important features was that it provided line-of-sight communications between Europe and North America. Line of sight in telecommunications refers to *a clear straight path for transmitting between sender and receiver*. Because satellites travel many miles above Earth they can pick up signals from over a massive area on the planet. An antenna placed on a mountain cannot reach nearly as far.

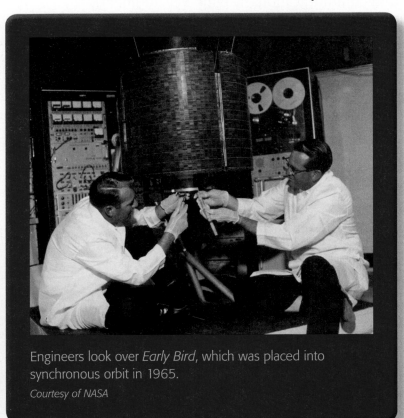

Engineers look over *Early Bird*, which was placed into synchronous orbit in 1965.
Courtesy of NASA

Furthermore, some signals, such as television signals, must travel in a straight line. A television signal cannot curve around the Earth. A satellite can receive and transmit signals in a straight line from its orbit in space to stations on Earth. But ground antennas don't have that advantage.

COMSAT placed *Early Bird* on reserve status in January1969 but brought it back briefly during the summer to help with the *Apollo 11* landing on the Moon. It remains in orbit today, on inactive status.

The Commercial Use of Communications Satellites

COMSAT's contract for *Early Bird* paved the way for more private companies to enter the satellite business. In 1964, about a year before *Early Bird's* launch, COMSAT added telecommunications agencies from 17 other countries to form one large commercial satellite company. It was called the International Telecommunications Satellite Organization (INTELSAT). COMSAT remained this agency's main manager, however.

Also by the time *Early Bird* launched, countries had placed ground stations around the globe to transmit signals to and from satellites. These Earth stations sat in the United Kingdom, France, Germany, Italy, Brazil, and France. INTELSAT formed, in part, to manage this new global telecommunications system.

In the beginning, COMSAT/INTELSAT mostly supported NASA operations. But as time went on, it took on a more commercial aspect as well. For instance, in 1969 the organization's *Intelsat III* satellite series gave the Indian Ocean coverage for the first time. When this happened the world was for the first time globally connected via satellite. Only days later nearly 500 million people across the globe watched the *Apollo 11* landing on the Moon because of these *Intelsat* satellites.

Furthermore, in 1976, COMSAT launched a new satellite series, *Marisat. Marisat* provided communication services to the US Navy, as well as to commercial shipping and other offshore businesses. In 1979 a United Nations body organized the International Maritime Satellite Organization (INMARSAT) to support communications on the seas as INTELSAT did on land. In the 1980s the Europeans launched satellites called *Marecs*, which were comparable to *Marisat*. In the beginning INMARSAT used both the US and European satellites before eventually launching its own in 1990.

Telesat Canada, a private Canadian company, produced *the first domestic communications satellites*. It named these satellites Anik. The first launched in 1972. RCA leased circuits on *Anik* to provide the same service to the United States until it developed its own satellite in 1975. Western Union developed the first American domestic communications satellite, *Westar 1*, which it launched in 1974. These satellites were all used at first for voice and data, but they quickly developed into prime carriers for television.

The costs of building and launching satellites have dramatically dropped over the years for both operators and consumers. For example, the circuits themselves, which are housed in satellites and carry telephone calls, once cost operators about $100,000 each. Now they have fallen to a few thousand dollars. And while customers used to pay about $10 per minute for a call, they now pay just pennies.

A Thor-Delta-68 launches Intelsat III into space in 1969. This satellite series gave the Indian Ocean telecommunications coverage for the first time. When this happened the world was for the first time globally connected via satellite.

Courtesy of NASA

Antennas on the ground have also fallen in price and size. More-powerful satellite transmissions mean that Earth stations no longer need 100-foot (30.5-meter) dish reflectors at $10 million each (in 1960 dollars). By 1990, such dishes were usually about 15 feet (4.6 meters) across and cost only $30,000. In 2010, you could buy a 10-foot (three-meter) dish for as little as $879. And home antennas for satellite TV and radio reception cost even less.

Commercial Launches for NASA

In the past, NASA has been the owner and operator of spacecraft designed to transport humans and cargo to and from the space station. Now it's looking for help from private companies to provide those services.

NASA's Commercial Crew and Cargo Program (C3PO) provides both funding and technology to help private companies develop safe and reliable space transportation. Currently its Commercial Orbital Transportation Services (COTS) has partnerships with US industry totaling $500 million for transporting commercial cargo. Likewise, it has invested $50 million for commercial crew transportation.

One company, Space Exploration Technologies Corp. (SpaceX), has developed two partially reusable launch vehicles, the *Falcon 1*, and the larger *Falcon 9*. The *Falcon 1* became the first privately funded rocket to reach orbit in 2008. It carried its first commercial payload into orbit the next year. *Falcon 9* completed a successful test firing at Cape Canaveral in March 2010. SpaceX is also developing the *Dragon* space capsule, which could carry astronauts and supplies to the International Space Station.

Another company involved is Orbital, which has developed several launch vehicles. The small *Pegasus*, itself launched from an airplane, has launched more than 80 satellites. *Taurus* has conducted six successful missions out of eight attempts, launching 13 satellites. The *Minotaur* has delivered 30 satellites into orbit. The *Taurus II* launch vehicle is scheduled to conduct its first mission in 2011.

Other companies developing space systems and technology include:

- PlanetSpace, which is developing launch vehicles and the *Silver Dart* orbital glider
- SpaceDev, which has supplied 2,500 devices on some 250 space missions
- Bigelow Aerospace, which has launched the *Genesis I* and *Genesis II* experimental spacecraft into Earth orbit
- Well-known aviation and space technology companies such as Boeing, Lockheed Martin, and Northrop Grumman.

Taurus II is a privately developed launch vehicle.

Courtesy of Orbital Sciences Corporation

Unit 4 Space Technology

The Possibility of Space Tourism

One price tag that's out of this world and probably won't drop anytime soon is a ride into space as a private individual. Space tourism is actually quite a new phenomenon. For those with enough money, paying a hefty fee to be a space tourist can be the way to go. As with satellites, the price will come down as technology improves. And this in turn will make space tourism accessible to more people.

The Growing Interest in Space Travel

Movies such as *2001: A Space Odyssey* fascinated ordinary people. It gave them a desire to travel to outer space. Until recently this was not an option. Those who traveled beyond the Earth's boundaries were almost all career astronauts. Teacher Christa McAuliffe, who died in the *Challenger* explosion, would have been a notable exception—perhaps the first of many shuttle passengers who did not work full-time as astronauts.

American Dennis Tito, the first space tourist to travel to the International Space Station

© Mikhial Metzel/AP Photos

When the Soviet Union broke up, its space program took a tremendous hit. To help raise revenue, Russia began selling seats on its *Soyuz* capsules traveling to the International Space Station (ISS). So far seven people have paid upwards of $20 million each to travel to the ISS. The first was American Dennis Tito in 2001.

These seven people have performed experiments while at the International Space Station. Many object to the term "space tourist," instead preferring "private astronaut." The six men and one woman have been of various nationalities, including an Iranian-American, a Canadian, and a Malaysian, among others.

In March 2010 Russia announced that because NASA was retiring its space shuttles later in the year, it would suspend its space tourist program indefinitely. Without the space shuttles to carry crews back and forth to the International Space Station, Russia would need all available seats on *Soyuz* for cosmonauts and other astronauts.

Besides Russia, several private companies are developing the technology to take private citizens to outer space. The US government's Federal Aviation Administration, tasked with ensuring safety in the air and enforcing aviation rules, will issue space launch permits to those companies taking off from the United States.

Plans for Space Tourism

Even if the Russians stop sending tourists into space, those who wish to reach toward the stars may soon have other options. A joint venture between American Burt Rutan and British billionaire Sir Richard Branson aims to offer the first commercial travel to space.

Aside from Rutan and Branson, however, other companies and groups are looking to expand service to space, much the way airlines shuttle people around the globe.

Instead of traveling to the ISS, which is too expensive except for the mega-wealthy, these companies are working on vehicles to transport paying customers on suborbital flights. These trips would cost about $200,000 and last a matter of hours at most, including just a few minutes in microgravity.

Spacecraft designer Burt Rutan (*left*) and British billionaire Sir Richard Branson (*right*) are behind efforts to offer commercial travel to space.
© Jae C. Hong/AP Photos

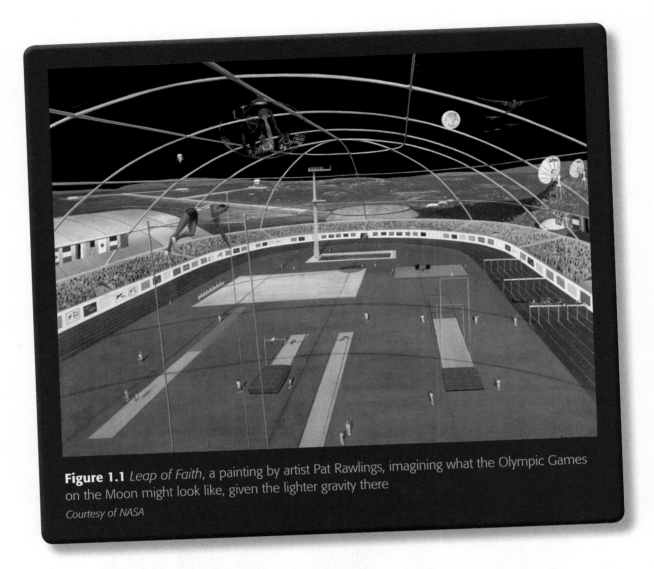

Figure 1.1 *Leap of Faith*, a painting by artist Pat Rawlings, imagining what the Olympic Games on the Moon might look like, given the lighter gravity there

Courtesy of NASA

Plans even include a commercial hotel in space at some point, where travelers could stop over. According to Reuters, the Galactic Suite Space Resort, based in Barcelona, Spain, says a three-night stay at its hotel will cost $4.4 million. The price includes an eight-week training course. During their stay, guests would see the Sun rise 15 times a day and travel around the world every 18 minutes. They would wear Velcro suits so they can crawl around their rooms by sticking themselves to the walls. The first pod in space would hold four guests and two astronaut-pilots.

The VSS *Enterprise*

If you Google Branson, you might wonder if there is anything he doesn't do. The British billionaire has a successful airline, a mobile communications firm, a music line, and countless other ventures. At one time he even sold a cola he hoped would rival Coke and Pepsi.

He has teamed up with Rutan to develop the first private commercial spaceship. It will travel from New Mexico to space. Rutan is well known as the designer of *Voyager*, the first aircraft to circle the globe without refueling. He also won a $10 million prize for his development of *SpaceShipOne*, the first privately built and manned spacecraft to reach outer space twice in two weeks.

The two men have joined to bring space tourism to non-astronauts. Branson and Rutan announced the joint venture in 2005. Since then more than 300 people have signed up to take part—and 80,000 more are on a waiting list. The first trip for VSS (Virgin Space Ship) *Enterprise* may launch as early as 2011. The spacecraft made its first successful test flight on 22 March 2010, and spent about three hours at 45,000 feet (15,000 meters). When it is ready for passengers, its voyage will begin in the Mojave Desert, where an airplane will lift off with the spacecraft. The two will travel together, reaching an altitude of 50,000 feet (15,240 meters). There they will separate, with a rocket motor thrusting VSS *Enterprise* to an altitude of 60 miles (97 km).

The tourists will experience weightlessness for a period of about four minutes. They will also be able to see the Earth curving below their windows.

The VSS *Enterprise* (*center*) takes its first "captive carry" flight with mothership *WhiteNightTwo* on 22 March 2010 over the Mojave Desert.

© 2010 Mark Greenberg/Virgin Galactic

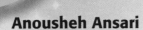

Anousheh Ansari

Anousheh Ansari came to the United States from Iran as a teenager in 1984. She did not speak a word of English. But she had big dreams. She graduated from George Mason University and later received a master's degree from George Washington University. She became a successful businesswoman and, along with her brother-in-law, set up the $10 million Ansari X Prize competition. The contest encouraged private development of a manned, reusable spacecraft. The winner was Burt Rutan.

Ansari didn't settle for letting others go into space. She became the fourth private space traveler to go to the International Space Station in conjunction with Russia. It's unclear how much she paid, but other space travelers paid between $12 million and $35 million for their trips. In 2006, she became the first Muslim woman, the first self-funded woman, and the first person of Iranian descent to travel into space.

Anousheh Ansari, first woman to travel to space as a tourist

Courtesy of Roscosmos/Gagarin Cosmonaut Training Center/NASA

During her trip, she sent videos back to Earth. She described her time on the International Space Station as "floating like a feather throughout the station." And she said that the way everyone on the station worked well together, despite their diverse backgrounds, should be an example to people on Earth. It could show them how people can work for common goals.

The Potential of Mining Asteroids and Moons

Another commercial avenue for space exploration is mining asteroids and moons for their natural resources. These resources include everything from precious metals to water. Companies could make a lot of money extracting these materials if only a safe and cost-effective way could be found. These metals and other supplies could also support colonies on the Moon and perhaps even Mars.

In Chapter 12 you read about NASA's NEAR spacecraft, which landed on the asteroid Eros in 2001. NASA scientists fired its engines for a controlled descent at only 5.28 feet (1.6 meters) per second. Engineers expected that the spacecraft would crash-land, but it survived and its instruments sent back details on Eros for more than two weeks. So human beings now know that a gentle landing on an asteroid is possible. But just how to get a manned mission safely to one of these orbiting objects is the next great leap.

The Mineral Composition of Asteroids

A virtually endless supply of asteroids orbits in the Solar System. While most remain in the asteroid belt between Mars and Jupiter, a few now and then fly near Earth. In fact, sometimes they approach nearer than the Moon.

Scientists have a number of ways to study the composition of asteroids. They've orbited and landed on Eros. Back in 1991 NASA's *Galileo* probe took dozens of photos of the asteroid Gaspara. Astronomers analyze in their labs meteorites that came from asteroids. And they rely on instruments that can read the way asteroids reflect sunlight. This reflected light helps scientists determine their makeup.

Scientists group asteroids into three categories. The most common are C-Types. These include 75 percent of the Solar System's asteroids. The same elements make up these asteroids as make up the Sun, except that they don't contain such unstable elements as hydrogen or helium. The S-Type includes 17 percent of the asteroids. These are made of nickel, iron, and magnesium. Finally, the remaining are M-Type, made of nickel and iron, but mostly iron.

Scientists also speculate that some asteroids may contain water and oxygen. Human colonies in space could use these necessities of life to survive. Furthermore, a number of asteroids might also be rich in such precious metals as gold or platinum.

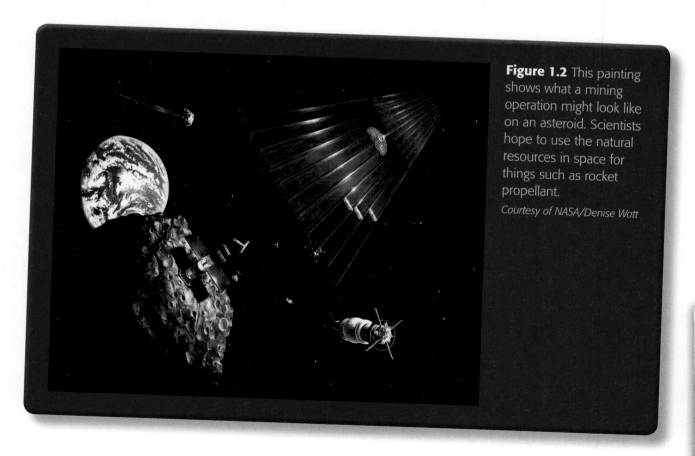

Figure 1.2 This painting shows what a mining operation might look like on an asteroid. Scientists hope to use the natural resources in space for things such as rocket propellant.

Courtesy of NASA/Denise Watt

The Estimated Mineral Wealth of Asteroids

It would not be practical at this point to collect the asteroids' precious metals and other bounties and bring them back to Earth for use. Instead scientists hope to use the natural resources in space for things such as rocket propellant or developing space structures. The ability to use resources available in space would save the millions of dollars it would cost to transport them there from Earth. This would allow for greater exploration activity.

In any event, acquiring these metals would be challenging. You might wonder why anyone would want to go to such extremes to mine metals (Figure 1.2). The explanation is financial: Scientists estimate that the amount of minerals and resources asteroids contain would be enough to give every single man, woman, and child on Earth $100 billion. According to one estimate, an asteroid 0.62 miles (1 km) across might have about 30 million tons (27 million metric tons) of nickel and 7,500 tons (6,803 metric tons) of platinum, among other metals. This amount of platinum, which is a good deal more precious than gold, would be worth about $150 billion.

Private industry's activity in space appears certain to grow in the future. Many people argue that having the private sector foot the bill for some space activities will save NASA and the US government money. They also argue that industry can carry out space missions more efficiently and at less cost.

It seems likely that the number of people employed by companies involved in space activities will also grow. The day may not be far off when people other than astronauts will routinely travel to space. If that comes true, you might even be one of them.

CHAPTER 13   Commercial Use of Space

CHECK POINTS

Lesson 1 Review

Using complete sentences, answer the following questions on a sheet of paper.

1. Why did NASA never use *Relay 3*?

2. How many people watched the *Apollo 11* landing thanks to a satellite?

3. What is INTELSAT?

4. How many satellites has the *Pegasus* launch vehicle carried into space?

5. Why will Russia no longer take space tourists to the ISS?

6. What's the expected price tag that space tourists will pay when private companies begin taking them to outer space in the coming years?

7. Where will the VSS *Enterprise* launch from? At what altitude will it separate from the airplane launching it?

8. What elements make up an S-Type asteroid?

9. How much per person on Earth do scientists estimate the natural resources of the Solar System's asteroids are worth?

APPLYING YOUR LEARNING

10. In the early days of satellite development, the US government requested and funded several spacecraft. But AT&T went it alone with its *Telstar* satellite. Do you think governments should fund space activities, or should they leave it to private industry? Why or why not?

LESSON 2 Space in Your Daily Life

Quick Write

_____

_____

Think of a space-related technology that you use every day, and describe what your life would be like without it.

Learn About

- how people use satellites every day
- the uses of a global positioning system
- how NASA shares its inventions with the private sector

Sometimes the hardest part of a trip is the journey home. That was certainly true for the early astronauts in the US space program. When they returned to Earth, the process was described as a splashdown. It perhaps sounds like fun, but it could be difficult and dangerous.

The capsule would plunge through the atmosphere. Then a series of parachutes would open to slow the capsule so that it could hit the water without too much impact. A US Navy ship would be standing by. It would dispatch a helicopter to retrieve the space travelers. As the copter made its way to the astronauts, they would climb out of their capsule onto a life raft. Just exiting the capsule could be a task. Virgil I. Grissom got entangled in some lines attached to _Liberty 7_, his Mercury capsule. He could have gone to the bottom of the ocean in his flooded ship before his helicopter rescued him.

But even once astronauts were out of their capsule and into their life raft, the very aircraft that came to rescue them could endanger their lives. The problem was the rotor downdraft from the helicopters. Reaching as much as 100 knots per hour, this wind was enough to flip a flat-bottomed life raft.

NASA scientists and engineers needed a solution. They didn't want to lose their crew on the last mile of the journey. So they went to work designing a raft that was hydrodynamically stabilized. Hydrodynamics refers to the study of the motion of water or other fluids. The raft would inflate instantly and take on water. The weight of this water would then stabilize the raft, the way a heavy keel stabilizes a sailboat. This system would allow the raft to keep from tipping in choppy seas and fierce winds. But if the craft did flip, its design meant that the weight of the water would help right the raft. It wasn't a smooth ride, but NASA was sure it would ultimately prove a safe one.

The technology of these rafts has been licensed for commercial use. It's estimated that since these rafts have been introduced, they have saved the lives of 450 sailors. In August 1980, for instance, four sailors found themselves caught in Hurricane Allen, at the time the second-worst storm ever recorded on the Atlantic. Their 30-ton ketch capsized, and the crew headed for the safety of their Givens Buoy Life Raft. They spent 42 hours in it before they were rescued. The raft flipped and righted itself again and again. At times they were under several feet of water. But they made it safely to shore. As one of them said afterward, "We didn't feel comfortable, but we did feel secure."

Vocabulary

- satellite cell phone
- direct-broadcast satellite television service
- technology transfer

How People Use Satellites Every Day

Since the Soviet launch of *Sputnik 1* in 1957, satellites have transformed the way the world communicates. People once had to book overseas telephone calls through special operators. The process could take hours. Nowadays you can call most of the world directly—and fairly cheaply. Calling overseas just means punching in a longer number.

When you travel in the United States or abroad, to give another example, you can generally get the cash you need, in the local currency, just by sticking a plastic card into the slot of a bank's cash dispenser. You may take this for granted. But it hasn't always been this way. And satellite technology helps make it all possible.

The Use of Satellite Cell Phones

Satellite cell phones are an important development from satellite technology. "Satphones," as they are known, are different from the phones that most people carry. A satellite cell phone is *a mobile telephone that connects to a network of satellites, rather than one of land-based cell towers*. This means that satphones work in remote locations where ordinary cell phones don't. For example, ships carry satphones for staying in touch on the high seas.

Air Force members operate a mobile satellite phone in South Carolina.

Courtesy of U.S. Air Force/Senior Master Sgt. Edward E. Snyder

Tweeting From Space

Not all high-tech communications take place in war zones and disaster areas. Much communication is just a matter of friends staying in touch. Social media services like Twitter are one way for people to do this. Your grandparents may remember a time when ordinary people made long-distance phone calls only in case of really big news—a birth or a death in the family, for instance. Nowadays, though, people can "tweet" messages such as "Am at coffee shop. Drinking skinny cinnamon dolce latte." Twitter lets users send messages of up to 140 characters. People call it a "microblogging" service.

Mission specialist Mike Massimino tweets from *Atlantis*'s mid-deck during STS-125.
Courtesy of NASA

And so as NASA has worked to make life in space more "normal" for astronauts, it's not surprising that they have learned to tweet from space.

On 12 May 2009 astronaut Mike Massimino connected with his Twitter followers from space as he was on his way to meet up with the Hubble Space Telescope. He was on an 11-day mission on the space shuttle.

He wrote: "From orbit: Launch was awesome!! I am feeling great, working hard, & enjoying the magnificent views, the adventure of a lifetime has begun!"

Word got around, and by the next day, he had more than a quarter-million followers. It wasn't, however, a real "tweet," strictly speaking. "Astro Mike's" message was an e-mail to the Johnson Space Center, which then posted his message as a tweet.

On 22 January 2010, though, the astronaut corps took the next step into the Twitterverse. Flight Engineer T. J. Creamer, aboard the International Space Station, was the first to make a "live" post to his Twitter account, "@Astro_TJ," from the space station. Unlike "Astro Mike," he didn't need an assist from ground crews.

"Hello Twitterverse! We r now LIVE tweeting from the International Space Station -- the 1st live tweet from Space! :) More soon, send your ?s"

Twitter in space may seem a bit frivolous. But NASA says that having one more way to keep in touch with people on Earth will help make space a bit less lonely for the astronauts.

Satphones cost much more than ordinary cell phones, both to buy and to use. But military personnel, journalists, diplomats, aid workers, and countless others now couldn't do without them. Satphones help cut through—somewhat— the "fog of war." They play a crucial role in disaster relief work and in helping explorers keep in touch while on expedition in remote areas.

The Use of Satellite Images During Evening Weather Reports

Since 1975 the weather forecasts that so many millions of Americans watch every day on television have depended on a network of special satellites. These are the Geostationary Operational Environmental Satellites (GOES), which you read about in Chapter 10. They collect data used to build extensive weather maps of the whole planet. NASA launches these satellites. And then, once they're safely in orbit, the agency turns them over to the control of the National Oceanic and Atmospheric Administration (NOAA).

GOES does not come cheap. The latest in the series, GOES-P, cost nearly half a billion dollars. But they last a long time—15 years or more. GOES continuously monitors the weather of about 90 percent of the planet.

NASA is also working on a system for forecasting ocean weather. This effort draws on satellite data, computer models, and on-site measurements of the ocean. Scientists meld these data to produce "three-dimensional" forecasts of ocean conditions. The forecasts are three-dimensional in that they cover the whole ocean, from the surface down to the ocean floor. Oceangoing vessels of all kinds, coastal managers, and marine rescuers will eventually benefit from this system.

Satellite dishes on home rooftops receive TV programs directly from telecommunications satellites orbiting Earth.

© foto.fritz/ShutterStock, Inc.

The Use of Direct-Broadcast Satellites

How do the television shows you watch at home make it to your screen? Over the air? Via cable? On your computer? What about by satellite? So-called direct-broadcast satellite television service (DBS) refers to *broadcasts sent via satellite directly to consumers.* The service is "direct" in that the signal goes directly to people's homes. This is in contrast with other ways the television industry uses satellites. For instance, they also carry "feeds" from network headquarters to local stations.

DBS programming is very much like what cable television offers. But instead of getting signals over cables in the ground, satellite subscribers get their programs directly from high-powered telecommunications satellites in geosynchronous orbit some 22,000 miles (35,046 km) above the Earth. Like cable providers, DBS providers offer different packages of services or "channels" and market them to customers, typically for a monthly fee.

DBS has been one of the more successful commercial uses of satellite communications. It's become more attractive to consumers as receiving dishes have gotten smaller. Early dishes had to be several feet across. Newer ones measure less than a couple of feet. Most people attach them to their roof or the side of their home.

DBS television got started in Britain in the late 1980s. And Europeans are generally ahead of Americans in adopting satellite TV. But by 2003 a trade association claimed that 1 in 5 US households was receiving programming by DBS. And satellite TV has long been especially popular in rural areas poorly served by broadcast or cable TV.

The Uses of a Global Positioning System

When older adults tell you that the world has changed since they were your age, Global Positioning System (GPS) technology may well be one of the things they have in mind. The US Department of Defense originally developed this technology. But like satellites now supporting television broadcasts and weather forecasts, GPS has come into wide civilian use. It's changed how people everywhere find their way.

How a Global Positioning System Uses Space Technology

During the Cold War, the Defense Department needed precise navigation for the nuclear-armed aircraft and submarines meant to protect against the Soviet Union. The original concept was navigation with reference to a system of atomic clocks on satellites orbiting the Earth.

Today's GPS consists of three parts: a constellation of satellites, the ground stations that control the satellites, and then countless individual GPS devices. These devices are used by bush pilots, taxi drivers, ambulance crews, pizza delivery people, and legions of others. The 24 or more satellites orbit Earth at about 12,000 miles (20,000 km) up. Each satellite circles the Earth every 12 hours. With this many satellites orbiting, any location on Earth is usually within range of at least four of them (Figure 2.1).

A GPS device is a type of radio receiver. It calculates a position by measuring its distance from the satellites. Actually, GPS measures distance by measuring time—the time the signal takes to travel from the satellite to the device. The signal moves at the speed of light. But it goes slightly more slowly when passing through Earth's atmosphere. (Remember the satellites are thousands of miles up. This puts them well outside the thin envelope of the atmosphere. And they are moving, too. So even if a GPS device on Earth is stationary, its distance from all the satellites in space is constantly changing.)

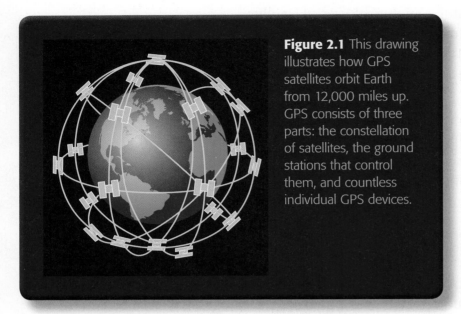

Figure 2.1 This drawing illustrates how GPS satellites orbit Earth from 12,000 miles up. GPS consists of three parts: the constellation of satellites, the ground stations that control them, and countless individual GPS devices.

A given device communicates with typically at least four satellites at a time. By calculating the distance from each, the device pinpoints location more precisely. It's like the triangulation that navigators have practiced for centuries.

Military GPS devices, more precise than their civilian counterparts, can pinpoint locations within two inches (five centimeters). Most consumer devices promise accuracy to within about 50 feet (15 meters). A more advanced system, differential GPS, is accurate to within three feet (one meter) most of the time.

How Internet Mapping Programs Use GPS Images

All the "pings" back and forth between satellites and GPS devices are in a very specialized language. It's nothing that ordinary people can read very well. But GPS data can be turned into familiar imagery such as road maps or street maps. GPS data is behind the popular mapping programs widely available on the Internet—Google Maps, Yahoo! Maps, and MapQuest. You've probably had some experience with them. You enter the addresses of your starting point and your destination, and then they calculate a route for you.

GPS technology has led to a number of hobbies as well. For instance, it allows amateur photographers to "tag" their digital images with information about the latitude, longitude, and altitude at which they took them. This is known as GPS photo tagging, or geotagging. An individual photographer can use geotagging to organize his or her own photo collection. And when tagged photos are uploaded onto a photo-sharing Web site, others can search for and identify them by location. Tagged photos can also be displayed on interactive maps. Other specialized software lets users combine GPS data with their own interests. They can map hiking trails in remote areas, for example.

How People Use GPS Technology While Driving

People have gotten used to having GPS in their cars—their own vehicles as well as rental cars. Sales reps, repair technicians, and many others have come to rely on their devices to get around in unfamiliar parts of town. In recent years, some people have begun to list their GPS coordinates as part of their address, just as they do their ZIP code. Like the online mapping services, onboard GPS in cars relies on the same satellite data stream.

Many mobile phones also include a GPS function. These systems are not without controversy. If drivers try to program their devices while behind the wheel, this can lead to accidents because they've taken their eyes off the road. But if drivers program their devices before they leave home, this step can actually make the roads safer for everyone. A pre-programmed GPS means drivers don't have to fiddle with an unwieldy paper map with one eye on the map and one on the road, or drive around lost. When people are lost, they too often make risky moves such as turning suddenly or braking to make an exit.

Engineers have designed some new technologies that come with a GPS function to prevent risky behavior. For instance, smart phones put the built-in GPS function to work to shut down other dangerous practices: talking on the phone and texting while driving. One new program uses GPS to determine the speed at which a phone is moving. If the phone is changing position fast enough for the system to conclude it must be in a moving car, the new program automatically places all e-mail and text messages on hold. It also sends calls to voice mail until the GPS indicates the vehicle is stopped once again.

Star POINTS

A study reported by the Canadian Broadcasting Corporation found that nearly half of all drivers using GPS admitted to programming their devices while they were driving.

Star POINTS

On 1 September 1983, when a South Korean airliner strayed into the airspace of the Soviet Union, Moscow responded by shooting the jet down, killing 269 passengers and crew. This was widely seen as a horrifying overreaction to an innocent mistake by the Koreans. The episode prompted President Reagan to order that GPS data be made freely available to civilian users such as airlines. The idea was that this could reduce navigational errors like the one responsible for the Korean tragedy.

Many people from truck drivers to vacationers mount GPS devices on their dashboards to guide them to their destinations.

Courtesy of Garmin International Inc.

How NASA Shares Its Inventions With the Private Sector

Ordinary people benefit from space programs in many other ways as well. As one of NASA's reports says: "Space exploration acts as a lens that sharply focuses the development of key technologies through the rigorous scientific demands that arise from pursuit of the near-impossible." In other words, the space program fosters the invention of new devices and technologies.

Once NASA has developed them, the agency shares them with the public. The term for this is technology transfer—*the movement of new technology from its creators to secondary users.* GPS is an excellent example of this. A technology developed for the military—using the military's considerable budget—ultimately enters civilian life.

Technology transfer has been part of NASA's mandate from the start. The act that established NASA in 1958 required the agency to share its discoveries as widely as practicable and appropriate. And the agency saw the need to do this in a structured, orderly way. Otherwise, technology transfer can be too much a matter of luck, chance, and happy coincidence.

After 1980 Congress spoke even more clearly on the need to share products of government-funded research. At that point, Americans had begun to worry they were losing their competitive edge to other countries, especially Japan. And so Congress required NASA and other federal agencies to take an even more active role in sharing new discoveries with the business community. But, as you will read, NASA not only shares new technologies that it develops. It also seeks to buy new technologies it needs from the private sector.

The Goals of NASA's Innovative Partnerships Program

NASA creates new technologies to meet the challenging goals of the aeronautics and space programs. Once these technologies are proven, they often turn out to be useful in many different fields, often very different from aerospace. Making the connections between aerospace and health care, or environmental engineering, or transportation, or whatever field, and then getting the technologies to the public is a primary goal of NASA's Innovative Partnerships Program, or IPP.

It works the other way, too. IPP helps find answers to problems NASA hasn't solved yet. IPP seeks solutions to some of NASA's technical challenges by, among other things, funding small-business research. The goal is to create new technologies by joining up with small businesses and sharing the cost of developing something new.

How the NASA Publication *Spinoff* Contributes to Technology Transfer

Not too many years after the space program started, new products began to emerge from NASA. These were known as *spinoffs*; you'll read more about them later in this lesson. Soon people at NASA began to think about producing an annual report on these spinoffs. NASA could then present this report to members of Congress at each year's budget hearings.

CHAPTER 13   Commercial Use of Space

Cowboys Stadium in Arlington, Texas, sports a Teflon-coated fiberglass dome inspired by astronaut spacesuits.

Courtesy of Richard and Cindy Krauss

And so in 1973 NASA published a black-and-white document with a not terribly exciting title: "Technology Utilization Program Report." The next year, a sequel came out. Even in black and white, it created some buzz. NASA decided to recast the report in full color, with the more energetic title *Spinoff*. Every year since 1976, *Spinoff* has highlighted the transfer of NASA technology to the private sector.

The report goes to elected officials, corporate executives, scholars, and specialists in technology transfer, as well as to the public and the press. Since its first black-and-white effort, NASA has published nearly 2,000 stories of technology-transfer success.

Spinoff accomplishes several goals:

- It helps justify NASA's budget
- It helps educate the press and the public by telling them about what NASA is accomplishing
- It helps dispel the myth of "wasted taxpayer dollars"
- It builds public interest in space exploration
- It shows how aerospace technology can be applied in other environments, such as health care
- It spotlights American inventors, engineers, and entrepreneurs
- It also spotlights the willingness of a government entity to assist them
- And it helps ensure that the United States keeps its global edge and its technological leadership.

How NASA's Direct Research Has Benefited Society

So far in this lesson, much of the discussion has been about practical applications of science, and specific products. But you may be wondering about the role of NASA's own direct research and its effect on society.

Michael DeBakey, the famous heart surgeon who worked on projects with NASA, had this to say about the kinds of research that NASA carries out: "NASA is engaged in very active research. It has as its goal to explore space. But to do so, you have to do all kinds of research—biological research, physical research, and so on. So it's really a very, very intensive research organization. And anytime you have any type of intensive research organization or activity going on, new knowledge is going to flow from it."

Dramatic space missions may get more attention than the quieter story of NASA's tangible impacts on Americans' daily lives. But the return on investment for society from NASA is significant. *USA Today* recently offered a list of the "Top 25 Scientific Breakthroughs" that have occurred since that newspaper's founding in 1982. Nine of them came from space exploration. Of these nine, eight came directly from NASA.

Star POINTS

NASA has received more than 6,300 patents in its history.

At NASA's 50th anniversary, administrator Michael Griffin said, "We see the transformative effects of the space economy all around us through numerous technologies and lifesaving capabilities."

How Products Developed for NASA Have Benefited Society

NASA spinoffs have improved people's quality of life as well as their economic growth. These spinoffs have been in the areas of:

- health and medicine
- transportation
- public safety
- consumer goods
- environmental and agricultural resources
- computer technology
- industrial productivity.

Here are some concrete examples of spinoffs from NASA research:

- In the 1970s NASA developed a new space suit out of Teflon-coated fiberglass. This new "fabric" has since been used around the world on roofs for stadiums and other buildings.
- In 1986 a joint project by NASA with the National Bureau of Standards led to a new lightweight breathing system for firefighters. Every major manufacturer of these systems now uses some form of the new technology. Using the new equipment, firefighters suffer significantly fewer smoke-inhalation injuries.

- In 1991 a Chicago company used three different NASA technologies in the design and testing of its school bus chassis. The result was a safer, more reliable advanced chassis. With this new product, the company captured nearly half the market within its first year of production.

- In 1994 a firm in Santa Barbara, California, used technologies created for servicing spacecraft to develop a mechanical arm for use in advanced laparoscopic surgery (procedures that call for a medical utensil that helps examine organs).

- In 1995 David Saucier of NASA's Johnson Space Center and Michael DeBakey of the Baylor College of Medicine developed a device that supplements the heart's pumping capacity. It can stabilize a patient in need of a heart transplant for up to a year. In some cases it can even make a transplant unnecessary.

NASA spinoffs have worldwide applications. In Pakistan, the Dominican Republic, and Iraq people are using a portable water filtration device based on technology developed for the space shuttle and the International Space Station. Scientists have used space-suit technologies to make balloons that serve as low-cost "satellites" to provide cell phone coverage in remote parts of Africa.

In Kosovo and Jordan, a technique using surplus NASA rocket fuels is in use to defuse land mines. A device developed to monitor astronaut health is now in use to track water quality in Vietnam and public health information in Ethiopia. Charities shipped "space blankets" in mass quantities to Pakistan in 2005 after earthquakes there. Scientists have used techniques developed to clean up

Runners wrap themselves in space blankets while cooling down after a 10K run. Space blankets are a spinoff of NASA technology.

© Martin Cameron/Alamy Images

Some NASA Myths

A fair bit of what people "know" about NASA simply isn't so. Many people think NASA developed Tang, a sweet, orange-flavored soft drink. Not so. General Foods developed Tang in 1957. It first went onto the supermarket shelves in 1959. But Tang was on the menu when John Glenn performed eating experiments as he orbited Earth in *Friendship 7* in 1962. This got Tang noticed by the public, and led to much improved sales.

Likewise, NASA didn't invent Teflon, the nonstick coating widely used in cookware; DuPont did, in 1938. But NASA has used lots of Teflon in such things as heat shields and space suits.

STS-90 astronaut Richard Searfoss prepares a meal for the crew by attaching food packets to trays with Velcro.

NASA often uses Velcro to deal with microgravity, but the agency didn't invent it. A Swiss engineer did.

Courtesy of NASA

You might think of Velcro as the opposite of Teflon. After all, Velcro is about sticking, and Teflon is all about not sticking. But like Teflon, Velcro is an invention mistakenly credited to NASA. A Swiss engineer invented Velcro in the 1940s. But the space program has used it so much—notably for anchoring equipment aboard spacecraft in low-gravity environments—that people think NASA invented it.

groundwater near the Kennedy Space Center's launch sites to reclaim other badly polluted industrial sites. And the list goes on.

The economic effect of NASA's technology has been enormous. In 2003, the agency's Inventions and Contributions Board report estimated that since 1958, NASA technologies have contributed more than *half a trillion dollars* in wealth to the world economy. And this is only part of the positive impact NASA has had on people's lives.

★ ★ ★

This textbook has introduced you to humanity's efforts to learn about space and noted the landmarks along the way. In the last 2,000 years, humanity has gone from thinking that the Sun revolves around the Earth to sending space probes beyond the Solar System. Its understanding of the universe has grown from the work of deep thinkers—people who understood the relationship of mathematics and physics here on Earth to what happens beyond Earth's atmosphere. These scientists made careful measurements and observations that others could use to further advance understanding.

Travel into Earth orbit seems so routine today that it's hard to remember the time in which Alan Shepard's and John Glenn's flights riveted Americans to their black-and-white TVs. Yet as the *Challenger* and *Columbia* tragedies showed, getting safely into orbit and back again is not as simple as it sometimes seems.

Space exploration has greatly increased scientists' understanding of the universe around us. It's raised new questions as well. But as you've read in this lesson, it has also brought many direct benefits to people around the world. They include inventions that improve people's health, the quality of their water, their ability to navigate, and their ability to predict life-threatening storms. Space exploration has also increased people's understanding of Earth, humanity's home planet. For example, satellites orbiting Earth allow scientists to observe significant changes in climate and the ozone layer.

When will people return to the Moon? Will it be possible to journey to Mars? What new discoveries await mankind regarding the nature of the universe? The answers to these questions lie with you and your fellow students. You are the voters of the future whose representatives will decide the space program's goals and funding. You are the engineers and technicians of the future who will create and build the technologies to travel and live in space. You are the scientists of the future who will interpret the data from probes that go "where no one has gone before."

If you are interested in space, now is the time to start thinking about a possible career in space exploration. In addition to astronauts and scientists, careers with NASA include engineers, mathematicians, technicians, accountants, historians, writers, computer support personnel, project and public relations managers, artists, educators, human resource personnel, physicians, lawyers, doctors, and many more. To learn more about these and other careers in exploring space, the NASA website is a good place to start. It contains information about the qualifications you need for various jobs with the agency. You can also find information about NASA's many programs and internships for students. In any event, you'll want to prepare by studying one of the sciences (chemistry, physics, biology, astronomy, and so on), mathematics, or technology and engineering. Most importantly, in preparing to find a job at NASA, take the necessary courses in high school as recommended above. Study what you're interested in, pursue your goals, and work very hard to achieve those goals. You've already started your journey by completing this course.

The future of space exploration will soon be in your hands. Like Newton, you stand on the shoulders of giants. Are you ready to take up the challenge?

CHECK POINTS

Lesson 2 Review

Using complete sentences, answer the following questions on a sheet of paper.

1. What is the difference between a satellite cell phone and the phones most people carry?

2. What is the mission of GOES?

3. What specific development has helped make satellite television more attractive to consumers?

4. How precisely can GPS pinpoint exact locations?

5. What are some examples of Internet mapping programs with GPS data behind them, beyond just simple driving directions?

6. How do GPS devices make the roads safer for everyone? How do they increase risk?

7. What is a primary goal of NASA's Innovative Partnerships Program?

8. What became of the NASA publication originally titled "Technology Utilization Program Report"?

9. Space exploration and NASA accounted for how many of the "Top 25 Scientific Breakthroughs" that the newspaper *USA Today* listed at its own 25th anniversary?

10. How did a joint project by NASA and the National Bureau of Standards benefit firefighters?

APPLYING YOUR LEARNING

11. Of all the adaptations of space technology to daily life that you've read about in this lesson, which seems most interesting to you? Why?

References

CHAPTER 1, Lesson 1   Prehistoric and Classical Astronomy

Koupelis, T., & Kuhn, K. F. (2007). *In quest of the universe (5th ed.).* Sudbury, Massachusetts: Jones and Bartlett Publishers.

Koupelis, T., & Kuhn, K. F. (in press). *In quest of the universe (6th ed.).* Sudbury, Massachusetts: Jones and Bartlett Publishers.

Raymo, C. (2001). *An intimate look at the night sky.* New York: Walker & Company.

CHAPTER 1, Lesson 2   Astronomy and the Renaissance

The Curious Team. (2005). *How do you measure the distance between Earth and the Sun?* Cornell University: Astronomy Department. Retrieved 16 October 2009 from http://curious.astro.cornell.edu/question.php?number=400

Imagine the universe! Kepler's Laws I, II, and III. (2004). NASA. Retrieved 16 October 2009 from http://imagine.gsfc.nasa.gov/docs/features/movies/kepler.html

Kepler's Laws of Planetary Motion. (2000). University of Illinois at Urbana-Champaign. Retrieved 16 October 2009 from http://www.astro.illinois.edu/projects/data/KeplersLaws/

Koupelis, T., & Kuhn, K. F. (2007). *In quest of the universe (5th ed.).* Sudbury, Massachusetts: Jones and Bartlett Publishers.

Koupelis, T., & Kuhn, K. F. (in press). *In quest of the universe (6th ed.).* Sudbury, Massachusetts: Jones and Bartlett Publishers.

O'Connor, J. J., & Robertson, E. F. (2003). *Tycho Brahe.* School of Mathematics and Statistics: University of St. Andrews, Scotland. Retrieved 19 October 2009 from http://www.gap-system.org/~history/Biographies/Brahe.html

Tycho's life. (n.d.). Tychobrahe.com, a Universal Website. Retrieved 19 October 2009 from http://www.tychobrahe.com/UK/tychos_liv.html

Van Helden, A. (1995). *The Galileo Project.* Rice University. Retrieved 19 October 2009 from http://galileo.rice.edu/sci/kepler.html

CHAPTER 1, Lesson 3   The Enlightenment and Modern Astronomy

Koupelis, T., & Kuhn, K. F. (2007). *In quest of the universe (5th ed.).* Sudbury, Massachusetts: Jones and Bartlett Publishers.

Koupelis, T., & Kuhn, K. F. (in press). *In quest of the universe (6th ed.).* Sudbury, Massachusetts: Jones and Bartlett Publishers.

Van Helden, A. (1995). *The Galileo Project.* Rice University. Retrieved 23 October 2009 from http://galileo.rice.edu/sci/lipperhey.html

CHAPTER 2, Lesson 1   Earth, Inside and Out

Ask an astrobiologist: What makes Earth so unique in the Solar System? (2001). NASA. Retrieved 26 October 2009 from http://astrobiology.nasa.gov/ask-an-astrobiologist/question/?id=22

Ask an astrophysicist: What causes the SAA? (2005). NASA. Retrieved 26 October 2009 from http://imagine.gsfc.nasa.gov/docs/ask_astro/answers/970307a.html

Johnson, J., Jr. (2006, August 10). James Van Allen, 91; Physicist whose work helped boost space exploration. *Los Angeles Times*. Retrieved 26 October 2009 from http://articles.latimes .com/2006/aug/10/local/me-vanallen10?pg=3

Koupelis, T., & Kuhn, K. F. (2007). *In quest of the universe (5th ed.)*. Sudbury, Massachusetts: Jones and Bartlett Publishers.

Koupelis, T., & Kuhn, K. F. (in press). *In quest of the universe (6th ed.)*. Sudbury, Massachusetts: Jones and Bartlett Publishers.

Layers of the atmosphere. (2009). National Weather Service: JetStream—Online School for Weather. Retrieved 4 November 2009 from http://www.srh.noaa.gov/jetstream/atmos/layers.htm

NASA satellite data show progress of 2009 Antarctic ozone hole. (2009). NASA. Retrieved 4 November 2009 from http://www.nasa.gov/topics/earth/features/2009_anta.html

Snowden, S. L. (2002, August 16). *South Atlantic Anomaly*. High Energy Astrophysics Science Archive Research Center, NASA. Retrieved 26 October 2009 from http://heasarc.gsfc.nasa.gov/ docs/rosat/gallery/misc_saad.html

Sullivan, W. (2006, August 10). James A. Van Allen, discoverer of Earth-circling radiation belts, is dead at 91. *The New York Times*. Retrieved 28 October 2009 from http://www.nytimes.com/ images/2006/08/10/obituaries/VanAllen.pdf

CHAPTER 2, Lesson 2 The Moon: Earth's Fellow Traveler

Kennedy, J. (1961). *Special message to the Congress on urgent national needs* (speech). John F. Kennedy Presidential Library and Museum. Retrieved 30 October 2009 from http://www.jfklibrary.org/ Historical+Resources/Archives/Reference+Desk/Speeches/JFK/003POF03NationalNeeds 05251961.htm

Koupelis, T., & Kuhn, K. F. (2007). *In quest of the universe (5th ed.)*. Sudbury, Massachusetts: Jones and Bartlett Publishers.

Koupelis, T., & Kuhn, K. F. (in press). *In quest of the universe (6th ed.)*. Sudbury, Massachusetts: Jones and Bartlett Publishers.

Williams, D. R. (2005). *30th anniversary of* Apollo 11: *1969–1999*. NASA Goddard Space Flight Center. Retrieved 30 October 2009 from http://nssdc.gsfc.nasa.gov/planetary/lunar/ apollo_11_30th.html

CHAPTER 3, Lesson 1 The Sun and Its Domain

Do-it-yourself podcast: Sports demo. (2009). NASA. Retrieved 7 December 2009 from http://www .nasa.gov/audience/foreducators/diypodcast/sportsdemo-index-diy.html

Historical aspects of our Sun. (n.d.). NASA. Retrieved 4 November 2009 from http://www.nasa.gov/ vision/universe/solarsystem/sun_history.html

Koupelis, T., & Kuhn, K. F. (2007). *In quest of the universe (5th ed.)*. Sudbury, Massachusetts: Jones and Bartlett Publishers.

Koupelis, T., & Kuhn, K. F. (in press). *In quest of the universe (6th ed.)*. Sudbury, Massachusetts: Jones and Bartlett Publishers.

CHAPTER 3, Lesson 2 The Terrestrial Planets

Brown, D., & Campbell, P. (2009, November 3). MESSENGER *spacecraft reveals more hidden territory on Mercury*. John Hopkins University Applied Physics Laboratory. Retrieved 13 November 2009 from http://messenger.jhuapl.edu/news_room/details.php?id=138

Koupelis, T., & Kuhn, K. F. (2007). *In quest of the universe (5th ed.)*. Sudbury, Massachusetts: Jones and Bartlett Publishers.

Koupelis, T., & Kuhn, K. F. (in press). *In quest of the universe (6th ed.)*. Sudbury, Massachusetts: Jones and Bartlett Publishers.

StarChild question of the month for August 2002. (2002). The StarChild Team. NASA/Goddard Space Flight Center. Retrieved 18 November 2009 from http://starchild.gsfc.nasa.gov/docs/StarChild/questions/question48.html

Venus flagship mission study. (n.d.). NASA: Jet Propulsion Laboratory. Retrieved 13 November 2009 from http://vfm.jpl.nasa.gov/othervenusmissions/flybymissionsmarinermessenger

Young, C. (Ed.). (1990). *The* Magellan *Venus explorer's guide.* Pasadena, California: Jet Propulsion Laboratory, California Institute of Technology. Retrieved 14 December 2009 from http://www2.jpl.nasa.gov/magellan/guide.html

CHAPTER 3, Lesson 3   The Outer Planets

Buratti, B., Hillier, J., & Hicks, M. (n.d.). *Triton and Pluto: A comparison of changes in ground-based photometry* [abstract]. Pasadena, California: Jet Propulsion Laboratory, California Institute of Technology. Retrieved 14 December 2009 from http://trs-new.jpl.nasa.gov/dspace/bitstream/2014/8669/1/02-1202.pdf

Historical background of Saturn's rings. (n.d.). NASA. Retrieved 13 November 2009 from http://www2.jpl.nasa.gov/saturn/back.html

Koupelis, T., & Kuhn, K. F. (2007). *In quest of the universe (5th ed.).* Sudbury, Massachusetts: Jones and Bartlett Publishers.

Koupelis, T., & Kuhn, K. F. (in press). *In quest of the universe (6th ed.).* Sudbury, Massachusetts: Jones and Bartlett Publishers.

CHAPTER 3, Lesson 4   Dwarf Planets, Comets, Asteroids, and Kuiper Belt Objects

Koupelis, T., & Kuhn, K. F. (2007). *In quest of the universe (5th ed.).* Sudbury, Massachusetts: Jones and Bartlett Publishers.

Koupelis, T., & Kuhn, K. F. (in press). *In quest of the universe (6th ed.).* Sudbury, Massachusetts: Jones and Bartlett Publishers.

Pluto downgraded to "dwarf planet" status; Solar System now has eight planets. (2006, August 24). *Science Daily.* Retrieved 23 November 2009 from http://www.sciencedaily.com/releases/2006/08/060825003742.htm

Wilford, J. N. (1985, October 29). Sir Edmund Halley: Orbiting forever in Newton's shadow. *The New York Times.* Retrieved 25 November 2009 from http://www.nytimes.com/1985/10/29/science/sir-edmund-halley-orbiting-forever-in-newton-s-shadow.html

CHAPTER 4, Lesson 1   The Milky Way Galaxy

Freudenrich, C. (n.d.). *How planet hunting works.* HowStuffWorks, Inc. Retrieved 1 December 2009 from http://www.howstuffworks.com/planet-hunting2.htm

Koupelis, T., & Kuhn, K. F. (2007). *In quest of the universe (5th ed.).* Sudbury, Massachusetts: Jones and Bartlett Publishers.

Koupelis, T., & Kuhn, K. F. (in press). *In quest of the universe (6th ed.).* Sudbury, Massachusetts: Jones and Bartlett Publishers.

Searching for Earthlike worlds. (n.d.). NASA. Retrieved 30 November 2009 from http://planetquest.jpl.nasa.gov/overview/overview_index.cfm

CHAPTER 4, Lesson 2   What Lies Beyond

Koupelis, T., & Kuhn, K. F. (2007). *In quest of the universe (5th ed.).* Sudbury, Massachusetts: Jones and Bartlett Publishers.

Koupelis, T., & Kuhn, K. F. (in press). *In quest of the universe (6th ed.).* Sudbury, Massachusetts: Jones and Bartlett Publishers.

Link, J. (2007). *RE: what percentage of the electromagnetic spectrum is visible light?* MadSci Network. Retrieved 13 January 2010 from http://www.madsci.org/posts/archives/2007-08/1188407794 .Ph.r.html

CHAPTER 5, Lesson 1   Why Explore Space?

Achenbach, J. (2009, September 9). Mars and Moon are out of NASA's reach for now, review panel says. *The Washington Post.* Retrieved 13 December 2009 from http://www.washingtonpost.com/ wp-dyn/content/article/2009/09/08/AR2009090802464.html

Boorstin, D. J. (1983). *The Discoverers.* New York, NY: Random House, Inc.

Corps of Discovery: United States Army. (2006). US Army Center of Military History. Retrieved 13 January 2010 from http://www.history.army.mil/lc/index.htm

Derbyshire, J. (2009, June 15). End of an extravaganza: Manned space travel always was, and still is, a pointless project. *National Review Online.* Retrieved 18 December 2009 from http://article .nationalreview.com/?q=Nzc2NTYxMzVjNWRkMzc0YzQ0Y2VhNGI1ZGRkMTc2N2I

Griffin, M. (2007, January 18). *Why explore space?* NASA. Retrieved 18 December 2009 from http://www.nasa.gov/exploration/home/griffin_why_explore.html

Hale, W. (2009, April 10). *Why explore space?* NASA: Wayne Hale's Blog. Retrieved 18 December 2009 from http://blogs.nasa.gov/cm/blog/waynehalesblog/posts/post_1239387201344.html

Hale, W. (2009, June 29). *Point, counterpoint.* NASA: Wayne Hale's Blog. Retrieved 18 December 2009 from http://blogs.nasa.gov/cm/blog/waynehalesblog.blog/posts/post_1246303539257.html

Leonard, T. (2009, July 6). NASA experts scale back Moon and Mars plans in face of Obama funding cut fears. *The Daily Telegraph.* Retrieved 13 December 2009 from http://www.telegraph .co.uk/science/space/5760285/Nasa-experts-scale-back-moon-and-Mars-plans-in-face-of -Obama-funding-cut-fears.html

President Bush announces new vision for space exploration program. (2004, January 14). The White House. Retrieved 18 December 2009 from http://history.nasa.gov/SEP%20Press%20Release.htm

Tyson, N. deG. (2007, August 5). Why America needs to explore space. *Parade.* Retrieved 18 December 2009 from http://www.parade.com/articles/editions/2007/edition_08-05-2007/Space

Walker, R. (2004, August 3). A spicy history of humanity: The quest for spices drove exploration around the world. *The Christian Science Monitor.* Retrieved 18 December 2009 from http://www .csmonitor.com/2004/0803/p15s02-bogn.html

Why explore space? (2005, March 11). European Space Agency. Retrieved 18 December 2009 from http://www.esa.int/esaSC/SEMC3VZO4HD_index_0.html

CHAPTER 5, Lesson 2   Assembling a Space Mission

About the Deep Space Network. (2009, November 5). NASA. Retrieved 21 December 2009 from http://deepspace.jpl.nasa.gov/dsn/

Astronaut requirements. (2004, January 29). NASA. Retrieved 19 December 2009 from http://www .nasa.gov/audience/forstudents/postsecondary/features/F_Astronaut_Requirements.html

Astronaut selection. (2008, March 12). NASA. Retrieved 19 December 2009 from http://nasajobs .nasa.gov/ASTRONAUTS/

Astronaut training timeline. (2009, April 10). NASA. Retrieved 20 December 2009 from http:// www.nasa.gov/audience/foreducators/k-4/features/F_Astronaut_Training_Timeline.html

Aviation Systems Division: Division overview. (2008, December 3). NASA. Retrieved 22 December 2009 from http://www.aviationsystemsdivision.arc.nasa.gov/about/overview.shtml

Fowler, W. T. (n.d.). *An introduction to space mission planning.* The University of Texas at Austin: Cockrell School of Engineering. Retrieved 20 December 2009 from http://www.ae.utexas.edu/ design/mission_planning/mission_resources/mission_planning/Intro_to_Mission_Planning.pdf

IBEX launch. (2008, October 24). NASA. Retrieved 20 December 2009 from http://www.nasa.gov/mission_pages/ibex/launch/index.html

IBEX launch puts Telemetry and Communications Group to the test. (2008, October 10). NASA. Retrieved 20 December 2009 from http://www.nasa.gov/centers/kennedy/launchingrockets/comm_telem.html

In their own words: Gregory R. (Reid) Wiseman. (2009, June 29). NASA. Retrieved 21 December 2009 from http://www1.nasa.gov/astronauts/2009_wiseman.html

JAXA history. (2009). Japan Aerospace Exploration Agency. Retrieved 19 January 2010 from http://www.jaxa.jp/about/history/index_e.html

Kulacki, G., & Lewis, J. G. (2009). *A place for one's mat: China's space program, 1956–2003* (occasional paper). Cambridge, Massachusetts: American Academy of Arts and Sciences.

Matson, J. (2009, July 15). Why does NASA launch space shuttles from such a weather-beaten place? *Scientific American*. Retrieved 20 December 2009 from http://www.scientificamerican.com/article.cfm?id=space-shuttle-weather-florida

NASA aeronautics research onboard. (n.d.). NASA. Retrieved 19 January 2010 from http://www.aeronautics.nasa.gov/aero_onboard/flash_index.html

Russia, Japan to draft space cooperation agreement. (2009, December 20). *RIA Novosti*. Retrieved 21 December 2009 from http://en.rian.ru/science/20091220/157306702.html

Science strategy. (n.d.). NASA. Retrieved 20 December 2009 from http://nasascience.nasa.gov/about-us/science-strategy

Welcome to the Aeronautics Research Mission Directorate. (2010). NASA. Retrieved 19 January 2010 from http://www.aeronautics.nasa.gov/

What does NASA do? (2008, March 9). NASA. Retrieved 19 December 2009 from http://www.nasa.gov/about/highlights/what_does_nasa_do.html

What is ESA? (2009, May 29). European Space Agency. Retrieved 19 December 2009 from http://www.esa.int/SPECIALS/About_ESA/SEMW16ARR1F_0.html

CHAPTER 5, Lesson 3   The Hazards for Spacecraft

The hazards. (2008, June 17). NASA. Retrieved 22 December 2009 from http://aerospacescholars.jsc.nasa.gov/HAS/Modules/Earth-to-Mars/9/4.cfm

NASA spacecraft take cover from the Leonids. (1998, November 16). NASA. Retrieved 22 December 2009 from http://science.nasa.gov/newhome/headlines/ast16nov98_2.htm

Not just another old flame. (2008, May 12). NASA. Retrieved 22 December 2009 from http://science.nasa.gov/headlines/y2000/ast12may_1.htm

Phillips, T. (n.d.). *Shielding spacecraft from dangers of space dust*. NASA. Retrieved 22 December 2009 from http://www.nasa.gov/vision/universe/watchtheskies/meteor_cloud.html

Radiation and long-term spaceflight. (n.d.). National Space Biomedical Research Institute. Retrieved 22 December 2009 from http://www.nsbri.org/Radiation/HumanAffects.html

Steigerwald, B. (2008, December 16). *Sun often "Tears out a wall" in Earth's solar storm shield*. NASA. Retrieved 22 December 2009 from http://www.nasa.gov/mission_pages/themis/news/themis_leaky_shield.html

Technical discipline area: Atomic oxygen. (2001). Hampton, Virginia: NASA Langley Research Center. Retrieved 25 January 2010 from http://setas-www.larc.nasa.gov/LDEF/ATOMIC_OXYGEN/ao_intro.html

To go where no spacecraft has gone before. (2008, December 23). NASA. Retrieved 22 December 2009 from http://www.nasa.gov/topics/moonmars/features/alhat2-20081223.html

What goes up doesn't always come down. (2004, December 2). NASA. Retrieved 22 December 2009 from http://www.nasa.gov/audience/forstudents/k-4/home/F_What_Goes_Up_K-4.html

Areas of study: Muscle function. (2009, November 5). NASA. Retrieved 28 December 2009 from http://www.nasa.gov/exploration/humanresearch/areas_study/physiology/physiology _muscle.html

Astronauts test Glenn exercise harnesses. (2009, December 21). NASA Glenn Research Center. Retrieved 27 January 2010 from http://www.nasa.gov/centers/glenn/shuttlestation/station/ harness.html

Barry, P. L., & Phillips, T. (2004, October 22). *Blinding flashes.* NASA. Retrieved 27 December 2009 from http://science.nasa.gov/headlines/y2004/22oct_cataracts.htm

Coulter, D. (2009, July 10). *The beating heart, minus gravity.* NASA. Retrieved 28 December 2009 from http://science.nasa.gov/headlines/y2009/10jul_cardio.htm

Get a leg up. (n.d.). NASA. Retrieved 1 February 2010 from http://education.jsc.nasa.gov/explorers/ pdf/p3_student.pdf

Hipschman, R. (1997). *Your weight on other worlds.* Exploratorium. Retrieved 28 December 2009 from http://www.exploratorium.edu/ronh/weight/

Human Research Program. (2009, November 12). NASA. Retrieved 28 December 2009 from http://www.nasa.gov/exploration/humanresearch/index.html

Long-duration spaceflight. (2008, June 17). NASA. Retrieved 27 December 2009 from http:// aerospacescholars.jsc.nasa.gov/HAS/Modules/Earth-to-Mars/9/3.cfm

Microgravity: Zero g and Mars g. (2008, June 17). NASA. Retrieved 27 December 2009 from http://aerospacescholars.jsc.nasa.gov/HAS/Modules/Earth-to-Mars/9/6.cfm

Miller, K, & Phillips, T. (2003, October 7). *DNA biosentinels.* NASA. Retrieved 27 December 2009 from http://science.nasa.gov/headlines/y2003/07oct_dna.htm

Mission to Mars. (2008, June 17). NASA. Retrieved 27 December 2009 from http:// aerospacescholars.jsc.nasa.gov/HAS/Modules/Earth-to-Mars/9/2.cfm

Mixed up in space. (2001). Science @ NASA. Retrieved 1 February 2010 from http://science.nasa .gov/headlines/y2001/ast07aug_1.htm

NASA has a mystery to solve: Can people go to Mars, or not? (2004). NASA manuscript. Retrieved 1 February 2010 from http://ciencia.nasa.gov/headlines/y2004/images/radiation/ story.doc

NASA STS-92. (n.d.). NASA. Retrieved 28 December 2009 from http://www.nasa.gov/mission _pages/shuttle/shuttlemissions/archives/sts-92.html

National Science Biomedical Research Institute. (n.d.). NSBRI. Retrieved 27 December 2009 from http://www.nsbri.org/

Phillips, T. (2004, February 17). *Can people go to Mars?* NASA. Retrieved 27 December 2009 from http://science.nasa.gov/headlines/y2004/17feb_radiation.htm

Radiation biology educator guide, revision 2. (2006). NASA. Retrieved 1 February 2010 from http://er.jsc.nasa.gov/seh/RB_Module3_10.pdf

Space medicine. (2008, June 17). NASA. Retrieved 28 December 2009 from http:// aerospacescholars.jsc.nasa.gov/HAS/Modules/Earth-to-Mars/9/5.cfm

STS-92 wakeup calls. (2002, April 9). NASA. Retrieved 28 December 2009 from http://spaceflight .nasa.gov/gallery/audio/shuttle/sts-92/html/ndexpage.html

The team. (2009, December 8). NASA. Retrieved 28 December 2009 from http://aerospacescholars .jsc.nasa.gov/HAS/Modules/Earth-to-Mars/9/10.cfm

Weak in the knees—the quest for a cure. (n.d.). NASA. Retrieved 28 December 2009 from http://weboflife.nasa.gov/currentResearch/currentResearchGeneralArchives/weakKnees.htm

CHAPTER 6, Lesson 1   The US Manned Space Program

Apollo 13 (29): *"Houston, we have a problem ..."* (2002). NASA: Kennedy Space Center. Retrieved 11 January 2010 from http://www-pao.ksc.nasa.gov/kscpao/history/apollo/apollo-13/apollo-13.htm

The Apollo program. (2009). NASA: Human Space Flight. Retrieved 11 January 2010 from http://spaceflight.nasa.gov/history/apollo/

Compton, W. D. (1989). *Where no man has gone before: A history of Apollo lunar exploration missions.* NASA Special Publication 4214. Retrieved 19 February 2010 from http://www.hq.nasa.gov/office/pao/History/SP-4214/contents.html

Flight summary. (2008). NASA: Kennedy Space Center. Retrieved 12 January 2010 from http://www-pao.ksc.nasa.gov/kscpao/history/apollo/flight-summary.htm

Gemini IV. (2003). NASA: Kennedy Space Center. Retrieved 8 January 2010 from http://www-pao.ksc.nasa.gov/kscpao/history/gemini/gemini-4/gemini4.htm

Gemini *overview.* (2000). NASA: Kennedy Space Center. Retrieved 8 January 2010 from http://www-pao.ksc.nasa.gov/kscpao/history/gemini/gemini-overview.htm

Glenn contributions to Apollo. (2008). NASA. Retrieved 11 January 2010 from http://www.nasa.gov/centers/glenn/about/history/apollew.html

Gray, T. (n.d.). *Alan B. Shepard, Jr.* NASA: 40th Anniversary of the Mercury 7. Retrieved 11 January 2010 from http://history.nasa.gov/40thmerc7/shepard.htm

Gray, T. (n.d.). *Donald K. "Deke" Slayton.* NASA: 40th Anniversary of the Mercury 7. Retrieved 11 January 2010 from http://history.nasa.gov/40thmerc7/slayton.htm

Gray, T. (n.d.). *John H. Glenn, Jr.* NASA: 40th Anniversary of the Mercury 7. Retrieved 11 January 2010 from http://history.nasa.gov/40thmerc7/glenn.htm

Gray, T. (n.d.). *L. Gordon Cooper, Jr.* NASA: 40th Anniversary of the Mercury 7. Retrieved 11 January 2010 from http://history.nasa.gov/40thmerc7/cooper.htm

Gray, T. (n.d.). *M. Scott Carpenter.* NASA: 40th Anniversary of the Mercury 7. Retrieved 11 January 2010 from http://history.nasa.gov/40thmerc7/carpenter.htm

Gray, T. (n.d.). *Walter M. Schirra, Jr.* NASA: 40th Anniversary of the Mercury 7. Retrieved 11 January 2010 from http://history.nasa.gov/40thmerc7/schirra.htm

Mercury Freedom 7 MR-3 (18). (2002). NASA: Kennedy Space Center. Retrieved 8 January 2010 from http://www-pao.ksc.nasa.gov/kscpao/history/mercury/mr-3/mr-3.htm

Project Mercury goals. (2000). NASA: Kennedy Space Center. Retrieved 8 January 2010 from http://www-pao.ksc.nasa.gov/kscpao/history/mercury/mercury-goals.htm

Swenson, L. S., Jr., Grimwood, J. M., & Alexander, C. C. (1989). *This new ocean: A history of Project Mercury.* NASA. Retrieved 8 January 2010 from http://history.nasa.gov/SP-4201/ch11-4.htm

Warhol was right about "15 minutes of fame." (2008, October 8). *National Public Radio.* Retrieved 8 January 2010 from http://www.npr.org/templates/story/story.php?storyId=95516647

White, M. C. (2006). *Detailed biographies of* Apollo I *crew—Ed White.* NASA. Retrieved 8 January 2010 from http://www.hq.nasa.gov/office/pao/History/Apollo204/zorn/white.htm

Zornio, M. C. (n.d.). *Virgil Ivan "Gus" Grissom.* NASA: 40th Anniversary of the Mercury 7. Retrieved 11 January 2010 from http://history.nasa.gov/40thmerc7/grissom.htm

CHAPTER 6, Lesson 2   The Soviet/Russian Manned Space Program

Apollo-Soyuz Test Project: Biographies. (2005). NASA History Division. Retrieved 12 January 2010 from http://history.nasa.gov/30thastp/bios.html#Leonov

Ezell, E. C., & Ezell, L. N. (1978). *The partnership: A history of the Apollo-Soyuz Test Project.* Published as NASA Special Publication-4209 in the NASA History Series, 1978. Retrieved 12 January 2010 from http://www.hq.nasa.gov/office/pao/History/SP-4209/toc.htm

Leonov. A. (2005, January 1). The nightmare of *Voskhod 2*. *Air and Space Magazine*. Retrieved 13 January 2010 from http://www.airspacemag.com/space-exploration/voskhod.html

Mellies, A. (2004). *Apollo-Soyuz Test Project*. NASA. Retrieved 12 January 2010 from http://history.nasa.gov/astp/index.html

Short tour: A quick overview and timeline of shuttle-Mir. (2004). NASA. Retrieved 19 January 2010 from http://spaceflight.nasa.gov/history/shuttle-mir/history/h-t-short.htm

Soyuz 1. (2009). NASA: National Space Science Data Center. Retrieved 12 January 2010 from http://nssdc.gsfc.nasa.gov/nmc/spacecraftDisplay.do?id=1967-037A

Soyuz 3. (2009). NASA: National Space Science Data Center. Retrieved 12 January 2010 from http://nssdc.gsfc.nasa.gov/nmc/spacecraftDisplay.do?id=1968-094A

Soyuz 4. (2009). NASA: National Space Science Data Center. Retrieved 12 January 2010 from http://nssdc.gsfc.nasa.gov/nmc/spacecraftDisplay.do?id=1969-004A

Space race: Racing to the Moon: Vostok *and* Voskhod. (2002). Smithsonian: National Air and Space Museum. Retrieved 12 January 2010 from http://www.nasm.si.edu/exhibitions/gal114/SpaceRace/sec300/sec330.htm

Space race: Racing to the Moon: Why land on the ground? (2002). Smithsonian: National Air and Space Museum. Retrieved 12 January 2010 from http://www.nasm.si.edu/exhibitions/gal114/SpaceRace/sec500/sec533.htm

Space: Self-control in *Soyuz 3*. (1968, November 8). *Time*. Retrieved 19 January 2010 from http://www.time.com/time/magazine/article/0,9171,902519,00.html

Space station: A rare inside view of the next frontier in space exploration. (1999). *Houston Public Television*. Retrieved 16 January 2010 from http://www.pbs.org/spacestation/station/russian.htm

Valentina Tereshkova. (n.d.). NASA/Goddard Space Flight Center: The StarChild Team. Retrieved 12 January 2010 from http://starchild.gsfc.nasa.gov/docs/StarChild/whos_who_level2/tereshkova.html

Voskhod 1. (2009). NASA: National Space Science Data Center. Retrieved 12 January 2010 from http://nssdc.gsfc.nasa.gov/nmc/spacecraftDisplay.do?id=1964-065A

West, J. B. (2000, July). Historical perspectives: Physiology in microgravity. *Journal of Applied Physiology, 89*(1), 379–384. Retrieved 12 January 2010 from http://jap.physiology.org/cgi/content/full/89/1/379

Yuri Gagarin. (n.d.). NASA/Goddard Space Flight Center: The StarChild Team. Retrieved 12 January 2010 from http://starchild.gsfc.nasa.gov/docs/StarChild/whos_who_level2/gagarin.html

CHAPTER 6, Lesson 3   Space Programs Around the World

About ISRO. (2008). Indian Space Research Organisation. Retrieved 21 January 2010 from http://www.isro.org/scripts/Aboutus.aspx

About JAXA. (2007). Japan Aerospace Exploration Agency. Retrieved 18 January 2010 from http://www.jaxa.jp/about/index_e.html

About the International Space Station. (2008, June 26). European Space Agency. Retrieved 18 January 2010 from http://www.esa.int/esaHS/ESA6NE0VMOC_iss_0.html

Adams, P. (2009, December 21). Japanese astronaut to serve first sushi in space. *Popular Science*. Retrieved 18 January 2010 from http://www.popsci.com/technology/article/2009-12/japanese-astronaut-brings-sushi-space

Cassini-Huygens *mission facts*. (2005, February 2). European Space Agency. Retrieved 17 January 2010 from http://www.esa.int/SPECIALS/Cassini-Huygens/SEMVOZ1VQUD_0.html

Chinese space program. (2009, August). NASA Headquarters Library. Retrieved 17 January 2010 from http://www.hq.nasa.gov/office/hqlibrary/pathfinders/china.htm

De Selding, P. B. (2009, December 17). ESA members approve two joint Mars missions with U.S. *Space News*. Retrieved 17 January 2010 from http://www.spacenews.com/civil/091217-esa-approves-collaborative-mars-program-with-nasa.html

ESA astronaut Frank De Winne safely back on Earth. (2009, December 1). European Space Agency. Retrieved 18 January 2010 from http://www.esa.int/esaCP/SEMJYW49J2G_index_0.html

Farooq, O. (2007, September 27). India plans manned space mission by 2015. *The Associated Press*. Retrieved 17 January 2010 from http://www.msnbc.msn.com/id/21019302/

Gonzalez, S. (2009, February 20). *Slow and steady: China takes a walk and Russia partners with Cuba*. JSC Advanced Planning Office Blog. Retrieved 17 January 2010 from http://blogs.nasa.gov/cm/blog/JSC%20Advanced%20Planning%20Office%20Blog.blog/posts/post_1235174436501.html

History of China in space. (2005). *Space Today Online*. Retrieved 17 January 2010 from http://www.spacetoday.org/China/ChinaHistory.html

INSAT 2. (2000, May 1). FAS Space Policy Project: World Space Guide. Retrieved 17 January 2010 from http://www.fas.org/spp/guide/india/earth/insat2_eo.htm

IRS (Indian Remote Sensing Satellite). (2000, April 20). FAS Space Policy Project: World Space Guide. Retrieved 17 January 2010 from http://www.fas.org/spp/guide/india/earth/irs.htm

ISRO METSAT satellite series named after Columbia *astronaut Kalpana Chawla*. (2003, February 6). SpaceRef.com. Retrieved 17 January 2010 from http://www.spaceref.com/news/viewnews.html?id=732

Japanese experiment module "Kibo." (2007). Japan Aerospace Exploration Agency. Retrieved 18 January 2010 from http://www.jaxa.jp/projects/iss_human/kibo/index_e.html

Japanese space program. (2010). Aerospaceguide.net. Retrieved 18 January 2010 from http://www.aerospaceguide.net/worldspace/japanesespaceprogram.html

Kaguya. (2009, November 23). NASA. Retrieved 21 January 2010 from http://nssdc.gsfc.nasa.gov/nmc/spacecraftDisplay.do?id=2007-039A

Kaguya. (2010). In *Encyclopædia Britannica*. Retrieved 18 January 2010 from http://www.britannica.com/EBchecked/topic/1386132/Kaguya

Kaguya (Selene). (2007). Japan Aerospace Exploration Agency. Retrieved 18 January 2010 from http://www.selene.jaxa.jp/index_e.htm

Laxman, S. (2009, December 26). Easy-on-the-pocket rocket. *The Times of India*. Retrieved 17 January 2010 from http://timesofindia.indiatimes.com/india/Indias-coming-out-party-/articleshow/5380665.cms

Malik, T. (2007, May 2). *Japan prepares space station's largest laboratory for flight*. Space.com. Retrieved 21 January 2010 from http://www.space.com/businesstechnology/070502_techwed_kibo.html

Mars Express. (2010). European Space Agency. Retrieved 18 January 2010 from http://sci.esa.int/science-e/www/area/index.cfm?fareaid=9

Moskowitz, C. (2008, April 8). New station crew, Korean astronaut rocket into space. *Space News*. Retrieved 18 January 2010 from http://www.space.com/missionlaunches/080408-exp17-launch.html

NASA Kennedy Space Center Safety and Mission Assurance Directorate: NASA range safety annual report 2005. (2005). NASA. Retrieved 17 January 2010 from http://kscsma.ksc.nasa.gov/Range_Safety/Annual_Report/2005/PrintPages/China.pdf

Profile of JAXA's astronauts. (2008, May 19). Japan Aerospace Exploration Agency. Retrieved 20 January 2010 from http://iss.jaxa.jp/astro/profile_e.html

Reeves, P. (2009, July 22). Despite glitches, India shoots for the Moon. *National Public Radio*. Retrieved 17 January 2010 from http://www.npr.org/templates/story/story.php?storyId=106876605&ps=rs

Rosetta—*mission to intercept a comet*. (2009, November 5). British National Space Centre. Retrieved 18 January 2010 from http://www.bnsc.gov.uk/Missions/Exploring-the-Solar-System/8118.aspx

Sang-Hun, C. (2008, February 22). Kimchi goes to space, along with first Korean astronaut. *The New York Times*. Retrieved 18 January 2010 from http://www.nytimes.com/2008/02/22/world/asia/22iht-kimchi.1.10302283.html?pagewanted=1&_r=1

Sappenfield, M. (2008, October 22). Moon mission takes India's space program in new direction. *The Christian Science Monitor*. Retrieved 17 January 2010 from http://www.csmonitor.com/ World/Asia-South-Central/2008/1022/p06s12-wosc.html

Smith, M. S. (2003). *China's space program: An overview*. Department of Defense. Retrieved 17 January 2010 from http://www.defense.gov/pubs/20030730chinaex.pdf

Soldier blasts off for mission on International Space Station. (2009, December 20). NASA. Retrieved 18 January 2010 from http://www.army.mil/-news/2009/12/20/32143-soldier-blasts-off-for -mission-on-international-space-station/

Space Flight 2007—The Year in Review. (2008, July 13). NASA. Retrieved 20 January 2010 from http://www.hq.nasa.gov/osf/2007/2007_asia.htm

STS-128 mission information. (2009, September 23). NASA. Retrieved 21 January 2010 from http://www.nasa.gov/mission_pages/shuttle/shuttlemissions/sts128/main/index.html

Two new ESA satellites successfully lofted into orbit. (2009, November 2). European Space Agency. Retrieved 18 January 2010 from http://www.esa.int/esaCP/SEMNEYAOE1G_index_0.html

Venus Express. (2010). European Space Agency. Retrieved 18 January 2010 from http://sci.esa.int/ science-e/www/area/index.cfm?fareaid=64

Wax, E. (2009, November 4). India's space ambitions taking off. *The Washington Post*. Retrieved 17 January 2010 from http://www.washingtonpost.com/wp-dyn/content/article/2009/11/03/ AR2009110303419.html?sid=ST2009110400142

What is ESA? (2009, May 29). European Space Agency. Retrieved 18 January 2010 from http://www.esa.int/SPECIALS/About_ESA/SEMW16ARR1F_0.html

Wines, M. (2009, November 3). Qian Xuesen, father of China's space program, dies at 98. *The New York Times*. Retrieved 17 January 2010 from http://www.nytimes.com/2009/11/04/ world/asia/04qian.html?_r=2

CHAPTER 7, Lesson 1   The Shuttle Program

Astronaut selection and training. (2003, June 27). NASA. Retrieved 25 January 2010 from http://spaceflight.nasa.gov/shuttle/reference/factsheets/asseltrn.html

Background and Status. (2002, April 7). NASA. Retrieved 29 January 2010 from http://spaceflight .nasa.gov/shuttle/reference/shutref/sts/background.html

Biographical data: Eileen Marie Collins (Colonel, USAF, Ret.). (2006, May). NASA/Lyndon B. Johnson Space Center. Retrieved 29 January 2010 from http://www11.jsc.nasa.gov/Bios/ htmlbios/collins.html

Biographical data: Ellison S. Onizuka (Colonel, USAF). (2007, January). NASA/Lyndon B. Johnson Space Center. Retrieved 29 January 2010 from http://www.jsc.nasa.gov/Bios/htmlbios/ onizuka.html

Biographical data: Guion S. Bluford, Jr. (Colonel, USAF, Ret.). (2009, October). NASA/Lyndon B. Johnson Space Center. Retrieved 28 January 2010 from http://www.jsc.nasa.gov/Bios/ htmlbios/bluford-gs.html

Biographical data: John W. Young. (2005, May). NASA/Lyndon B. Johnson Space Center. Retrieved 29 January 2010 from http://www11.jsc.nasa.gov/Bios/htmlbios/young.html

Biographical data: Mae C. Jemison (M.D.). (1993, March). NASA/Lyndon B. Johnson Space Center. Retrieved 28 January 2010 from http://www.jsc.nasa.gov/Bios/htmlbios/jemison-mc.html

Biographical data: Sally K. Ride, Ph.D. (2006, July). NASA/Lyndon B. Johnson Space Center. Retrieved 28 January 2010 from http://www.jsc.nasa.gov/Bios/htmlbios/ride-sk.html

Biographical data: Sidney M. Gutierrez (Colonel, USAF, Ret.). (1996, July). NASA/Lyndon B. Johnson Space Center. Retrieved 29 January 2010 from http://www.jsc.nasa.gov/Bios/htmlbios/ gutierrez-sm.html

Columbia Accident Investigation Board: Report volume 1. (2003, August). Columbia Accident Investigation Board. Retrieved 22 January 2010 from http://caib.nasa.gov/news/report/pdf/ vol1/full/caib_report_volume1.pdf

Enterprise *(OV-101).* (1994, March 18). NASA/Kennedy Space Center. Retrieved 26 January 2010 from http://science.ksc.nasa.gov/shuttle/resources/orbiters/enterprise.html

Garber, S. J. (2001, April 25). *The flight of STS-1.* NASA. Retrieved 26 January 2010 from http://history.nasa.gov/sts1/index.html

Launch schedule. (2010, January 27). NASA. Retrieved 28 January 2010 from http://www.nasa.gov/missions/highlights/schedule.html

Solid rocket boosters. (2000, August 31). NASA/Kennedy Space Center. Retrieved 25 January 2010 from http://science.ksc.nasa.gov/shuttle/technology/sts-newsref/srb.html

Some space shuttle stories. (2003). *Space Today Online.* Retrieved 22 January 2010 from http://www.spacetoday.org/SpcShtls/SpcShtlStories.html

Space shuttle basics. (2008, February 27). NASA. Retrieved 27 January 2010 from http://spaceflight.nasa.gov/shuttle/reference/basics/history/index.html

Space Shuttle Main Engines. (2009, July 16). NASA. Retrieved 26 January 2010 from http://www.nasa.gov/returntoflight/system/system_SSME.html

Space shuttle: Mission archives: STS-8. (2007, November 23). NASA. Retrieved 28 January 2010 from http://www.nasa.gov/mission_pages/shuttle/shuttlemissions/archives/sts-8.html

Space shuttle overview. (2008, March 20). NASA. Retrieved 25 January 2010 from http://www.nasa.gov/mission_pages/shuttle/vehicle/index.html

Space shuttle overview: Challenger *(OV-099).* (2008, August 6). NASA/Kennedy Space Center. Retrieved 22 January 2010 from http://www.nasa.gov/centers/kennedy/shuttleoperations/orbiters/challenger-info.html

Space shuttle: The orbiter. (2006, March 5). NASA. Retrieved 25 January 2010 from http://www.nasa.gov/returntoflight/system/system_Orbiter.html

STS-1. (2001, June 29). NASA/Kennedy Space Center. Retrieved 22 January 2010 from http://science.ksc.nasa.gov/shuttle/missions/sts-1/mission-sts-1.html

STS-61. (2001, June 29). NASA/Kennedy Space Center. Retrieved 26 January 2010 from http://science.ksc.nasa.gov/shuttle/missions/sts-61/mission-sts-61.html

STS-114 return to flight. (2006, August 4). NASA. Retrieved 29 January 2010 from http://www.nasa.gov/returntoflight/main/index.html

STS-133 mission information. (2010, January 27). NASA. Retrieved 28 January 2010 from http://www.nasa.gov/mission_pages/shuttle/shuttlemissions/sts133/index.html

Team Hubble: Servicing missions. (n.d.). Space Telescope Science Institute. Retrieved 26 January 2010 from http://hubblesite.org/the_telescope/team_hubble/servicing_missions.php#sm1

CHAPTER 7, Lesson 2   Lessons Learned: *Challenger* and *Columbia*

Atkinson, N. (2008, December 30). *New report details* Columbia *accident, recommends improvements.* Universe Today. Retrieved 9 February 2010 from http://www.universetoday.com/2008/12/30/new-report-details-columbia-accident-recommends-improvements/

Biographical data: David M. Brown (Captain, USN). (2004, May). NASA/Lyndon B. Johnson Space Center. Retrieved 5 February 2010 from http://www.jsc.nasa.gov/Bios/htmlbios/brown.html

Biographical data: Ellison S. Onizuka (Colonel, USAF). (2007, January). NASA/Lyndon B. Johnson Space Center. Retrieved 1 February 2010 from http://www.jsc.nasa.gov/Bios/htmlbios/onizuka.html

Biographical data: Francis R. (Dick) Scobee (Mr.). (2003, December). NASA/Lyndon B. Johnson Space Center. Retrieved 1 February 2010 from http://www.jsc.nasa.gov/Bios/htmlbios/scobee.html

Biographical data: Gregory B. Jarvis (Mr.). (2003, December). NASA/Lyndon B. Johnson Space Center. Retrieved 1 February 2010 from http://www.jsc.nasa.gov/Bios/htmlbios/jarvis.html

Biographical data: Ilan Ramon (Colonel, Israel Air Force). (2004, May). NASA/Lyndon B. Johnson Space Center. Retrieved 5 February 2010 from http://www11.jsc.nasa.gov/Bios/PS/ramon.html

Biographical data: Judith A. Resnik (Ph.D.). (2003, December). NASA/Lyndon B. Johnson Space Center. Retrieved 1 February 2010 from http://www.jsc.nasa.gov/Bios/htmlbios/resnik.html

Biographical data: Kalpana Chawla (Ph.D.). (2004, May). NASA/Lyndon B. Johnson Space Center. Retrieved 5 February 2010 from http://www.jsc.nasa.gov/Bios/htmlbios/chawla.html

Biographical data: Laurel Blair Salton Clark, M.D. (Captain, USN). (2004, May). NASA/Lyndon B. Johnson Space Center. Retrieved 5 February 2010 from http://www.jsc.nasa.gov/Bios/htmlbios/clark.html

Biographical data: Michael J. Smith (Captain, USN). (2003, December). NASA/Lyndon B. Johnson Space Center. Retrieved 1 February 2010 from http://www.jsc.nasa.gov/Bios/htmlbios/smith-michael.html

Biographical data: Michael P. Anderson (Lieutenant Colonel, USAF). (2004, May). NASA/Lyndon B. Johnson Space Center. Retrieved 5 February 2010 from http://www.jsc.nasa.gov/Bios/htmlbios/anderson.html

Biographical data: Rick Douglas Husband (Colonel, USAF). (2004, May). NASA/Lyndon B. Johnson Space Center. Retrieved 5 February 2010 from http://www.jsc.nasa.gov/Bios/htmlbios/husband.html

Biographical data: Ronald E. McNair (Ph.D.). (2003, December). NASA/Lyndon B. Johnson Space Center. Retrieved 1 February 2010 from http://www.jsc.nasa.gov/Bios/htmlbios/mcnair.html

Biographical data: S. Christa Corrigan McAuliffe (Teacher in Space Participant). (2006, May). NASA/Lyndon B. Johnson Space Center. Retrieved 1 February 2010 from http://www.jsc.nasa.gov/Bios/htmlbios/mcauliffe.html

Biographical data: William C. McCool (Commander, USN). (2004, May). NASA/Lyndon B. Johnson Space Center. Retrieved 5 February 2010 from http://www.jsc.nasa.gov/Bios/htmlbios/mccool.html

Columbia Accident Investigation Board: Report volume 1. (2003, August). Columbia Accident Investigation Board. Retrieved 5 February 2010 from http://caib.nasa.gov/news/report/pdf/vol1/full/caib_report_volume1.pdf

External tank. (2000, August 31). NASA/Kennedy Space Center. Retrieved 10 February 2010 from http://science.ksc.nasa.gov/shuttle/technology/sts-newsref/et.html

Gleick, J. (1988, February 17). Richard Feynman dead at 69; leading theoretical physicist. *The New York Times.* Retrieved 11 February 2010 from http://www.nytimes.com/books/97/09/21/reviews/feynman-obit.html

Report of the Presidential Commission on the space shuttle Challenger *accident.* (1986, June 6). Retrieved 9 February 2010 from http://history.nasa.gov/rogersrep/genindex.htm

Roberts, J. (n.d.). Langley employees assist in debris collection. *Researcher News.* Retrieved 8 February 2010 from http://researchernews.larc.nasa.gov/archives/2003/050903/Search.html

Rumerman, J. (n.d.). *The* Challenger *accident.* US Centennial of Flight Commission. Retrieved 9 February 2010 from http://www.centennialofflight.gov/essay/SPACEFLIGHT/challenger/SP26.htm

Space shuttle Columbia *accident.* (n.d.). NASA: The StarChild Team. Retrieved 11 February 2010 from http://starchild.gsfc.nasa.gov/docs/StarChild/space_level2/columbia.html

STS-51L. (2007, November 23). NASA. Retrieved 10 February 2010 from http://www.nasa.gov/mission_pages/shuttle/shuttlemissions/archives/sts-51L.html

CHAPTER 8, Lesson 1   From *Salyut* to the International Space Station

Biographical data: Clayton C. Anderson. (2009, October). NASA/Lyndon B. Johnson Space Center. Retrieved 15 February 2010 from http://www.jsc.nasa.gov/Bios/htmlbios/anderson-c.html

Biographical data: David M. Brown (Captain, USN). (2004, May). NASA/Lyndon B. Johnson Space Center. Retrieved 19 February 2010 from http://www.jsc.nasa.gov/Bios/htmlbios/mcarthur.html

Biographical data: Edward Michael "Mike" Fincke (Colonel, USAF). (2009, August). NASA/Lyndon B. Johnson Space Center. Retrieved 15 February 2010 from http://www.jsc.nasa.gov/Bios/htmlbios/fincke.html

Biographical data: Sandra H. Magnus (PhD). (2009, August). NASA/Lyndon B. Johnson Space Center. Retrieved 15 February 2010 from http://www.jsc.nasa.gov/Bios/htmlbios/magnus.html

Biographical data: Tracy Caldwell Dyson (PhD). (2009, November). NASA/Lyndon B. Johnson Space Center. Retrieved 15 February 2010 from http://www.jsc.nasa.gov/Bios/htmlbios/caldwell.html

Biographical data: William Surles "Bill" McArthur, Jr., (Colonel, USA, Ret.). (2009, June). NASA/Lyndon B. Johnson Space Center. Retrieved 19 February 2010 from http://www.jsc.nasa.gov/Bios/htmlbios/mcarthur.html

Brooks, C. G., Ertel, I. D., & Newkirk, R. W. (n.d.). Skylab: *A chronology.* SP-4011. NASA. Retrieved 10 March 2010 from http://history.nasa.gov/SP-4011/contents.htm

Culbertson, F. L. (1996). *What's in a name?* NASA. Retrieved 14 February 2010 from http://history.nasa.gov/SP-4225/documentation/mirmeanings/meanings.htm

Dean, B. (2007). *Clay Anderson: The little engine that flew.* NASA. Retrieved 15 February 2010 from http://www.nasa.gov/astronauts/anderson_profile.html

Esperance to mark *Skylab* anniversary. (2009, July 10). *ABC News* (Australian Broadcasting Corporation). Retrieved 12 February 2010 from http://www.abc.net.au/news/stories/2009/07/10/2622503.htm

Greetings, earthlings: Ed's musings from space: Expedition 7: *Watching the world go by.* (2003, July 17). NASA. Retrieved 18 February 2010 from http://spaceflight.nasa.gov/station/crew/exp7/luletters/lu_letter5.html

International Space Station. (n.d.). *Internet Encyclopedia of Science.* Retrieved 15 February 2010 from http://www.daviddarling.info/encyclopedia/I/ISS.html

The International Space Station. (1999, June 3). NASA. Retrieved 15 February 2010 from http://www.shuttlepresskit.com/ISS_OVR/index.htm

Mir *space station.* (n.d.). NASA. Retrieved 14 February 2010 from http://history.nasa.gov/SP-4225/mir/mir.htm

Mir's *15 years.* (2004, April 3). NASA. Retrieved 17 February 2010 from http://spaceflight.nasa.gov/history/shuttle-mir/spacecraft/s-mir-15yrs-main.htm

Phillips, T. (2001). *The end is* Mir. Science and Technology Directorate, NASA. Retrieved 13 February 2010 from http://science.nasa.gov/headlines/y2001/ast10mar_1.htm

Salyut 1. (2009, November 23). NASA/Goddard Space Flight Center, NSSDC Master Catalog. Retrieved 12 February 2010 from http://nssdc.gsfc.nasa.gov/nmc/masterCatalog.do?sc=1971-032A

Satellite orbits. (2009, September 30). NASA. Retrieved 18 February 2010 from http://science-edu.larc.nasa.gov/SCOOL/orbits.html

*Shuttle-*Mir *overview.* (2007). NASA. Retrieved 13 February 2010 from http://www.nasa.gov/mission_pages/shuttle-mir/index.html

Skylab: *America's first space station: History.* (2009). NASA. Retrieved 13 February 2010 from http://www.nasa.gov/mission_pages/skylab/missions/skylab_summary.html

Skylab: *America's first space station: Overview.* (2009). NASA. Retrieved 12 February 2010 from http://www.nasa.gov/mission_pages/skylab/

Space factoids: Lost tidbits recovered. (2003). *Space Today Online.* Retrieved 14 February 2010 from http://www.spacetoday.org/History/SpaceFactoids/SpaceFactoids2.html

Space shuttle Canadarm robotic arm marks 25 years in space. (2006). NASA. Retrieved 15 February 2010 from http://www.nasa.gov/mission_pages/shuttle/behindscenes/rms_anniversary.html

Space station: The station: Russian space history. (1999). *Houston Public Television.* Retrieved 12 February 2010 from http://www.pbs.org/spacestation/station/russian.htm

Space station under construction: Building a ship outside a shipyard. (2006). NASA. Retrieved 15 February 2010 from http://www.nasa.gov/mission_pages/station/main/iss_construction.html

To the galaxies and beyond—Three astronauts land at tech. (n.d.). Georgia Institute of Technology, College of Engineering. Retrieved 15 February 2010 from http://www.coe.gatech.edu/feature/6_nasa.php

World Book *at NASA: International Space Station*. (2007). NASA. Retrieved 15 February 2010 from http://www.nasa.gov/worldbook/intspacestation_worldbook.html

World Book *at NASA: Space exploration*. (2007, November 29). NASA. Retrieved 17 February 2010 from http://www.nasa.gov/worldbook/space_exploration_worldbook.html

CHAPTER 8, Lesson 2   The Future in Space

Biographical data: Sunita L. Williams (Captain, USN). (2009, December). NASA/Lyndon B. Johnson Space Center. Retrieved 28 February 2010 from http://www.jsc.nasa.gov/Bios/htmlbios/williams-s.html

A crewed mission to Mars. (2005, January 6). NASA Goddard Space Flight Center. Retrieved 28 February 2010 from http://nssdc.gsfc.nasa.gov/planetary/mars/mars_crew.html

Dean, B. (2006, October 10). *Astronaut Suni Williams: As chance would have it*. NASA. Retrieved 28 February 2010 from http://www.nasa.gov/astronauts/s_williams_profile.html

Exploration Technology Development Program. (2007, October 16). NASA. Retrieved 28 February 2010 from http://www.nasa.gov/directorates/esmd/aboutesmd/acd/technology_dev.html

Hitt, D. (2008, July 29). *What is Altair?* NASA Educational Technology Services. Retrieved 3 March 2010 from http://www.nasa.gov/audience/forstudents/k-4/stories/what-is-altair-k4.html

Lunar exploration themes and objectives development process. (2006, December 4). NASA. Retrieved 19 February 2010 from http://www.nasa.gov/exploration/home/why_moon_process.html

Lunar outpost plans taking shape. (2007, October 1). NASA. Retrieved 19 February 2010 from http://www.nasa.gov/exploration/lunar_architecture.html

Lunar Reconnaissance Orbiter: LRO launch information. (2009, June 25). NASA. Retrieved 28 February 2010 from http://www.nasa.gov/mission_pages/LRO/launch/index.html

Lunar Reconnaissance Orbiter, mission overview. (n.d.). NASA. Retrieved 19 February 2010 from http://lunar.gsfc.nasa.gov/mission.html

President Bush offers new vision for NASA. (2004, January 14). NASA. Retrieved 28 February 2010 from http://www.nasa.gov/missions/solarsystem/bush_vision.html

Spotts, P. (2009). Star trek: This generation. *The Christian Science Monitor, 101*(113), 13–18.

Why the Moon? (2009, August 18). NASA. Retrieved 19 February 2010 from http://www.nasa.gov/exploration/home/why_moon.html

Zaroulis, N. (2008, April 27). The man who invented Mars. *The Boston Globe*. Retrieved 2 March 2010 from http://www.boston.com/bostonglobe/magazine/articles/2008/04/27/the_man_who_invented_mars/?page=1

CHAPTER 9, Lesson 1   Missions to the Sun, Moon, Venus, and Mars

Christian, E. R., & Davis, A. J. (2008, April 15). ACE *mission overview*. California Technical University. Retrieved 12 January 2010 from http://www.srl.caltech.edu/ACE/

Exploring Mars. (2007). *Space Today Online*. Retrieved 11 January 2010 from http://www.spacetoday.org/SolSys/Mars/MarsThePlanet/MarsStats.html

Factsheet: Ulysses. (2008, November 7). European Space Agency. Retrieved 11 January 2010 from http://www.esa.int/esaSC/SEMQ035KXMF_0_spk.html

Gamma ray bursts: Introduction to a mystery. (2008, July 22). NASA. Retrieved 6 January 2010 from http://imagine.gsfc.nasa.gov/docs/science/know_l1/bursts.html

Helios 1. (2009, January 23). NASA. Retrieved 22 January 2010 from https://sse.jpl.nasa.gov/missions/profile.cfm?Sort=Target&Target=Sun&MCode=Helios_01&Display=ReadMore

Helios 2. (2009, January 23). NASA. Retrieved 22 January 2010 from https://sse.jpl.nasa.gov/missions/profile.cfm?Sort=Target&Target=Sun&MCode=Helios_02&Display=ReadMore

Hinode *mission delves into solar mysteries*. (2007, December 6). American Association for the Advancement of Science. Retrieved 11 January 2010 from http://www.eurekalert.org/pub_releases/2007-12/aaft-hmd113007.php

Hinode's first images from our violent Sun. (2006, December 22). European Space Agency. Retrieved 11 January 2010 from http://www.esa.int/esaSC/SEMII1R08ZE_index_0.html

IBEX *explores galactic frontier, releases first-ever all-sky map*. (2009, October 15). NASA. Retrieved 12 January 2010 from http://www.nasa.gov/mission_pages/ibex/allsky_map.html

The interstellar medium. (n.d.). The University of New Hampshire, Experimental Space Plasma Group. Retrieved 11 January 2010 from http://www-ssg.sr.unh.edu/ism/what1.html

Inward spirals. (n.d.). NASA. Retrieved 12 January 2010 from http://marsprogram.jpl.nasa.gov/mgs/overvu/mplan/ab/ab.html

LCROSS *science briefing November 13th 2009*. (2009, November 13). NASA. Retrieved 12 January 2010 from http://www.youtube.com/watch?v=5xVlBa6YKH4

Luna 1. (2009, November 23). NASA. Retrieved 14 January 2010 from http://nssdc.gsfc.nasa.gov/nmc/spacecraftDisplay.do?id=1959-012A

Luna 2. (2009, November 23). NASA. Retrieved 12 January 2010 from http://nssdc.gsfc.nasa.gov/nmc/masterCatalog.do?sc=1959-014A

Lunar exploration timeline. (2009, June 22). NASA. Retrieved 12 January 2010 from http://nssdc.gsfc.nasa.gov/planetary/lunar/lunartimeline.html

Lunar Reconnaissance Orbiter. (2008, October). NASA. Retrieved 12 January 2010 from http://www.nasa.gov/pdf/359938main_LRO_factsheet.pdf

Lunar Reconnaissance Orbiter. (n.d.). NASA Goddard Space Flight Center. Retrieved 3 March 2010 from http://lunar.gsfc.nasa.gov/moonfacts.html

Lure of Mars. (n.d.). NASA. Retrieved 2 February 2010 from http://www.nasa.gov/externalflash/m2k4/whymars.html

Magellan *summary sheet*. (n.d.). NASA. Retrieved 12 January 2010 from http://www2.jpl.nasa.gov/magellan/fact1.html

Mars Reconnaissance Orbiter. (2009, January 23). NASA. Retrieved 11 January 2010 from http://solarsystem.nasa.gov/missions/profile.cfm?Sort=Target&Target=Mars&MCode=MRO

Mitchell, C. (2006, May 4). *Fahrenheit vs. Celsius*. NASA. Retrieved 2 February 2010 from http://asd-www.larc.nasa.gov/GLOBE/resources/lesson_plans/Fahrenheit_vs_Celsius.html

NASA Phoenix results point to Martian climate cycles. (2009, July 2). NASA. Retrieved 13 January 2010 from http://phoenix.lpl.arizona.edu/07_02_09_pr.php

NASA science missions: ACE. (n.d.). NASA. Retrieved 12 January 2010 from http://nasascience.nasa.gov/missions/ace

NASA science missions: TRACE. (n.d.). NASA. Retrieved 12 January 2010 from http://nasascience.nasa.gov/missions/trace

NASA thesaurus machine aided indexing. (n.d.). NASA. Retrieved 11 January 2010 from http://mai.larc.nasa.gov/ct?t=micrometeorites

NASA unveils latest results from lunar missions, helps prepare for next stage of scientific discovery. (2009, December 15). NASA. Retrieved 12 January 2010 from http://www.nasa.gov/mission_pages/LRO/news/agu-results-2009.html

A new solar dynamo mechanism discovered by the Hinode *satellite*. (2009, April 7). National Astronomical Observatory of Japan. Retrieved 11 January 2010 from http://solar-b.nao.ac.jp/news_e/20090407_press_e/index_e.shtml

Phillips, T. (2001, January 6). *Shields up: Heliosphere-helium*. FirstScience.com. Retrieved 11 January 2010 from http://www.firstscience.com/home/articles/space/shields-up-the-heliosphere-helium_1424.html

Photospheric features. (2007, January 18). NASA. Retrieved 22 January 2010 from http://solarscience.msfc.nasa.gov/feature1.shtml

Pioneer 7. (2009, August 17). NASA. Retrieved 22 January 2010 from http://solarsystem.nasa.gov/missions/profile.cfm?Sort=Target&Target=Comets&MCode=Pioneer_07&Display=ReadMore

The Pioneer *missions*. (2007, March 26). NASA. Retrieved 22 January 2010 from http://www.nasa.gov/centers/ames/missions/archive/pioneer.html

Pioneer Venus *overview*. (n.d.). NASA. Retrieved 12 January 2010 from http://www.nasa.gov/mission_pages/pioneer-venus/index.html

Solar physics UToV space missions. (2009). Physics Department, Università degli Studi di Roma "Tor Vergata." Retrieved 18 March 2010 from http://www.fisica.uniroma2.it/~solare/english/index.php?option=com_content&view=article&id=24:utov-space-missions

Solar System exploration: Pioneer Venus 1 *(comets)*. (2009, October 16). NASA. Retrieved 12 January 2010 from http://sse.jpl.nasa.gov/missions/profile.cfm?Sort=Nation&Target=Comets&MCode=Pioneer_Venus_01&Nation=USA&Display=ReadMore

Solar System exploration: Ulysses. (2009, January 23). NASA. Retrieved 6 January 2010 from http://solarsystem.nasa.gov/missions/profile.cfm?Sort=Target&Target=Sun&MCode=Ulysses

Solar System exploration: Venera 4. (2009, August 20). NASA. Retrieved 18 January 2010 from http://solarsystem.nasa.gov/missions/profile.cfm?MCode=Venera_04

Solar System exploration: Venera 16. (2009, October 14). NASA. Retrieved 12 January 2010 from http://solarsystem.nasa.gov/missions/profile.cfm?Sort=Target&Target=Venus&MCode=Venera_16

Spacecraft reveals new insights about origins of solar wind. (2007, December 6). NASA. Retrieved 11 January 2010 from http://www.nasa.gov/home/hqnews/2007/dec/HQ_07264_Hinode_Waves.html

Spirit *and* Opportunity: *Mars exploration rovers*. (2010). NASA. Retrieved 11 January 2010 from http://www.nasa.gov/mission_pages/mer/index.html

Sunspots and magnetic fields. (2005, August 11). University Corporation for Atmospheric Research. Retrieved 11 January 2010 from http://www.windows.ucar.edu/tour/link=/sun/atmosphere/sunspot_magnetism.html&edu=high

TRACE *educational resources*. (n.d.). Lockheed Martin Solar and Astrophysics Labs. Retrieved 12 January 2010 from http://trace.lmsal.com/Public/eduprodu.htm

CHAPTER 9, Lesson 2 The Hubble Space Telescope and Missions to Comets and Outer Planets

Astronaut John Grunsfeld appointed STScI deputy director. (2010, January 4). NASA. Retrieved 31 January 2010 from http://hubblesite.org/newscenter/archive/releases/2010/04/full/

Cassini *mission overview*. (n.d.). NASA. Retrieved 20 January 2010 from http://saturn.jpl.nasa.gov/mission/introduction/

Cassini: *Saturn*. (2009, September 28). NASA. Retrieved 20 January 2010 from http://solarsystem.nasa.gov/missions/profile.cfm?Sort=Alpha&Alias=Cassini&Letter=C&Display=ReadMore

Cassini's *earthly benefits*. (1995, May). NASA. Retrieved 20 January 2010 from http://saturn.jpl.nasa.gov/multimedia/products/pdfs/earthly.pdf

Celebrating the fifth anniversary of Huygen's Titan touchdown. (2010, January 14). European Space Agency. Retrieved 20 January 2010 from http://www.esa.int/SPECIALS/Cassini-Huygens/SEM5KSLJ74G_0.html

Deep Space 1. (2009, November 4). NASA. Retrieved 20 January 2010 from http://solarsystem.nasa.gov/missions/profile.cfm?Sort=Target&Target=Asteroids&MCode=DS1&Display=ReadMore

Deep Space *technology brochure*. (2005, December 7). NASA. Retrieved 20 January 2010 from http://deepspace.jpl.nasa.gov/dsn/brochure/technology2.html

Discovery *highlights*. (2007, August 9). NASA. Retrieved 19 January 2010 from http://solarsystem.nasa.gov/galileo/discovery.cfm

The Earth's magnetosphere. (2006, April 25). NASA. Retrieved 5 February 2010 from http://www.nasa.gov/mission_pages/themis/auroras/magnetosphere.html

Edwin P. Hubble. (2008, June 16). NASA. Retrieved 19 January 2010 from http://hubble.nasa.gov/overview/hubble_bio.php

Edwin Powell Hubble. (2002, May 3). Edwinhubble.com. Retrieved 19 January 2010 from http://www.edwinhubble.com/hubble_bio_001.htm

Far-flung supernovae shed light on dark universe. (2003, April 10). NASA. Retrieved 20 January 2010 from http://hubblesite.org/newscenter/archive/releases/2003/12/

Galilean moons: Callisto. (2001, October 1). NASA. Retrieved 20 January 2010 from http://www2.jpl.nasa.gov/galileo/moons/callisto.html

Galilean moons: Ganymede. (2001, October 1). NASA. Retrieved 20 January 2010 from http://www2.jpl.nasa.gov/galileo/moons/ganymede.html

Galileo *crosses boundary into Jupiter's environment.* (1995, December 1). NASA/Jet Propulsion Laboratory. Retrieved 5 February 2010 from http://www2.jpl.nasa.gov/sl9/gll36.html

Galileo: *Jupiter.* (2009, April 30). NASA. Retrieved 19 January 2010 from http://solarsystem.nasa.gov/missions/profile.cfm?Sort=Alpha&Letter=G&Alias=Galileo

Galileo *makes new discoveries at Ganymede.* (n.d.). NASA. Retrieved 20 January 2010 from http://www2.jpl.nasa.gov/galileo/status960710.html

Galileo's *Jupiter journey began two decades ago.* (2009, October 16). NASA/Jet Propulsion Laboratory. Retrieved 3 February 2010 from http://www.jpl.nasa.gov/news/features.cfm?feature=2338

Harper, J. (2010, February 3). Inside the Beltway: The unkindest cut. *The Washington Times*, p. A2.

The history of the Hubble Space Telescope. (2007, June). NASA Quest. Retrieved 16 January 2010 from http://quest.nasa.gov/hst/about/history.html

Hubble completes eight-year effort to measure expanding universe. (1999, May 25). NASA. Retrieved 20 January 2010 from http://hubblesite.org/newscenter/archive/releases/1999/19/

Hubble finds hidden exoplanet in archival data. (2009, April 1). Space Telescope Science Institute. Retrieved 18 January 2010 from http://hubblesite.org/newscenter/archive/releases/2009/15/results/50/

Hubble finds smallest Kuiper belt object ever seen. (2009, December 16). NASA. Retrieved 18 January 2010 from http://www.nasa.gov/mission_pages/hubble/science/hst_img_kuiper-smallest.html

Hubble makes first direct measurements on world around another star. (2001, November 27). Space Telescope Science Institute. Retrieved 20 January 2010 from http://hubblesite.org/newscenter/archive/releases/2001/38/

Hubble Space Telescope. (2009, January 23). NASA. Retrieved 16 January 2010 from http://solarsystem.nasa.gov/missions/profile.cfm?Sort=Alpha&Letter=H&Alias=Hubble%20Space%20Telescope

The Hubble Space Telescope. (2009, August 13). NASA. Retrieved 26 January 2010 from http://hubble.nasa.gov/

Hubble Space Telescope: Eyes in the sky. (2010). National Geographic Society. Retrieved 17 January 2010 from http://science.nationalgeographic.com/science/space/space-exploration/hubble.html

Hubble Space Telescope servicing mission 3A. (1999, June). NASA. Retrieved 31 January 2010 from http://hubble.nasa.gov/a_pdf/news/facts/FS12.pdf

Hubble's top ten discoveries. (2005, April). National Geographic Society. Retrieved 19 January 2010 from http://news.nationalgeographic.com/news/2005/04/photogalleries/hubble/

Jupiter and its moons. (n.d.). The Open University. Retrieved 20 January 2010 from http://openlearn.open.ac.uk/mod/resource/view.php?id=307700

Jupiter: Moons: Io. (2008, December 8). NASA. Retrieved 20 January 2010 from http://solarsystem.nasa.gov/planets/profile.cfm?Object=Jup_Io

Kuiper belt. (2009, January 23). NASA. Retrieved 20 January 2010 from http://solarsystem.nasa.gov/planets/profile.cfm?Object=KBOs&Display=OverviewLong

Lemonick, M. A. (1999, March 29). Astronomer Edwin Hubble. *Time*. Retrieved 19 January 2010 from http://www.time.com/time/magazine/article/0,9171,990615,00.html

Missions to Jupiter: Voyager 2. (2009, January 23). NASA. Retrieved 19 January 2010 from http://solarsystem.nasa.gov/missions/profile.cfm?Sort=Target&Target=Jupiter&MCode=Voyager_2

Missions to Saturn: Voyager 1. (2009, January 23). NASA. Retrieved 19 January 2010 from http://solarsystem.nasa.gov/missions/profile.cfm?Sort=Target&Target=Saturn&MCode=Voyager_1

NASA Facts: Galileo mission to Jupiter. (n.d.). NASA. Retrieved 20 January 2010 from http://www.jpl.nasa.gov/news/fact_sheets/galileo.pdf

NASA finds "big baby" galaxies in newborn universe. (2005, September 27). NASA. Retrieved 18 January 2010 from http://solarsystem.nasa.gov/news/display.cfm?News_ID=12019

NASA's Optical Verification Program. (1993, November). NASA. Retrieved 26 January 2010 from http://hubble.nasa.gov/a_pdf/news/facts/OpticalVerification.pdf

New Horizons *mission timeline.* (2010, January 20). NASA. Retrieved 20 January 2010 from http://pluto.jhuapl.edu/mission/mission_timeline.php

New Horizons: *Pluto.* (2009, September 5). NASA. Retrieved 20 January 2010 from http://solarsystem.nasa.gov/missions/profile.cfm?Sort=Alpha&Alias=New%20Horizons&Letter=N&Display=ReadMore

Okolski, G. (2008, April 18). *A brief history of the Hubble Space Telescope.* NASA History Division. Retrieved 16 January 2010 from http://history.nasa.gov/hubble/index.html

Overbye, D. (2009, April 13). Last voyage for keeper of the Hubble. *The New York Times.* Retrieved 31 January 2010 from http://www.nytimes.com/2009/04/14/science/space/14prof.html?pagewanted=1&_r=1

Pioneering NASA spacecraft mark thirty years of flight. (2007, August 20). NASA. Retrieved 19 January 2010 from http://solarsystem.nasa.gov/news/display.cfm?News_ID=22575

Rosetta *at a glance.* (2010, January 21). European Space Agency. Retrieved 21 January 2010 from http://www.esa.int/SPECIALS/Rosetta/SEMYMF374OD_0.html

Saturn: Moon: Titan. (2007, September 7). NASA. Retrieved 19 January 2010 from http://solarsystem.nasa.gov/planets/profile.cfm?Object=Titan

Science: Look upward. (1948, February 9). *Time.* Retrieved 19 January 2010 from http://www.time.com/time/magazine/article/0,9171,856024-2,00.html

A science odyssey: People and discoveries: Edwin Hubble 1889–1953. (1998). WGBH. Retrieved 19 January 2010 from http://www.pbs.org/wgbh/aso/databank/entries/bahubb.html

Spacecraft lifetime. (2010, January 19). NASA. Retrieved 19 January 2010 from http://voyager.jpl.nasa.gov/spacecraft/spacecraftlife.html

Stardust-NExT. (2009, September 3). NASA. Retrieved 20 January 2010 from http://solarsystem.nasa.gov/missions/profile.cfm?Sort=Target&Target=Asteroids&MCode=STARDUST&Display=ReadMore

Stathopoulos, V. (2010, January 16). *Hubble Space Telescope history.* Aerospaceguide.net. Retrieved 16 January 2010 from http://www.aerospaceguide.net/spacehistory/hubble-history.html

The telescope: Hubble essentials. (n.d.). Space Telescope Science Institute. Retrieved 19 January 2010 from http://hubblesite.org/the_telescope/hubble_essentials/

The telescope: Hubble essentials: About Edwin Hubble. (n.d.). Space Telescope Science Institute. Retrieved 19 January 2010 from http://hubblesite.org/the_telescope/hubble_essentials/edwin_hubble.php

Voyager 1: *Beyond our Solar System.* (2009, January 23). NASA. Retrieved 19 January 2010 from http://solarsystem.nasa.gov/missions/profile.cfm?Sort=Alpha&Alias=Voyager%201&Letter=V&Display=ReadMore

Voyager 1: *"The spacecraft that could" hits new milestone.* (2006, August 15). NASA. Retrieved 19 January 2010 from http://solarsystem.nasa.gov/news/display.cfm?News_ID=16038

Voyager *fast facts.* (2010, January 19). NASA. Retrieved 19 January 2010 from http://voyager.jpl.nasa.gov/mission/fastfacts.html

Voyager: *Golden record.* (2010, January 19). NASA. Retrieved 19 January 2010 from http://voyager.jpl.nasa.gov/spacecraft/goldenrec.html

Voyager: *Planetary Voyage.* (2010, January 19). NASA. Retrieved 19 January 2010 from http://voyager.jpl.nasa.gov/science/planetary.html

Voyager: *Saturn.* (2010, January 19). NASA. Retrieved 19 January 2010 from http://voyager.jpl.nasa.gov/science/saturn.html

Voyager: *Uranus.* (2010, January 19). NASA. Retrieved 19 January 2010 from http://voyager.jpl.nasa.gov/science/uranus.html

Weinstock, M. (2000, September 6). *Fallen Galileo probe uncovers sheets of Jupiter's hotspots*. Space.com. Retrieved 20 January 2010 from http://www.space.com/scienceastronomy/ solarsystem/jupiter_hotspots.html

What is the Hubble Space Telescope? (2009, December 30). NASA. Retrieved 19 January 2010 from http://www.nasa.gov/audience/forstudents/5-8/features/what-is-the-hubble-space -telescope-58.html

What is the Hubble Space Telescope? (n.d.). Space Telescope Science Institute. Retrieved 17 January 2010 from http://hubblesite.org/reference_desk/faq/answer.php.id=76&cat=topten

World Book *at NASA: Hubble Space Telescope*. (2007, November 29). NASA. Retrieved 19 January 2010 from http://www.nasa.gov/worldbook/hubble_telescope_worldbook.html

CHAPTER 10, Lesson 1   Orbits and How They Work

Artificial satellites. (2007, November 29). NASA. Retrieved 9 February 2010 from http://www.nasa .gov/worldbook/artificial_satellites_worldbook.html

Boyd, P. (2005, December 1). NASA: Ask an astrophysicist. Retrieved 9 February 2010 from http://imagine.gsfc.nasa.gov/docs/ask_astro/answers/970613a.html

Cain, F. (2009, March 6). *How fast does the Earth rotate?* Universe Today. Retrieved 8 February 2010 from http://www.universetoday.com/guide-to-space/earth/how-fast-does-the-earth-rotate/

Elert, G. (2005). *Speed needed to escape the Earth (escape velocity)*. Hypertextbook.com. Retrieved 23 January 2010 from http://hypertextbook.com/facts/2005/LeoTam.shtml

Europe's spaceport. (2009, July 20). European Space Agency. Retrieved 8 February 2010 from http://www.esa.int/esaMI/Launchers_Europe_s_Spaceport/index.html

Geostationary satellites. (n.d.). NOAA Satellite and Information Service. Retrieved 24 January 2010 from http://www.oso.noaa.gov/goes/

The International Space Station. (1999, June 3). NASA/The United Space Alliance/The Boeing Company. Retrieved 26 January 2010 from http://www.shuttlepresskit.com/ISS_OVR/index.htm

Launch a "rocket" from a spinning "planet." (2005, September 8). NASA. Retrieved 8 February 2010 from http://www.spaceplace.nasa.gov/en/kids/ds1_mgr.shtml

Lu, E. (2003, August 1). *Greetings, earthlings: Ed's musings from space*. NASA. Retrieved 8 February 2010 from http://spaceflight.nasa.gov/station/crew/exp7/luletters/lu_letter6.html

NASA extends the World Wide Web out into space. (2010, January 22). NASA. Retrieved 8 February 2010 from http://www.nasa.gov/home/hqnews/2010/jan/HQ_M10-011_Hawaii221169.html

Orbits "R" us. (2005, September 8). NASA. Retrieved 23 January 2010 from http://spaceplace.nasa .gov/en/kids/goes/goes_poes_orbits.shtml

Polar orbiting satellites. (n.d.). NOAA Satellite and Information Service. Retrieved 26 January 2010 from http://www.oso.noaa.gov/poes/

Riebeek, H. (2009, September 4). *Catalog of Earth satellite orbits*. NASA: Earth Observatory. Retrieved 26 January 2010 from http://earthobservatory.nasa.gov/Features/OrbitsCatalog/

Satellite orbits. (2006, July 17). NASA. Retrieved 24 January 2010 from http://asd-www.larc.nasa .gov/SCOOL/orbits.html

Shoot a cannonball into orbit! (2005, September 8). NASA. Retrieved 8 February 2010 from http://spaceplace.nasa.gov/en/kids/orbits1.shtml

Types of orbits. (2004, June 22). European Space Agency. Retrieved 9 February 2010 from http://www.esa.int/SPECIALS/Launchers_Home/ASEHQOI4HNC_0.html

Watson, T. (2009, February 25). Failed launch dumps NASA climate satellite into ocean. *USA Today*. Retrieved 25 January 2010 from http://www.usatoday.com/weather/climate/ globalwarming/2009-02-24-global-warming-satellite-NASA_N.htm

What is NPOESS? (2009, February 2). NOAA Satellite and Information Service. Retrieved 26 January 2010 from http://www.ipo.noaa.gov/

What is orbit? (2009, April 9). NASA. Retrieved 23 January 2010 from http://www.nasa.gov/ audience/forstudents/5-8/features/orbit_feature_5-8.html

CHAPTER 10, Lesson 2   Maneuvering and Traveling in Space

Aqua: Project science. (2010, February 2). NASA/Goddard Space Flight Center. Retrieved 2 February 2010 from http://aqua.nasa.gov/

Aqua's instruments. (2009, June 1). NASA/Goddard Space Flight Center. Retrieved 16 February 2010 from http://aqua.nasa.gov/about/instruments.php

Basics of space flight: Interplanetary trajectories. (2001). NASA/Jet Propulsion Laboratory. Retrieved 2 February 2010 from http://www2.jpl.nasa.gov/basics/bsf4-1.php

Basics of space flight: Reference systems. (2001). NASA/Jet Propulsion Laboratory. Retrieved 16 February 2010 from http://www2.jpl.nasa.gov/basics/bsf2-2.php

Basics of space flight: Spacecraft navigation. (2001). NASA/Jet Propulsion Laboratory. Retrieved 31 January 2010 from http://www2.jpl.nasa.gov/basics/bsf13-1.php

Biographical data: Ellen Ochoa (Ph.D.). (2008, January). NASA. Retrieved 3 February 2010 from http://www.jsc.nasa.gov/Bios/htmlbios/ochoa.html

Doppler shift. (2007, September 26). NASA/Goddard Space Flight Center. Retrieved 16 February 2010 from http://imagine.gsfc.nasa.gov/YBA/M31-velocity/Doppler-shift-2.html

Gravity assist maneuvers. (n.d.). NASA. Retrieved 15 February 2010 from http://science.nasa.gov/newhome/headlines/ast24jun99_1.htm#gravityassist

A gravity assist mechanical simulator. (2004, August 9). NASA/Jet Propulsion Laboratory. Retrieved 31 January 2010 from http://www2.jpl.nasa.gov/basics/grav/index.php

A gravity assist primer. (n.d.). NASA/Jet Propulsion Laboratory. Retrieved 15 February 2010 from http://www2.jpl.nasa.gov/basics/grav/primer.php

Greeting earthlings: Ed's Musings from Space: Expedition 7: Progress. (2003, July 29). NASA. Retrieved 16 February 2010 from http://spaceflight.nasa.gov/station/crew/exp7/luletters/lu_letter4.html

Hispanic heritage: Meet Ellen Ochoa. (1999). Scholastic. Retrieved 3 February 2010 from http://teacher.scholastic.com/activities/hispanic/ochoatscript.htm

Hohmann transfer & plane changes. (1995, September 21). NASA. Retrieved 15 February 2010 from http://science.nasa.gov/realtime/rocket_sci/satellites/hohmann.html

Mariner 10: (Mercury). (2009, June 30). NASA. Retrieved 31 January 2010 from http://solarsystem.nasa.gov/missions/profile.cfm?Sort=Alpha&Letter=M&Alias=Mariner%2010

Riebeek, H. (2009, August 18). *Flying steady: Mission control tunes up Aqua's orbit.* NASA. Retrieved 2 February 2010 from http://earthobservatory.nasa.gov/Features/OrbitsManeuver/page1.php

Solar electric (ion) propulsion. (n.d). NASA. Retrieved 1 February 2010 from http://nmp.nasa.gov/ds1/tech/sep.html

Special delivery. (2004, October 18). NASA. Retrieved 17 February 2010 from http://www.nasa.gov/mission_pages/dart/rendezvous/pegasus_l1011.html

Steve Gauvain, shuttle rendezvous instructor. (2008, February 21). NASA. Retrieved 16 February 2010 from http://www.nasa.gov/audience/foreducators/stseducation/stories/Steve_Gauvain_Profile.html

Summary of DART accident report. (2007, December 18). NASA. Retrieved 2 February 2010 from http://www.nasa.gov/mission_pages/dart/main/index.html

Trajectories and orbits. (2004, October 22). NASA. Retrieved 16 February 2010 from http://www.hq.nasa.gov/pao/History/conghand/traject.htm

CHAPTER 11, Lesson 1   It *Is* Rocket Science: How Rockets Work

Air rockets. (2006, February 5). NASA. Retrieved 6 February 2010 from http://exploration.grc.nasa.gov/education/rocket/rktstomp.html

All about water rockets. (2005, September 9). NASA. Retrieved 8 February 2010 from http://exploration.grc.nasa.gov/education/rocket/BottleRocket/about.htm

Basic properties of the atmosphere. (n.d.). University of Wisconsin, Green Bay. Retrieved 6 February 2010 from http://74.125.113.132/search?q=cache:l-U44Dp-sKMJ:www.uwgb.edu/DutchS/EnvSC102Notes/102BasicAtmo.ppt+properties+of+atmosphere&cd=1&hl=en&ct=clnk&gl=us

Biographical data: Leland D. Melvin. (2009, December). NASA. Retrieved 10 February 2010 from http://www.jsc.nasa.gov/Bios/htmlbios/melvin.html

Corliss, W. R. (1971). *NASA sounding rockets, 1958–1968: A historical summary.* NASA publication SP-4401. Retrieved 6 April 2010 from http://history.nasa.gov/SP-4401/contents.htm

Determining center of gravity—cg. (2005, November 17). NASA. Retrieved 24 February 2010 from http://exploration.grc.nasa.gov/education/rocket/rktcg.html

First law of thermodynamics. (2005, November 18). NASA. Retrieved 24 February 2010 from http://exploration.grc.nasa.gov/education/rocket/thermo1.html

First space flight. (2009, August 22). Universe Today. Retrieved 19 February 2010 from http://www.universetoday.com/tag/v-2-rocket/

Forces on a rocket. (2005, November 17). NASA. Retrieved 6 February 2010 from http://exploration.grc.nasa.gov/education/rocket/rktfor.html

Gas properties definitions. (2005, December 16). NASA. Retrieved 25 February 2010 from http://exploration.grc.nasa.gov/education/rocket/gasprop.html

Gas temperature. (2008, July 11). NASA. Retrieved 11 February 2010 from http://www.grc.nasa.gov/WWW/K-12/rocket/temptr.html

Liquid engine rocket. (2008, July 11). NASA. Retrieved 8 February 2010 from http://www.grc.nasa.gov/WWW/K-12/rocket/lrockth.html

Model rocket launches. (n.d.). NASA. Retrieved 9 February 2010 from http://www.nasa.gov/centers/goddard/visitor/events/rocket_launch.html

Model rocket safety. (2008, July 11). NASA. Retrieved 9 February 2010 from http://www.grc.nasa.gov/WWW/K-12/rocket/rktsafe.html

Model rockets. (2008, July 11). NASA. Retrieved 9 February 2010 from http://www.grc.nasa.gov/WWW/K-12/rocket/rktparts.html

Newton's laws of motion. (2006, January 5). NASA. Retrieved 23 February 2010 from http://exploration.grc.nasa.gov/education/rocket/newton.html

Newton's Second Law: Definitions. (2006, January 5). NASA. Retrieved 23 February 2010 from http://exploration.grc.nasa.gov/education/rocket/newton2r.html

Preflight interview: Leland D. Melvin. (2007, November 21). NASA. Retrieved 10 February 2010 from http://www.nasa.gov/mission_pages/shuttle/shuttlemissions/sts122/interview_melvin.html

Propulsion system. (2005, December 20). NASA. Retrieved 11 February 2010 from http://exploration.grc.nasa.gov/education/rocket/rocket.html

Robert Goddard and his rockets. (2005, January 14). NASA. Retrieved 11 February 2010 from http://www-istp.gsfc.nasa.gov/stargaze/Sgoddard.htm

Rocket aerodynamics. (2005, December 20). NASA. Retrieved 10 February 2010 from http://exploration.grc.nasa.gov/education/rocket/rktaero.html

Rocket parts. (2005, November 18). NASA. Retrieved 23 February 2010 from http://exploration.grc.nasa.gov/education/rocket/rockpart.html

Rocket principles. (n.d.). NASA. Retrieved 7 February 2010 from http://quest.nasa.gov/space/teachers/rockets/principles.html

Rocket weight. (2005, December 20). NASA. Retrieved 22 February 2010 from http://exploration.grc.nasa.gov/education/rocket/rktwt1.html

Scalars and vectors. (2005, November 18). NASA. Retrieved 23 February 2010 from http://exploration.grc.nasa.gov/education/rocket/vectors.html

Shearer, D., & Vogt, G. L. (2008). *Rockets.* NASA. Retrieved 19 February 2010 from http://www.nasa.gov/pdf/280754main_Rockets.Guide.pdf

Shuttle basics. (2006, March 5). NASA. Retrieved 23 February 2010 from http://www.nasa.gov/returntoflight/system/system_STS.html

Solid rocket boosters. (2000, August 31). NASA. Retrieved 9 February 2010 from http://science.ksc.nasa.gov/shuttle/technology/sts-newsref/srb.html

Solid rocket engines. (2008, July 11). NASA. Retrieved 10 February 2010 from http://www.grc.nasa.gov/WWW/K-12/airplane/srockth.html

Space exploration. (2007, November 29). NASA. Retrieved 19 February 2010 from http://www.nasa .gov/worldbook/space_exploration_worldbook.html

Sputnik 1. (2009, November 23). NASA. Retrieved 19 February 2010 from http://nssdc.gsfc.nasa .gov/nmc/spacecraftDisplay.do?id=1957-001B

Water rocket. (2009, August 20). NASA. Retrieved 8 February 2010 from http://www.grc.nasa.gov/ WWW/K-12/rocket/rktbot.html

The weight equation. (2010, January 20). NASA. Retrieved 24 February 2010 from http://www.grc .nasa.gov/WWW/K-12/airplane/wteq.html

Welcome to the beginner's guide to rockets. (2009, August 21). NASA. Retrieved 8 February 2010 from http://www.grc.nasa.gov/WWW/K-12/rocket/bgmr.html

When was the first rocket launched into space? (n.d.). NASA. Retrieved 5 February 2010 from http://coolcosmos.ipac.caltech.edu/cosmic_kids/AskKids/v2rocket.shtml

World Book *at NASA: Rocket*. (2007, November 29). NASA. Retrieved 5 February 2010 from http://www.nasa.gov/worldbook/rocket_worldbook.html

Work done by a gas. (2008, July 11). NASA. Retrieved 11 February 2010 from http://www.grc.nasa .gov/WWW/K-12/rocket/work2.html

CHAPTER 11, Lesson 2   Propulsion and Launch Vehicles

Ares: *NASA's new rockets get names*. (2008, November 26). NASA. Retrieved 15 February 2010 from http://www.nasa.gov/mission_pages/constellation/ares/ares_naming.html

Astronomy picture of the day. (2000, July 18). NASA. Retrieved 15 February 2010 from http://apod .nasa.gov/apod/ap000718.html

Biographical data: Stephanie D. Wilson. (2010, March). NASA. Retrieved 9 March 2010 from http://www.jsc.nasa.gov/Bios/htmlbios/wilson.html

Brief history of rockets. (n.d.). NASA. Retrieved 13 February 2010 from http://quest.nasa.gov/space/ teachers/rockets/history.html

Chapter 14: Launch phase. (n.d.). NASA/Jet Propulsion Laboratory. Retrieved 16 February 2010 from http://www2.jpl.nasa.gov/basics/bsf14-1.php

Countdown 101. (2009). NASA. Retrieved 7 April 2010 from http://www.nasa.gov/mission_pages/ shuttle/launch/countdown101.html

Custom-made blankets for the world-class observatory. (2007, December 17). NASA. Retrieved 10 March 2010 from http://www.nasa.gov/mission_pages/hubble/servicing/series/hst _blankets.html

Delta IV *overview*. (2010). Boeing. Retrieved 9 March 2010 from http://www.boeing.com/defense -space/space/delta/delta4/delta4.htm

Delta *history*. (n.d.). Boeing. Retrieved 16 February 2010 from http://www.boeing.com/defense -space/space/bls/deltaHistory.html

Dr. Wernher von Braun: First center director, July 1, 1960–Jan. 27, 1970. (n.d.). MSFC History Office: Marshall Space Flight Center. Retrieved 4 March 2010 from http://history.msfc.nasa.gov/ vonbraun/bio.html

From a different view. (2009, September 15). NASA. Retrieved 11 March 2010 from http://www .nasa.gov/audience/forstudents/9-12/features/from-a-different-view.html

Guillemette, R. (n.d.). Atlas. U.S. Centennial of Flight Commission. Retrieved 16 February 2010 from http://www.centennialofflight.gov/essay/SPACEFLIGHT/Atlas/SP10.htm

Guillemette, R. (n.d.). *The* Titan *launch vehicle*. U.S. Centennial of Flight Commission. Retrieved 16 February 2010 from http://www.centennialofflight.gov/essay/SPACEFLIGHT/titan/SP11.htm

Heppenheimer, T. A. (n.d.). Thor, Agena, *and* Delta. U.S. Centennial of Flight Commission. Retrieved 16 February 2010 from http://www.centennialofflight.gov/essay/SPACEFLIGHT/delta/SP9.htm

ILS Proton *successfully launches DIRECTV 12 satellite for DIRECTV; 7th ILS* Proton *mission of 2009*. (2009, December 29). International Launch Services. Retrieved 15 February 2010 from http://www.ilslaunch.com/news-122909

Initial Soviet reaction to Sputnik 1 *launch*. (n.d.). NASA. Retrieved 15 February 2010 from http://history.nasa.gov/sputnik/harford.html

Konstantin E. Tsiolkovsky. (2004, March 12). ALLSTAR Network. Retrieved 15 February 2010 from http://www.allstar.fiu.edu/aero/tsiolkovsky.htm

Konstantin E. Tsiolkovsky. (2010). New Mexico Museum of Space History. Retrieved 4 March 2010 from http://www.nmspacemuseum.org/halloffame/detail.php?id=27

Launch vehicle parts. (2009, January 23). NASA. Retrieved 10 March 2010 from http://marsrover.nasa.gov/mission/launch_vehicle.html

Launch vehicle: Payload fairing. (2009, January 23). NASA. Retrieved 10 March 2010 from http://marsrover.nasa.gov/mission/launch_payload.html

Launch vehicles. (2009, July 21). European Space Agency. Retrieved 15 February 2010 from http://www.esa.int/esaMI/Launchers_Access_to_Space/index.html

Lethbridge, C. (2000). *History of rocketry: Ancient times through the 17th century*. Spaceline, Inc. Retrieved 14 February 2010 from http://www.spacearium.com/special/spaceline/spaceline.org/history/1.html

NASA's new gamma ray satellite currently lodging in a comfortable "clean room." (2007, June 26). NASA. Retrieved 10 March 2010 from http://www.nasa.gov/vision/universe/starsgalaxies/gamma_cleanroom.html

R-36 / SS-9 SCARP. (2000, July 29). Federation of American Scientists. Retrieved 8 March 2010 from http://www.fas.org/nuke/guide/russia/icbm/r-36.htm

R-36M / SS-18 Satan. (2000, July 29). Federation of American Scientists. Retrieved 8 March 2010 from http://www.fas.org/nuke/guide/russia/icbm/r-36m.htm

Robert H. Goddard: American rocket pioneer. (n.d.). NASA Goddard Space Flight Center. Retrieved 4 March 2010 from http://www.gsfc.nasa.gov/gsfc/service/gallery/fact_sheets/general/goddard/goddard.htm

Rocket history. (2004, March 12). ALLSTAR Network. Retrieved 15 February 2010 from http://www.allstar.fiu.edu/aero/Rock_Hist1.html

Rocket, solid-fuel, Hale, 24-pounder. (n.d.). Smithsonian: National Air and Space Museum. Retrieved 13 February 2010 from http://www.nasm.si.edu/collections/artifact.cfm?id=A19790727000

Rockets for the Moon race (1960s). (n.d.). MSFC History Office: Marshall Space Flight Center. Retrieved 9 March 2010 from http://history.msfc.nasa.gov/rocketry/tl7.html

Rockets for warfare (18th through 19th centuries). (n.d.). MSFC History Office: Marshall Space Flight Center. Retrieved 13 February 2010 from http://history.msfc.nasa.gov/rocketry/tl2.html

Russian launch vehicles. (n.d.). Century of Flight. Retrieved 15 February 2010 from http://www.century-of-flight.net/Aviation%20history/space/Soviet%20Russian%20Launch%20Vehicles.htm

Saturn V: *America's Moon rocket*. (n.d.). Smithsonian: National Air and Space Museum. Retrieved 16 February 2010 from http://www.nasm.si.edu/exhibitions/gal114/SpaceRace/sec300/sec384.htm

Shearer, D., & Vogt, G. L. (2008). *Rockets*. NASA. Retrieved 19 February 2010 from http://www.nasa.gov/pdf/280754main_Rockets.Guide.pdf

Siddiqi, A. (n.d.). *Foreign launch vehicles*. U.S. Centennial of Flight Commission. Retrieved 16 February 2010 from http://www.centennialofflight.gov/essay/SPACEFLIGHT/foreign_launch_vehicles/SP15.htm

Soviet/Russian launch vehicles. (n.d.). U.S. Centennial of Flight Commission. Retrieved 16 February 2010 from http://www.centennialofflight.gov/essay/SPACEFLIGHT/Soviet_launch_vehicles/SP14.htm

Space station assembly: Russian Soyuz TMA *spacecraft details*. (2007, November 23). NASA. Retrieved 15 February 2010 from http://www.nasa.gov/mission_pages/station/structure/elements/soyuz/spacecraft_detail.html

Stephanie Wilson: Becoming an astronaut kicking and swimming. (2006, November 20). NASA. Retrieved 9 March 2010 from http://www.nasa.gov/astronauts/s_wilson_profile.html

Stephanie Wilson, Soaring in Discovery. (2006, August 15). NASA. Retrieved 15 February 2010 from http://solarsystem.nasa.gov/people/profile.cfm?Code=WilsonS

Stern, D. P. (2004, September 24). *The evolution of the rocket*. NASA Goddard Space Flight Center. Retrieved 13 February 2010 from http://www-istp.gsfc.nasa.gov/stargaze/Srockhis.htm

Stern, D. P. (2005, January 14). *Robert Goddard and his rockets*. NASA Goddard Space Flight Center. Retrieved 15 February 2010 from http://www-istp.gsfc.nasa.gov/stargaze/Sgoddard.htm

STS-72: Endeavour: OAST-Flyer/Space Flyer Unit (SFU). (1995, December). Kennedy Space Center. Retrieved 10 March 2010 from http://www-pao.ksc.nasa.gov/kscpao/nasafact/72facts2.htm

Voskhod 1. (2009, November 23). NASA. Retrieved 15 February 2010 from http://nssdc.gsfc.nasa.gov/nmc/spacecraftDisplay.do?id=1964-065A

William Hale. (2010). New Mexico Museum of Space History. Retrieved 15 February 2010 from http://www.nmspacemuseum.org/halloffame/detail.php?id=150

CHAPTER 12, Lesson 1   Developing Robots for Space

Canada lends a mechanized hand for space station assembly. (2000, August 21). *Cable News Network*. Retrieved 23 February 2010 from http://archives.cnn.com/2000/TECH/space/08/21/canada.hand/index.html

A countdown of countdowns: The space shuttle's finale. (2010, March 12). NASA. Retrieved 18 March 2010 from http://www.nasa.gov/mission_pages/shuttle/behindscenes/shuttle_countdowns.html

Dick, S. J., & Launius, R. D. (Eds.). (2006). *Critical issues in the history of spaceflight*. NASA. Retrieved 23 February 2010 from http://history.nasa.gov/SP-2006-4702/frontmatter.pdf

Dr. Robert H. Goddard, American rocketry pioneer. (2009, August 14). NASA. Retrieved 20 February 2010 from http://www.nasa.gov/centers/goddard/about/dr_goddard.html

Educational Brief: Humans and Robots. (2001). NASA.

ERA: European Robotic Arm. (2009, January 16). European Space Agency. Retrieved 23 February 2010 from http://www.esa.int/esaHS/ESAQEI0VMOC_iss_0.html

Goddard information. (2009, October 1). NASA. Retrieved 20 February 2010 from http://www.nasa.gov/centers/goddard/about/info/faq.html

Harford, J. J. (n.d.). *Korolev's triple play: Sputniks 1, 2, and 3*. NASA. Retrieved 21 February 2010 from http://history.nasa.gov/sputnik/harford.html

Her time for Discovery. (2006, November 15). NASA. Retrieved 12 March 2010 from http://www.nasa.gov/astronauts/j_higginbotham_profile.html

History of the Explorers program. (n.d.). NASA. Retrieved 20 February 2010 from http://explorers.gsfc.nasa.gov/history.html

International Space Station. (2007, November 29). NASA. Retrieved 23 February 2010 from http://www.nasa.gov/worldbook/intspacestation_worldbook.html

The International Space Station: The making of an orbital outpost and Canada's role. (2008, February 6). *Canadian Broadcasting Corporation*. Retrieved 23 February 2010 from http://www.cbc.ca/news/background/space/iss.html

ISEE-3. (2009, November 23). NASA. Retrieved 18 March 2010 from http://nssdc.gsfc.nasa.gov/nmc/spacecraftDisplay.do?id=1978-079A

Mars 2. (2009, November 23). NASA. Retrieved 17 March 2010 from http://nssdc.gsfc.nasa.gov/nmc/spacecraftDisplay.do?id=1971-045A

Missions to our Solar System: ISEE-3/ICE. (2009, January 23). NASA. Retrieved 23 February 2010 from http://solarsystem.nasa.gov/missions/profile.cfm?MCode=ISEEICE

NASA TV's This Week @ NASA, December 4. (2009, December 4). NASA. Retrieved 20 February 2010 from http://www.nasa.gov/multimedia/podcasting/twan_12_04_09.html

NASA's WISE eye on the universe begins all-sky survey mission. (2009, December 14). NASA. Retrieved 23 February 2010 from http://www.nasa.gov/mission_pages/WISE/news/wise20091214.html

Preflight interview: Joan Higginbotham. (2006, November 3). NASA. Retrieved 12 March 2010 from http://www.nasa.gov/mission_pages/shuttle/shuttlemissions/sts116/interview_higginbotham.html

R2. (2008, March 13). NASA. Retrieved 12 March 2010 from http://robonaut.jsc.nasa.gov/

Robonaut 1. (2008, March 13). NASA. Retrieved 12 March 2010 from http://robonaut.jsc.nasa.gov/R1/index.asp

The robotic exploration of space: An interactive journey through time. (2008, August 19). NASA. Retrieved 20 February 2010 from http://sse.jpl.nasa.gov/history/index.cfm

Robotics. (2010, February 25). NASA. Retrieved 25 February 2010 from http://www.nasa.gov/audience/foreducators/robotics/home/index.html

Robotics—multimedia. (2009, November 24). NASA. Retrieved 16 March 2010 from http://www.nasa.gov/audience/foreducators/robotics/multimedia/index.html

Robotics: What is robotics? (2009, November 9). NASA. Retrieved 21 February 2010 from http://www.nasa.gov/audience/foreducators/robotics/home/what_is_robotics_58.html

Schneider, M. (2006, December 3). *Discovery's* 7 astronauts a diverse bunch. *USA Today.* Retrieved 20 February 2010 from http://www.usatoday.com/tech/science/space/2006-12-03-discovery-crew_x.htm

Smith, W. (2006, February 22). Explorer *series of spacecraft.* NASA: NASA History Division. Retrieved 16 March 2010 from http://history.nasa.gov/explorer.html

Solar System exploration: Explorer 1. (2009, March 14). NASA. Retrieved 20 February 2010 from http://solarsystem.nasa.gov/missions/profile.cfm?Sort=Alpha&Alias=Explorer%2001&Letter=E&Display=ReadMore

Solar System exploration: Mariner 9. (2009, September 11). NASA. Retrieved 23 February 2010 from http://solarsystem.nasa.gov/missions/profile.cfm?Sort=Alpha&Alias=Mariner%2009&Letter=M&Display=ReadMore

Solar System exploration: MESSENGER: *(Mercury).* (2009, September 10). NASA. Retrieved 19 March 2010 from http://sse.jpl.nasa.gov/missions/profile.cfm?Sort=Alpha&Alias=MESSENGER&Target=Mercury&CFID=54577241&CFTOKEN=3c01e8c921ecbaf1-7611B5C3-EEEE-CD28-C3C1FE4D7D55E8EE

Solar System exploration: New Horizons: *(Kuiper Belt & Oort Cloud).* (2009, September 5). NASA. Retrieved 19 March 2010 from http://sse.jpl.nasa.gov/missions/profile.cfm?Sort=Alpha&Alias=New%20Horizons&Target=KBOs&CFID=54577241&CFTOKEN=3c01e8c921ecbaf1-7611B5C3-EEEE-CD28-C3C1FE4D7D55E8EE

Solar System exploration: Ranger 3. (2009, March 31). NASA. Retrieved 16 March 2010 from http://solarsystem.jpl.nasa.gov/missions/profile.cfm?Sort=Chron&StartYear=1960&EndYear=1969&MCode=Ranger_03&CFID=54413663&CFTOKEN=e3810206a0eb1761-685CA7AB-09CC-CE39-A365B5E29EE439DF

Solar System exploration: Ranger 8. (2009, April 29). NASA. Retrieved 16 March 2010 from http://solarsystem.jpl.nasa.gov/missions/profile.cfm?Sort=Chron&StartYear=1960&EndYear=1969&MCode=Ranger_08&CFID=54413663&CFTOKEN=e3810206a0eb1761-685CA7AB-09CC-CE39-A365B5E29EE439DF

Solar System exploration: Ranger 9. (2010, March 9). NASA. Retrieved 16 March 2010 from http://solarsystem.jpl.nasa.gov/missions/profile.cfm?Sort=Chron&StartYear=1960&EndYear=1969&MCode=Ranger_09&CFID=54413663&CFTOKEN=e3810206a0eb1761-685CA7AB-09CC-CE39-A365B5E29EE439DF

Solar System exploration: Surveyor 1. (2009, August 17). NASA. Retrieved 21 February 2010 from http://solarsystem.nasa.gov/missions/profile.cfm?Sort=Alpha&Letter=S&Alias=Surveyor%2001

Solar System exploration: Surveyor 2. (2009, August 18). NASA. Retrieved 17 March 2010 from http://solarsystem.nasa.gov/missions/profile.cfm?Sort=Alpha&Letter=S&Alias=Surveyor%2002&CFID=54475574&CFTOKEN=5a4dcdb32632e186-6D7DFFD2-A605-5B7D-44922E580206AFA3

Solar System exploration: Surveyor 3. (2009, August 20). NASA. Retrieved 21 February 2010 from http://solarsystem.nasa.gov/missions/profile.cfm?Sort=Alpha&Letter=S&Alias=Surveyor%2003

Solar System exploration: Surveyor 4. (2009, October 10). NASA. Retrieved 21 February 2010 from http://solarsystem.nasa.gov/missions/profile.cfm?Sort=Alpha&Alias=Surveyor%2004&Letter=S&Display=ReadMore

Solar System exploration: Surveyor 5. (2009, October 1). NASA. Retrieved 21 February 2010 from http://solarsystem.nasa.gov/missions/profile.cfm?Sort=Alpha&Alias=Surveyor%2005 &Letter=S&Display=ReadMore

Solar System exploration: Surveyor 6. (2009, October 10). NASA. Retrieved 21 February 2010 from http://solarsystem.nasa.gov/missions/profile.cfm?Sort=Alpha&Alias=Surveyor%2006 &Letter=S&Display=ReadMore

Solar System exploration: Surveyor 7. (2009, October 11). NASA. Retrieved 17 March 2010 from http://solarsystem.nasa.gov/missions/profile.cfm?Sort=Alpha&Letter=S&Alias=Surveyor %2007&CFID=54475574&CFTOKEN=5a4dcdb32632e186-6D7DFFD2-A605-5B7D -44922E580206AFA3

Solar System exploration: T. Adrian Hill, software engineer. (2008, December 30). NASA. Retrieved 19 March 2010 from http://sse.jpl.nasa.gov/people/profile.cfm?Code=HillT

Solar System exploration: WISE *medley.* (2010, February 19). NASA. Retrieved 23 February 2010 from http://solarsystem.nasa.gov/news/display.cfm?News_ID=33537

Space station assembly: Canadarm2 and the Mobile Servicing System. (2007, November 23). NASA. Retrieved 23 February 2010 from http://www.nasa.gov/mission_pages/station/structure/ elements/mss.html

Space station assembly: Canadarm2 and the Mobile Servicing System: Subsystems. (2007, November 23). NASA. Retrieved 18 March 2010 from http://www.nasa.gov/mission_pages/station/ structure/elements/subsystems.html

Space station extravehicular activity: A new generation of space robotics. (2004, March 4). NASA. Retrieved 23 February 2010 from http://spaceflight.nasa.gov/station/eva/robotics.html

STS-127 MCC status report #17. (2009, July 23). NASA. Retrieved 18 March 2010 from http://www .nasa.gov/mission_pages/shuttle/shuttlemissions/sts127/news/STS-127-17.html

STS-132 mission information. (2010, March 4). NASA. Retrieved 18 March 2010 from http://www .nasa.gov/mission_pages/shuttle/shuttlemissions/sts132/index.html

Types of robots. (2003). ROVer Ranch. NASA. Retrieved 7 April 2010 from http://prime.jsc.nasa.gov/ ROV/types.html

CHAPTER 12, Lesson 2   The Mars Rover and Beyond

Amos, J. (2010, January 26). NASA accepts *Spirit* Mars rover "stuck for good." *British Broadcasting Corporation.* Retrieved 1 March 2010 from http://news.bbc.co.uk/2/hi/8481798.stm

Back to the future on Mars. (2000, July 28). NASA. Retrieved 1 March 2010 from http://science.nasa .gov/headlines/y2000/ast28jul_2m.htm

Bedrock in Mars' Gusev Crater hints at watery past. (2004, August 18). NASA. Retrieved 1 March 2010 from http://marsrover.nasa.gov/newsroom/pressreleases/20040818a.html

Curriculum vitae: Spirit, Mars Exploration Robot A. (2010, January 26). NASA. Retrieved 3 March 2010 from http://marsrovers.jpl.nasa.gov/spotlight/20100126a.html

Design reference missions and the decadal survey: Guiding the future of NASA missions. (2009, March 27). NASA. Retrieved 3 March 2010 from http://sse.jpl.nasa.gov/missions/future2.cfm

Destination: Gusev Crater. (2003, December 30). NASA. Retrieved 1 March 2010 from http:// science.nasa.gov/headlines/y2003/30dec_gusevcrater.htm

Exploration at NASA. (2010, March 3). NASA. Retrieved 3 March 2010 from http://www.nasa.gov/ exploration/home/index.html

The giant planet story is the story of the Solar System. (2009, March 3). NASA. Retrieved 3 March 2010 from http://www.nasa.gov/mission_pages/juno/overview/index.html

Girl with dreams names Mars rovers "Spirit"and "Opportunity." (2003, June 8). NASA. Retrieved 1 March 2010 from http://marsrover.nasa.gov/newsroom/pressreleases/20030608a.html

Healthy Spirit *cleans a Mars rock;* Opportunity *rolls.* (2004, February 6). NASA. Retrieved 2 March 2010 from http://marsrovers.jpl.nasa.gov/newsroom/pressreleases/20040206a.html

Key vocabulary. (2010). Moorehead Planetarium and Science Center. University of North Carolina. Retrieved 13 April 2010 from http://www.moreheadplanetarium.org/index.cfm?fuseaction=page &filename=EITS_vocab.html

Mars Exploration Rover *mission: Goals*. (2007, July 12). NASA. Retrieved 1 March 2010 from http://marsrovers.jpl.nasa.gov/science/goals.html

Mars Exploration Rover *status report concern increasing about* Opportunity. (2007, July 31). NASA. Retrieved 3 March 2010 from http://marsrovers.jpl.nasa.gov/newsroom/pressreleases/ 20070731a.html

Mars Odyssey *Themis: Discoveries*. (n.d.). Arizona State University: School of Earth & Space Exploration. Retrieved 1 March 2010 from http://themis2.mars.asu.edu/discoveries-meridiani

Mars rover Spirit *mission status*. (2003, November 4). NASA. Retrieved 1 March 2010 from http://marsrover.nasa.gov/newsroom/pressreleases/20031104a.html

Mission to Mars: Opportunity. (2009, January 23). NASA. Retrieved 3 March 2010 from http:// solarsystem.nasa.gov/missions/profile.cfm?Sort=Target&Target=Mars&MCode=MER_B

NASA and ESA prioritize outer planet missions. (2009, February 18). NASA. Retrieved 29 March 2010 from http://www.nasa.gov/topics/solarsystem/features/20090218.html

NASA Facts: Mars Exploration Rover. (2004, October). NASA. Retrieved 1 March 2010 from http://marsrovers.jpl.nasa.gov/newsroom/

NASA mission classes. (2009, March 27). NASA. Retrieved 3 March 2010 from http://sse.jpl.nasa .gov/missions/Mission_Classes.cfm

NASA's durable Spirit *sends intriguing new images from Mars*. (2005, September 1). NASA. Retrieved 2 March 2010 from http://marsrovers.jpl.nasa.gov/newsroom/pressreleases/20050901a.html

NASA's Mars rovers continue to explore and amaze. (2005, December 5). NASA. Retrieved 2 March 2010 from http://marsrovers.jpl.nasa.gov/newsroom/pressreleases/20051205a.html

Orbital eyes picked Mars rover Opportunity's *landing site*. (2004, January 24). Arizona State University: School of Earth & Space Exploration. Retrieved 1 March 2010 from http://themis .asu.edu/news/orbital-eyes-picked-mars-rover-opportunitys-landing-site

Possible meteorite in "Columbia Hills" on Mars. (2006, June 9). NASA. Retrieved 2 March 2010 from http://www.jpl.nasa.gov/missions/mer/images.cfm?id=1962

Programs and missions: Spirit *and* Opportunity. (n.d.). Jet Propulsion Laboratory. Retrieved 1 March 2010 from http://marsprogram.jpl.nasa.gov/programmissions/missions/present/2003/

Public events mark Mars rovers' five-year anniversary. (2009, January 12). NASA. Retrieved 1 March 2010 from http://marsrover.nasa.gov/newsroom/pressreleases/20090112a.html

Rocks tell stories in reports of Spirit's *first 90 Martian days*. (2004, August 5). NASA. Retrieved 2 March 2010 from http://marsrovers.jpl.nasa.gov/newsroom/pressreleases/20040805a.html

Scientists thrilled to see layers in Mars rocks near Opportunity. (2004, January 27). NASA. Retrieved 3 March 2010 from http://marsrovers.nasa.gov/newsroom/pressreleases/20040127a.html

Solar System strategic exploration plans. (2009, March 27). NASA. Retrieved 3 March 2010 from http://sse.jpl.nasa.gov/missions/future1.cfm

Spirit *discovers "new" highest peak in "Columbia Hills."* (2006, March 2). NASA. Retrieved 2 March 2010 from http://marsrover.nasa.gov/spotlight/20060302.html

Spirit *finds multi-layer hints of past water at Mars' Gusev site*. (2004, April 1). NASA: Jet Propulsion Laboratory. Retrieved 2 March 2010 from http://www.jpl.nasa.gov/releases/2004/93.cfm

CHAPTER 13, Lesson 1   Private Industry Enters Space

181 things to do on the Moon. (2007, February 2). NASA. Retrieved 11 March 2010 from http:// science.nasa.gov/headlines/y2007/02feb_181.htm

1960s—1970s. (n.d.). Air Force Space Command. Retrieved 23 March 2010 from http://www.afspc .af.mil/heritage/1960s-1970s.asp

1962: Satellite transmission. (2010). AT&T. Retrieved 9 March 2010 from http://www.corp.att.com/ attlabs/reputation/timeline/62trans.html

The 1970s: A decade of expansion. (2010). Intelsat. Retrieved 9 March 2010 from http://www.intelsat
.com/about-us/history/intelsat-1970s.asp

Anousheh Ansari dreams of stars. (2010, March 6). *National Public Radio.* Retrieved 12 March 2010
from http://www.npr.org/blogs/sundaysoapbox/2010/03/anousheh_ansari_from_tehran_to_1.html

Anousheh Ansari: First female private space explorer and first space ambassador. (2007). Anousheh
Ansari. Retrieved 12 March 2010 from http://www.anoushehansari.com/about.php

Asteroids. (n.d.). NASA. Retrieved 25 March 2010 from http://nssdc.gsfc.nasa.gov/planetary/text/
asteroids.txt

Bigelow Aerospace website. (2010). Retrieved 30 March 2010 from http://www.bigelowaerospace
.com/

Bonsor, K. (n.d.). *How asteroid mining will work.* HowStuffWorks. Retrieved 11 March 2010
from http://science.howstuffworks.com/asteroid-mining1.htm

Comsat. (2009). Encyclopædia Britannica. In *Encyclopædia Britannica 2009 Deluxe Edition.*
Chicago: Encyclopædia Britannica.

Dunn, M. (2001, April 27). Space-bound millionaire ready to lead charge into orbit. *Ludington
(Michigan) Daily News.* Retrieved 25 March 2010 from http://news.google.com/newspapers?nid
=110&dat=20010427&id=Rd4OAAAAIBAJ&sjid=nFUDAAAAIBAJ&pg=6943,2640837

Early Bird: *World's first commercial communications satellite.* (2010). Boeing. Retrieved 6 March 2010
from http://www.boeing.com/defense-space/space/bss/factsheets/376/earlybird/ebird.html

Got $200K to fly above the Earth? (2010, March 23). *CBS News.* Retrieved 24 March 2010 from
http://www.cbsnews.com/stories/2010/03/23/tech/main6326361.shtml

Harwood, W. (2006, September 28). Soyuz *capsule returns from space with station crew.* Spaceflight
Now. Retrieved 24 March 2010 from http://spaceflightnow.com/station/exp13/060928landing
.html

Heppenheimer, T. A. (n.d.). *Comsat and Intelsat.* U.S. Centennial of Flight Commission. Retrieved
9 March 2010 from http://www.centennialofflight.gov/essay/SPACEFLIGHT/intelsat/SP44.htm

History of AT&T and television. (2010). AT&T. Retrieved 6 March 2010 from http://www.corp.att
.com/history/television/

Inside the mind of a space tycoon. (2001, May 6). *MSNBC.* Retrieved 11 March 2010 from
http://www.msnbc.msn.com/id/3077967/

Intelsat. (2009). Encyclopædia Britannica. In *Encyclopædia Britannica 2009 Deluxe Edition.*
Chicago: Encyclopædia Britannica.

Jets and Moon rockets: 1957—1970: Hughes Aircraft Co. … satellites and Surveyors. (2010). Boeing.
Retrieved 9 March 2010 from http://www.boeing.com/history/narrative/n069hug.html

Johnson, J., Jr. (2009, December 8). Rutan and Branson make a giant leap for space tourism.
Los Angeles Times. Retrieved 11 March 2010 from http://www.latimes.com/news/local/la-sci
-virgin8-2009dec08,0,5479695.story

Kruzel, J. J. (2009, November 5). General calls for focus on protecting satellites. *American
Forces Press Service.* Retrieved 9 March 2009 from http://www.afspc.af.mil/news/story_print
.asp?id=123176330

Lehman Brothers collection—Twentieth-century business archives: RCA Corporation. (2010). Harvard
Business School. Retrieved 6 March 2010 from http://www.library.hbs.edu/hc/lehman/chrono
.html?company=rca_corporation

McDill, S. (2009, November 2). Space hotel says it's on schedule to open in 2012. *Reuters.*
Retrieved 14 April 2010 from http://www.reuters.com/article/idUSTRE5A151N20091102

Milstein, M. (2008, May 12). NASA makes space U-turn, opening arms to private industry.
Popular Mechanics. Retrieved 11 March 2010 from http://www.popularmechanics.com/science/
air_space/4263233.html

Missions to asteroids: NEAR Shoemaker. (2009, March 27). NASA. Retrieved 25 March 2010
from http://solarsystem.nasa.gov/missions/profile.cfm?MCode=NEAR&Display=ReadMore

Near-Earth objects as future resources. (2010). NASA. Retrieved 11 March 2010 from http://neo.jpl
.nasa.gov/neo/resource.html

Orbital website. (2010). Retrieved 30 March 2010 from http://www.orbital.com/

Our history: Making history is our legacy. (2010). Intelsat. Retrieved 9 March 2010 from http://www
.intelsat.com/about-us/history/

Piazza, D. (2007, October). Telstar *covers*. Smithsonian: National Postal Museum. Retrieved
23 March 2010 from http://www.postalmuseum.si.edu/museum/1d_Telstar_Covers.html

Pickup, A. (2009, February 18). Spacewatch: *Iridium* and *Cosmos* crash. *The Guardian*. Retrieved
9 March 2010 from http://www.guardian.co.uk/science/2009/feb/18/spacewatch-satellite-collision

PlanetSpace website. (2010). Retrieved 30 March 2010 from http://www.planetspace.org/

Relay. (2010, January 5). NASA. Retrieved 6 March 2010 from http://www.nasa.gov/centers/
goddard/missions/relay.html

Schwartz, J. (2008, December 29). With U.S. help, private space companies press their case:
Why not us? *The New York Times*. Retrieved 11 March 2010 from http://www.nytimes.com/
2008/12/30/science/30spacside.html?_r=2

Solovyov, D. (2010, March 3). Russia halts space tours as U.S. retires shuttle. *Reuters*. Retrieved
9 March 2010 from http://www.reuters.com/article/idUSTRE6223VF20100303

SpaceDev website. (2010). Retrieved 30 March 2010 from http://www.spacedev.com/

Space Exploration Technologies website. (2010). Retrieved 30 March 2010 from http://www
.spacex.com/

Spaceflight participant Anousheh Ansari. (n.d.). NASA. Retrieved 12 March 2010 from http://www
.nasa.gov/pdf/157366main_ansari.pdf

Space tourism a reality by 2012. (2009, November 5). *Fox News*. Retrieved 11 March 2010 from
http://www.foxnews.com/scitech/2009/11/05/space-tourism-reality/

Telstar 1. (2009, November 23). NASA. Retrieved 9 March 2010 from http://nssdc.gsfc.nasa.gov/
nmc/spacecraftDisplay.do?id=1962-029A

Tyler, P. E. (2001, May 7). Space tourist, back from 'Paradise,' lands on steppes. *The New York
Times*. Retrieved 11 March 2010 from http://www.nytimes.com/2001/05/07/world/space-tourist
-back-from-paradise-lands-on-steppes.html

Whalen, D. J. (2007, July 27). *Communications satellites: Making the global village possible.* NASA.
Retrieved 6 March 2010 from http://history.nasa.gov/satcomhistory.html

CHAPTER 13, Lesson 2   Space in Your Daily Life

Apollo-era life rafts save hundreds of sailors. (2010, February 15). NASA *Spinoff*. Retrieved
12 March 2010 from http://www.sti.nasa.gov/tto/Spinoff2009/ps_3.html

A brief history of satellite navigation. (1995, June 13). Stanford University News Service. Retrieved
7 March 2010 from http://news.stanford.edu/pr/95/950613Arc5183.html

Comstock, D.A., & Lockney, D. (2007). *NASA's legacy of technology transfer and prospects for future
benefits.* NASA. Retrieved 6 March 2010 from http://www.sti.nasa.gov/tto/hist_techtransfer.pdf

DePriest, D. (2002). *How your GPS works.* Retrieved 7 March 2010 from http://www.gpsinformation
.org/dale/theory.htm

Dunn, C. A. (2008, Fall). NASA's Inventions and Contributions Board: A historical perspective.
ASK Magazine, (32), 68–69. Retrieved 14 April 2010 from http://askmagazine.nasa.gov/pdf/
pdf_whole/NASA_APPEL_ASK_32_Fall_2008.pdf

Garay, R. (n.d.). *Direct Broadcast Satellite: Satellite delivery technology.* The Museum of Broadcast
Communications. Retrieved 6 March 2010 from http://www.museum.tv/eotvsection.
php?entrycode=directbroadc

GPS distraction: Convenience over safety? An investigation into GPSs. (2010, January 15).
Canadian Broadcasting Corporation: Marketplace. Retrieved 7 March 2010 from http://www.cbc.ca/
marketplace/2010/gps_distraction/main.html

Kendrick, J. (2009, September 30). *iZUP—Using phone GPS to stop cell phone use while driving.*
JKOntheRun. Retrieved 7 March 2010 from http://jkontherun.com/2009/09/30/izup-using
-phone-gps-to-stop-cell-phone-use-while-driving/

Mike Massimino becomes the first to "tweet" from space. (2009, May 13). NASA. Retrieved 11 March 2010 from http://www.nasa.gov/topics/people/features/massimino_tweet.html

Moskowitz, C. (2010, March 5). *NASA launches new high-tech weather satellite.* Space.com. Retrieved 6 March 2010 from http://www.space.com/missionlaunches/goes-p-weather-satellite-launch-100304.html

NASA extends the World Wide Web out into space. (2010, January 22). NASA. Retrieved 11 March 2010 from http://www.nasa.gov/home/hqnews/2010/jan/HQ_M10-011_Hawaii221169.html

NASA satellites help improve ocean weather forecasts. (2004, January 29). NASA: Goddard Space Flight Center. Retrieved 6 March 2010 from http://www.nasa.gov/centers/goddard/news/topstory/2004/0113forecastca.html

Ronca, D. (2010). *How space blankets work.* HowStuffWorks. Retrieved 29 March 2010 from http://adventure.howstuffworks.com/survival/gear/space-blanket.htm

Roos, D. (2010). *How GPS photo taggers work.* HowStuffWorks. Retrieved 29 March 2010 from http://electronics.howstuffworks.com/gadgets/travel/gps-photo-taggers.htm/printable

Spinoff frequently asked questions. (2010, February 15). NASA *Spinoff.* Retrieved 9 March 2010 from http://www.sti.nasa.gov/tto/spinfaq.htm

US satellite television subscribers top 20 million mark. (2003, August 20). SpaceRef.com. Retrieved 7 March 2010 from http://www.spaceref.com/news/viewpr.html?pid=12360

Wilson, J. R. (2008, August 27). *Space program benefits: NASA's positive impact on society.* NASA. Retrieved 7 March 2010 from http://www.nasa.gov/50th/50th_magazine/benefits.html

Woodford, C. (2009, June 2). *"Sat Nav" (GPS satellite navigation).* Explain that Stuff! Retrieved 7 March 2010 from http://www.explainthatstuff.com/howgpsworks.html

Glossary

A

accelerate—to change the speed and/or the direction of the motion of. (p. 38)

accretion disk—a rotating disk of gas orbiting a star, formed by material falling toward the star. (p. 157)

aerobraking—the use of a planet's atmosphere to slow an orbit and thereby lower a satellite closer to a planet. (p. 386)

aeronautical—anything related to the science, design, or operation of aircraft. (p. 199)

aft—the rear of a spacecraft or any other ship. (p. 309)

air lock—an airtight chamber, usually located between two regions of unequal pressure, in which air pressure can be regulated. (p. 274)

albedo—a celestial body's reflecting power, expressed as a ratio of reflected light to the total amount falling on the surface. (p. 364)

altitude—height measured as an angle above the horizon. (p. 57)

ALTO—assembly, test, and launch operations. (p. 470)

angularity—the sharpness of edges and corners. (p. 498)

Anik—the first domestic communications satellites. (p. 517)

aphelion—the point in a planet's orbit when it is farthest from the Sun. (p. 99)

apoapsis—when an orbiting body is farthest from the object it is orbiting. (p. 429)

apogee—farthest distance from the Earth. (p. 65)

apparent magnitude—the amount of light received from a celestial object. (p. 167)

asteroid belt—the region between Mars and Jupiter where most asteroids orbit. (p. 136)

astrometry—the branch of astronomy dealing with measurement of the positions and motions of celestial objects. (p. 155)

astronomical unit—the mean distance between the Earth and Sun, about 93 million miles (150 million km). (p. 134)

atrophy—a wasting away or shrinking of a body part, typically from lack of use. (p. 234)

attitude—a spacecraft's position, or angle, relative to the direction in which it is traveling. (p. 346)

aurora—light radiated in the upper atmosphere because of impacts from charged particles. (p. 59)

B

barred spiral galaxy—a spiral galaxy in which the spiral arms come from the ends of a bar through the nucleus, rather than from the nucleus itself. (p. 163)

bearing strength—the ability to support a certain amount of weight. (p. 485)

big bang—the theoretical initial explosion that began the universe's expansion. (p. 174)

binary star system—a pair of stars that revolve around each other. (p. 155)

black hole—an object whose escape velocity exceeds the speed of light. (p. 157)

brown dwarf—a star-like object that gives off light but lacks sufficient mass for nuclear reactions in its core. (p. 154)

budget—a sum of money set aside to spend for a specific purpose. (p. 189)

C

capture theory—the Moon is made up of Solar System debris captured by Earth. (p. 71)

cataract—a clouding of the lens of the eye. (p. 240)

celestial sphere—an imaginary sphere of heavenly objects that seems to center on the observer. (p. 9)

Cepheid variable stars—a special class of pulsating star used for accurate distance measurements. (p. 399)

chemical differentiation—the sinking of denser material toward the center of planets or other objects. (p. 52)

chromosphere—the region between the photosphere and the corona. (p. 82)

clean room—a workspace with a constant temperature and humidity and low levels of contaminants such as dust. (p. 470)

coma—the part of a comet's head made up of diffuse gas and dust. (p. 139)

combustible—flammable. (p. 263)

conduction—the transfer of energy in a solid by collisions between atoms and/or molecules. (p. 80)

conjunction—when two or more celestial bodies appear to be close to one another. (p. 489)

conservation of angular momentum—a law that says an object will spin more slowly as resistance increases and spin faster as resistance decreases. (p. 90)

constellation—an area of the sky containing a group of stars in a pattern. (p. 9)

continental drift—the gradual motion of the continents relative to one another. (p. 52)

contour map—a map that shows an area's different elevations. (p. 364)

convection—when the atoms of a warm liquid or gas move from one place to another. (p. 81)

core of the Earth—the central part of the Earth, made up of a solid inner core surrounded by a liquid outer core. (p. 52)

corona—the outermost portion of the Sun's atmosphere. (p. 82)

cosmonaut—astronauts in the space program of the Soviet Union and its successor state, Russia. (p. 210)

crust—the outer layer of Earth. (p. 51)

D

dark energy—an exotic form of energy whose negative pressure speeds up the expansion of the universe. (p. 172)

dark matter—matter that can be detected only by its gravitational interactions. (p. 172)

dark nebula—an interstellar molecular cloud whose dust blocks light from stars on the other side of it. (p. 169)

declination—the celestial sphere's equivalent to latitude on Earth. (p. 438)

deconditioning—the weakening of heart and lung function in microgravity. (p. 234)

density—the ratio of an object's mass to its volume. (p. 51)

deorbit—to cause to go out of orbit. (p. 348)

differential rotation—the phenomenon of different parts of a planet having different periods of rotation. (p. 117)

direct-broadcast satellite television service—broadcasts sent via satellite directly to consumers. (p. 532)

double planet theory—the Moon was formed at the same time as the Earth. (p. 70)

dynamo effect—the generation of magnetic fields due to circulating electric charges, such as in an electric generator. (p. 98)

E

ecliptic—the Sun's apparent path among the stars around the Earth. (p. 10)

ecliptic plane—the plane of a planet's orbit around the Sun. (p. 377)

electromagnetic spectrum—the entire array of electromagnetic waves. (p. 173)

ellipse—a geometrical shape of which every point (P) is the same total distance from two fixed points, or foci. (p. 28)

elliptical galaxy—a galaxy with a smooth spheroidal shape. (p. 165)

emission nebula—a cloud of interstellar gas receiving ultraviolet radiation and fluorescing as visible light. (p. 169)

end effector—a device located at the end of a robotic arm that can grasp and snare objects much as human hands do. (p. 480)

epicycle—the circular orbit of a planet, the center of which revolves around the Earth in another circle. (p. 16)

equilibrium—balance, neither noticeably contracting nor expanding. (p. 79)

equinox—the points when the Sun crosses the celestial equator. (p. 57)

exoplanet—planets that orbit stars outside the Solar System. (p. 148)

extravehicular activity—a spacewalk. (p. 258)

F

fireball—an extremely bright meteor. (p. 143)

fission—the splitting of an atom's nucleus. (p. 370)

fission theory—the Moon formed from material spun off from the Earth. (p. 70)

flight simulator—machines that duplicate what it's like to operate an airplane or a spacecraft. (p. 207)

fluorescence—the process of absorbing radiation of one frequency and re-emitting it at a lower frequency. (p. 169)

foreign national—someone who owes allegiance to a foreign country. (p. 313)

G

galactic corona—outer halo. (p. 153)

galaxy disk—the large, flat part of a spiral galaxy, rotating around its center. (p. 152)

gamma-ray burst—a short-lived burst of gamma-ray photons, the most energetic form of light. (p. 377)

geocentric—Earth-centered. (p. 33)

geodetic—related to measuring the Earth's shape, or the shape of another celestial body. (p. 364)

geostationary—orbiting at a speed and altitude that keeps satellites in the same place above the Earth at all times. (p. 289)

geosynchronous Earth orbit—an orbit around a planet or moon that places the satellite in the same place in the sky over a particular point on the surface each day. (p. 421)

globular cluster—a spherical group of up to hundreds of thousands of stars, found primarily in a galaxy's halo. (p. 152)

gyroscope—a device with a spinning wheel at its center that helps objects retain their balance. (p. 303)

H

halo—the outermost part of a spiral galaxy, nearly spherical and lying beyond the spiral. (p. 152)

heliocentric—Sun-centered. (p. 21)

heliosheath—the final frontier in the Solar System where the solar wind slows and meets the approaching wind in interstellar space. (p. 402)

hematite—an iron oxide that on Earth usually forms in an environment containing liquid water. (p. 500)

high-Earth orbit—an orbit at an altitude of about 22,300 miles (35,900 km). (p. 424)

Hohmann transfer orbit—a trajectory that travels exactly 180 degrees around the central body. (p. 429)

hot spot—a cloud-free area where warmer thermal heat from elsewhere on the planet emerges. (p. 405)

hydrostatic equilibrium—the equilibrium conditions in the Sun. (p. 80)

I

incendiary—able to set fire to something. (p. 285)

inclination—the angle an orbit takes as it circles the Earth. (p. 420)

inertia—the tendency of an object to resist a change in its motion. (p. 38)

intercontinental ballistic missile—a missile designed to deliver a payload to another spot on Earth several thousand miles away. (p. 460)

ionize—to gain either a positive or negative electric charge as a result of gaining or losing electrons. (p. 238)

irregular galaxy—a galaxy of irregular shape that cannot be classified as spiral or elliptical. (p. 166)

K

Kuiper belt—a disk-shaped region beyond Neptune's orbit, 30 AU to 1,000 AU from the Sun and the presumed source of short-period comets. (p. 143)

L

large impact theory—the Moon formed as the result of an impact between a large (Mars-sized) object and the Earth. (p. 71)

launch vehicle—rocket. (p. 204)

launch vehicle adapter—a physical structure used to connect a spacecraft to a launch vehicle. (p. 471)

launch window—the specific timeframe during which a launch can take place. (p. 469)

lenticular galaxy—a galaxy with a flat disk like a spiral galaxy, but with little spiral structure, and a large bulge in the nucleus. (p. 165)

light-year—the distance that light travels in a vacuum in one year (about 5.9 trillion miles or 9.5 trillion km). (p. 149)

line of sight—a clear straight path for transmitting between sender and receiver. (p. 516)

low-Earth orbit—an orbit up to about 1,240 miles (2,000 km) above the Earth. (p. 424)

luminosity—the rate at which electromagnetic energy is emitted from a celestial object. (p. 167)

M

magnetic field—what exists in a region of space where magnetic forces can be detected. (p. 57)

magnetosphere—the region in which a celestial body's magnetic field interacts with charged particles from the Sun. (p. 405)

magnitude—an amount, size, speed, or degree that can be measured. (p. 445)

mantle—a thick, solid layer between Earth's crust and its core. (p. 52)

maria—lunar lowlands that resemble seas when viewed from Earth. (p. 69)

medium-Earth orbit—an orbit with an altitude of about 12,400 miles (20,000 km). (p. 424)

meridian—an imaginary line drawn on the sky from south to north. (p. 22)

meteor—a streak of light in the sky caused when a rock particle falling to Earth is so heated by friction with the atmosphere that it emits light. (p. 143)

meteorite—interplanetary chunks of stone or matter that have crashed into a planet or moon from space. (p. 69)

meteoroid—an interplanetary chunk of matter that becomes a meteorite once it strikes the surface of a planet or moon. (p. 69)

microgravity—a condition of gravity so low that weightlessness results. (p. 226)

microlensing—temporary brightening of the light from a distant star orbited by an object such as a planet. (p. 157)

micrometeorites—meteorites that have a diameter of less than a meter. (p. 376)

missing mass—the difference between the mass of clusters of galaxies as calculated from Keplerian motions and the amount of visible mass. (p. 172)

mission directorate—the four main organizations through which NASA carries on its work. (p. 200)

monsoon—an annual period of heavy rainfall. (p. 288)

N

nanosecond—one billionth of a second. (p. 436)

neap tides—the tides that occur when the difference between high and low tides is least. (p. 67)

neutrino—an uncharged, or electrically neutral, particle believed to have very little or no mass. (p. 153)

neutrons—particles with no electrical charge. (p. 79)

normalization of deviance—process of reclassifying defects as acceptable. (p. 330)

north celestial pole—the point on the celestial sphere directly above Earth's North Pole. (p. 9)

nova—a star that grows brighter than usual for a time and then returns to its original state. (p. 159)

nozzle—a rocket's end that releases gas, smoke, and flame to produce thrust. (p. 445)

nuclear bulge—a spiral galaxy's central region. (p. 152)

nuclear fusion—two nuclei combine to form a larger nucleus. (p. 79)

nucleus—the solid core of a comet. (p. 139)

O

oblate—flattened at the poles. (p. 117)

occultation—the passing of one astronomical object in front of another. (p. 126)

Oort cloud—a theoretical sphere, between 10,000 AU and 100,000 AU from the Sun, containing billions of comet nuclei. (p. 141)

opposition—when a planet is directly opposite the Sun in the sky. (p. 103)

optical double—two stars that, from Earth, appear to be very close but are not actually gravitationally bound. (p. 168)

orbital decay—a gradual reduction in the height of a satellite's orbit. (p. 344)

orbital velocity—the speed an object must maintain to stay in orbit. (p. 419)

organizational culture—the values, norms, and shared experiences of an organization. (p. 332)

O-ring—a flat ring of rubber or plastic, used as a gasket or seal. (p. 320)

oxidant—a chemical substance that mixes with oxygen. (p. 226)

oxidizer—a substance that includes oxygen to aid combustion. (p. 450)

P

parallax—the apparent shifting of nearby objects with respect to distant ones as the position of the observer changes. (p. 16)

parsec—an astronomical unit equal to 3.26 light-years. (p. 150)

payload—the cargo the rocket is to carry aloft. (p. 204)

payload shroud—the thin metal cover, or nose cone, that protects a spacecraft and upper stages during a launch when aerodynamic forces can batter the rocket. (p. 471)

perchlorate—a chemical that attracts water. (p. 388)

periapsis—the point where an orbiting body is closest to the object it is orbiting. (p. 429)

perigee—closest distance from the Earth. (p. 65)

perihelion—when a planet is closest to the Sun. (p. 99)

photosphere—the visible part of the Sun and that part of the solar atmosphere that emits light. (p. 82)

physiological—having to do with the internal organs and how they work. (p. 237)

plasma—an electrically charged gaseous form of matter distinct from solids, liquids, and normal gases. (p. 220)

plate tectonics—the motion of sections (plates) of the Earth's crust across the underlying mantle. (p. 53)

posthumously—after death. (p. 328)

Glossary

precession—the conical shifting of the axis of a rotating object. (p. 68)

principle of equivalence—the statement that effects of acceleration are indistinguishable from gravitational effects. (p. 41)

proper motion—the angular velocity of a star as measured from the Sun. (p. 167)

protons—positively charged particles. (p. 79)

protoplanet—a hypothetical whirling gaseous mass within a giant cloud of gas and dust that rotates around a sun and becomes a planet. (p. 118)

pulsar—a rotating neutron star that emits beams of radio waves that, like a lighthouse beacon, are observed as pulses of radio waves with a regular period. (p. 155)

R

radiation—the transfer of energy by electromagnetic waves. (p. 81)

redundancy—duplication of each critical part. (p. 514)

reflection nebula—a cloud of interstellar dust that becomes visible because it refracts and reflects light from a nearby star. (p. 169)

regolith—top layer of silt-fine dust. (p. 368)

resilient—capable of bouncing back to its original shape after being compressed. (p. 324)

retrofire—the ignition of a retrorocket. (p. 275)

retrograde motion—backward motion. (p. 12)

retrorocket—a small rocket used to slow or change the course of a spaceship. (p. 275)

rift zone—a line near the center of the Atlantic Ocean from which lava flows upward. (p. 53)

right ascension—the celestial sphere's equivalent of longitude on Earth. (p. 438)

Robonaut—a robot that has an upper body shaped like a human, with a head, chest, arms, and hands. (p. 482)

robotic—machine- or robot-based. (p. 187)

robotics—the study of robots. (p. 478)

robots—machines that operate automatically or by remote control to perform tasks. (p. 478)

S

satellite cell phone—a mobile telephone that connects to a network of satellites, rather than one of land-based cell towers. (p. 530)

Schwarzschild radius—the radius of a sphere around a black hole from within which no light can escape. (p. 158)

silica—a white compound made of crystals that occurs naturally on Earth as quartz, sand, or flint, and is the main ingredient of window glass. (p. 502)

solar flares—an explosion near or at the Sun's surface, seen as an increase in activity such as prominences (bulges). (p. 85)

solar radiation—solar flares. (p. 217)

solar storm—an episode of violent space weather resulting from particles streaming outward from the Sun. (p. 218)

solar wind—a continuous outflow of charged particles from the Sun. (p. 84)

solstice—either of the twice-yearly times when the Sun is at its greatest distance from the celestial equator. (p. 57)

south celestial pole—the point on the celestial sphere directly above Earth's South Pole. (p. 9)

spar—structural support. (p. 330)

spectral analysis—the identification of an object based on the spectrum of light that it reflects, absorbs, or emits. (p. 497)

spectrograph—an instrument that separates light from the cosmos into its component colors. (p. 398)

spectrometer—a special instrument equipped with devices for measuring the wavelengths of the radiation it observes. (p. 382)

spring tides—exceptionally high and low tides that occur at the time of the new moon or the full moon, when the Sun, Moon, and Earth are approximately aligned. (p. 67)

stellar spectrum—the characteristic pattern of light that a star emits. (p. 340)

suborbital—having a trajectory, or path through space, of less than a complete orbit. (p. 253)

sunspots—the dark spots appearing periodically in groups on the Sun's surface. (p. 84)

Sun-synchronous orbit—an orbit coordinated with Earth's rotation so that the satellite always crosses the equator at the same local time on Earth. (p. 422)

T

taikonauts—Chinese astronauts. (p. 286)

tail of a comet—the gas and/or dust swept away from the comet's head. (p. 139)

technology transfer—the movement of new technology from its creators to secondary users. (p. 536)

telemedicine—the delivery of medical support to remote locations. (p. 243)

Telstar—the first privately sponsored satellite launched into space. (p. 515)

tidal force—a gravitational force that varies in strength and/or direction over an object and causes it to deform. (p. 66)

tidal friction—the friction that results from tides on a rotating object. (p. 67)

Type 1 trajectory—a route that is less than 180 degrees around the central body. (p. 429)

Type 2 trajectory—a path that is more than 180 degrees. (p. 429)

V

variable star—a star that appears to brighten or dim either because of changes going on within the star itself or because something has moved between it and an observer on Earth. (p. 164)

vestibular system—the system that helps people maintain balance and a sense of which way is up. (p. 236)

volatile—a substance that readily changes from solid or liquid to a vapor. (p. 369)

Z

zodiac—the group of constellations the Sun passes through on its apparent path along the ecliptic. (p. 10)

Index